Lecture Notes on Mathematical Olympiad Courses

For Junior Section Vol. 2

Mathematical Olympiad Series

ISSN: 1793-8570

Series Editors: Lee Peng Yee *(Nanyang Technological University, Singapore)*
Xiong Bin *(East China Normal University, China)*

Published

Vol. 1 A First Step to Mathematical Olympiad Problems
by Derek Holton (University of Otago, New Zealand)

Vol. 2 Problems of Number Theory in Mathematical Competitions
by Yu Hong-Bing (Suzhou University, China)
translated by Lin Lei (East China Normal University, China)

Vol. 3 Graph Theory
by Xiong Bin (East China Normal University, China) &
Zheng Zhongyi (High School Attached to Fudan University, China)
translated by Liu Ruifang, Zhai Mingqing & Lin Yuanqing
(East China Normal University, China)

Vol. 4 Combinatorial Problems in Mathematical Competitions
by Yao Zhang (Hunan Normal University, P. R. China)

Vol. 5 Selected Problems of the Vietnamese Olympiad (1962–2009)
by Le Hai Chau (Ministry of Education and Training, Vietnam)
& Le Hai Khoi (Nanyang Technology University, Singapore)

Vol. 6 Lecture Notes on Mathematical Olympiad Courses:
For Junior Section (In 2 Volumes)
by Xu Jiagu

Vol. 7 A Second Step to Mathematical Olympiad Problems
by Derek Holton (University of Otago, New Zealand &
University of Melbourne, Australia)

Vol. 8 Lecture Notes on Mathematical Olympiad Courses:
For Senior Section (In 2 Volumes)
by Xu Jiagu

Xu Jiagu

Former Professor of Mathematics, Fudan University, China

Vol. 8 | Mathematical
Olympiad
Series

Lecture Notes on Mathematical Olympiad Courses

For Senior Section Vol. 2

World Scientific

Published by

World Scientific Publishing Co. Pte. Ltd.

5 Toh Tuck Link, Singapore 596224

USA office: 27 Warren Street, Suite 401-402, Hackensack, NJ 07601

UK office: 57 Shelton Street, Covent Garden, London WC2H 9HE

British Library Cataloguing-in-Publication Data
A catalogue record for this book is available from the British Library.

Mathematical Olympiad Series — Vol. 8
LECTURE NOTES ON MATHEMATICAL OLYMPIAD COURSES
For Senior Section
(In 2 Volumes)

Copyright © 2012 by World Scientific Publishing Co. Pte. Ltd.

ISBN-13 978-981-4368-94-0 (pbk) (Set)
ISBN-10 981-4368-94-6 (pbk)   (Set)

ISBN-13 978-981-4368-95-7 (pbk) (Vol. 1)
ISBN-10 981-4368-95-4 (pbk)   (Vol. 1)

ISBN-13 978-981-4368-96-4 (pbk) (Vol. 2)
ISBN-10 981-4368-96-2 (pbk)   (Vol. 2)

Printed in Singapore by World Scientific Printers.

Preface

Although Mathematical Olympiad competitions are carried out by solving problems, the system of Mathematical Olympiads and the related training courses cannot consist only of problem solving techniques. Strictly speaking, it is a system of mathematical advancing education. To guide students, who are interested in and have the potential to enter the world of Olympiad mathematics, so that their mathematical ability can be promoted efficiently and comprehensively, it is important to improve their mathematical thinking and technical ability in solving mathematical problems.

An excellent student should be able to think flexibly and rigorously. Here, the ability to perform formal logic reasoning is an important basic component. However, it is not the main one. Mathematical thinking also includes other key aspects, such as starting from intuition and entering the essence of the subject, through the processes of prediction, induction, imagination, construction and design to conduct their creative activities. In addition, the ability to convert the concrete to the abstract and vice versa is essential.

Technical ability in solving mathematical problems does not only involve producing accurate and skilled-computations and proofs using the standard methods available, but also the more unconventional, creative techniques.

It is clear that the standard syllabus in mathematical education cannot satisfy the above requirements. Hence the Mathematical Olympiad training books must be self-contained basically.

This book is based on the lecture notes used by the editor in the last 15 years for Olympiad training courses in several schools in Singapore, such as Victoria Junior College, Hwa Chong Institution, Nanyang Girls High School and Dunman High School. Its scope and depth significantly exceeds that of the standard syllabus provided in schools, and introduces many concepts and methods from modern mathematics.

The core of each lecture are the concepts, theories and methods of solving mathematical problems. Examples are then used to explain and enrich the lectures, as well as to indicate the applications of these concepts and methods. A number of questions are included at the end of each lecture for the reader to try. Detailed solutions are provided at the end of book.

The examples given are not very complicated so that the readers can understand them easily. However, many of the practice questions at the end of lectures are taken from actual competitions, which students can use to test themselves. These questions are taken from a range of countries, such as China, Russia, the United States of America and Singapore. In particular, there are many questions from China for those who wish to better understand Mathematical Olympiads there. The questions at the end of each lecture are divided into two parts. Those in Part A are for students to practise, while those in Part B test students' ability to apply their knowledge in solving real competition questions.

Each volume can be used for training courses of several weeks with a few hours per week. The test questions are not considered part of the lectures as students can complete them on their own.

Acknowledgments

My thanks to Professor Lee Peng Yee for suggesting the publication of this the book, and to Professor Phua Kok Khoo for his strong support. I would also like to thank my former excellent student, Mr Tay Jingyi Kenneth, for his corrections, as well as Zhang Ji and He Yue, the editors of this book at World Scientific Publishing Co. (WSPC). This book would not have been published without their efficient assistance.

Abbreviations and Notations

Abbreviations

AHSME	American High School Mathematics Examination
AIME	American Invitational Mathematics Examination
AMC	American Mathematics Examination
APMO	Asia Pacific Mathematics Olympiad
ASUMO	Olympics Mathematical Competitions of All the Soviet Union
AUSTRIA	Austria Mathematical Olympiad
AUSTRALIA	Australia Mathematical Olympiad
BALKAN	Balkan Mathematical Olympiad
BELARUS	Belarus Mathematical Olympiad
BMO	British Mathematical Olympiad
BOSNIA	Bosnia Mathematical Olympiad
BULGARIA	Bulgaria Mathematical Olympiad
CGMO	China Girl's Mathematical Olympiad
CHINA	China Mathematical Competition for Secondary Schools except for CHNMOL
CHNMOL	China Mathematical Competition for Secondary Schools
CHNMO	China Mathematical Olympiad
CMC	China Mathematical Competition (for High Schools)
CMO	Canada Mathematical Olympiad
CNMO	China Northern Mathematical Olympiad
CSMO	China Southeastern Mathematical Olympiad
CROATIA	Croatia Mathematical Olympiad
CWMO	China Western Mathematical Olympiad

CZECH-POLISH-SLOVAK	International Competitions Czech-Polish-Slovak Match
ESTONIA	Estonia Mathematical Olympiad
GERMANY	Germany Mathematical Olympiad
GREECE	Greece Mathematical Olympiad
High-School Mathematics	Journal of "High-School Mathematics", China.
HONG KONG	Hong Kong Mathematical Olympiad
HUNGARY	Hungary Mathematical Competition
HUNGARY-ISRAEL	Hungary-Israel Binational Mathematical Competition
IMO	International Mathematical Olympiad
INDIA	India Mathematical Olympiad
IRAN	Iran Mathematical Olympiad
IRE	Ireland Mathematical Olympiad
JAPAN	Japan Mathematical Olympiad
KOREA	Korea Mathematical Olympiad
MACEDONIA	Macedonia Mathematical Olympiad
MEDITERRANEAN MO	Mediterranean Mathematics Olympiad
POLAND	Poland Mathematical Olympiad
PUTNAM	Putnam Mathematical Competition
ROMANIA	Romania Mathematical Olympiad
RUSMO	All-Russia Olympics Mathematical Competitions
SAUDI ARABIA	Saudi Arabia Mathematical Olympiad
SERBIA	Serbia Mathematical Olympiad
SLOVENIA	Slovenia Mathematical Olympiad
SMO	Singapore Mathematical Olympiads
SSSMO	Singapore Secondary Schools Mathematical Olympiads
SWE	Sweden Mathematical Olympiads
TAIWAN	Taiwan Mathematical Olympiads
THAILAND	Thailand Mathematical Olympiads
TST	Team Selection Test (including related training tests)
TURKEY	Turkey Mathematical Olympiad
UKRAINE	Ukraine Mathematical Olympiad
USAMO	United States of America Mathematical Olympiad
USSR	Union of Soviet Socialist Republics
VIETNAM	Vietnam Mathematical Olympiad

Notations for Numbers, Sets and Logic Relations

$\mathbb{N}$ the set of positive integers (natural numbers)

$\mathbb{N}_0$ the set of non-negative integers

$\mathbb{Z}$ the set of integers

$\mathbb{Z}^+$ the set of positive integers

$\mathbb{Q}$ the set of rational numbers

$\mathbb{Q}^+$ the set of positive rational numbers

$\mathbb{Q}_0^+$ the set of non-negative rational numbers

$\mathbb{R}$ the set of real numbers

$[m, n]$ the lowest common multiple of the integers m and n

(m, n) the greatest common divisor of the integers m and n

$a \mid b$ a divides b

$|x|$ the absolute value of x

$\lfloor x \rfloor$ the greatest integer not greater than x

$\lceil x \rceil$ the smallest integer not less than x

$\{x\}$ the decimal part of x, i.e. $\{x\} = x - \lfloor x \rfloor$

$a \equiv b \pmod{c}$ a is congruent to b modulo c

$\binom{n}{k}$ the binomial coefficient n choose k

$n!$ n factorial, equal to the product $1 \cdot 2 \cdot 3 \cdots n$

$[a, b]$ the closed interval, i.e. all x such that $a \leq x \leq b$

(a, b) the open interval, i.e. all x such that $a < x < b$

$\Leftrightarrow$ iff, i.e. if and only if

$\Rightarrow$ implies

$A \subset B$ A is a subset of B

$A - B$ the set formed by all the elements in A but not in B

$A \cup B$ the union of the sets A and B

$A \cap B$ the intersection of the sets A and B

$a \in A$ the element a belongs to the set A

Contents

Lecture 16

Mathematical Induction

The **Mathematical Induction** is a very powerful tool for proving infinitely many propositions by using only a few steps. In particular, it can be used often when the propositions are involving the nonnegative integer n.

Theorem I. **(The First Induction)** *Let* $\{P_n\}, n \geq m$ *be a family of infinitely many propositions (or statements) involving the integer n, where m is a nonnegative integer. If (i) P_m can be proved to be true, and (ii) P_{k+1} must be true provided P_k is true for any $k \geq m$, then P_n is true for all $n \geq m$.*

Theorem II. **(The Second Induction)** *Let* $\{P_n\}, n \geq m$ *be a family of infinitely many propositions (or statements) involving the integer n, where m is a nonnegative integer. If (i) P_m can be proved to be true, and (ii) P_{k+1} must be true provided $P_m, P_{m+1}, \ldots, P_k$ are all true for any $k \geq m$, then P_n is true for all $n \geq m$.*

Note that it seems to be more difficult for using the second induction, since it require more conditions. However it is actually not so. We can use either one freely if necessary.

Variations of Mathematical Induction

(i)   For proving each $P_n, n = 1, 2, \ldots$, we can prove $P_{n_k}, k = 1, 2, \ldots$ first by induction, and then prove P_n for each n.

(ii)   For proving $\{P_n, n \in \mathbb{N}\}$ by induction, sometimes it is convenient to prove a stronger proposition P'_n.

(iii)   For proving $\{P_n, n \geq n_0\}$ by induction, sometimes it is convenient to prove P_n by induction starting from $n = n_1 > n_0$.

(iv)   The induction can be conducted backwards: (i) Prove each proposition in the subsequence $\{P_{n_k}, k \in \mathbb{N}\}$ first, and then (ii) prove that P_k is true provided P_{k+1} is true.

(v)   Cross-Active (or Spiral) Induction: For proving propositions $P_n, n = 0, 1, 2, \ldots$ and $Q_n, n = 0, 1, 2, \ldots$, it suffices to show that (i) P_0 and Q_0 are true; (ii) If P_k, Q_k are true, then $P_k \Rightarrow Q_{k+1} \Rightarrow P_{k+1} \Rightarrow P_{k+1}$.

(vi)   Multiple Induction: Let $\{P_{n,m}, n, m \in \mathbb{N}\}$ be a sequence of propositions with two independent parameters n and m. If (i) $P_{1,m}$ is true for all $m \in \mathbb{N}$ and $P_{n,1}$ are true for all $n \in \mathbb{N}$; (ii) $P_{n+1,m+1}$ is true provided $P_{n,m+1}$ and $P_{n+1,m}$ are true, then $P_{n,m}$ is true for any $n, m \in \mathbb{N}$.

Examples

Example 1. Prove by induction that $1^3 + 2^3 + \cdots + n^3 = \dfrac{n^2(n+1)^2}{4}$ for each $n \in \mathbb{N}$. Hence prove that $1^3 + 3^3 + 5^3 + \cdots + (2n-1)^3 = n^2(2n^2 - 1)$ for each $n \in \mathbb{N}$.

Solution   Let P_n be the statement

$$1^3 + 2^3 + \cdots + n^3 = \frac{n^2(n+1)^2}{4}, \qquad n = 1, 2, \ldots.$$

For $n = 1$, left hand side $= 1 = \dfrac{1^2(1+1)^2}{4} =$ right hand side, P_1 is proven.

Assume that P_k is true $(k \geq 1)$, i.e. $1^3 + 2^3 + \cdots + k^3 = \dfrac{k^2(k+1)^2}{4}$, then for $n = k + 1$,

$$1^3 + 2^3 + \cdots + (k+1)^3 = 1^3 + 2^3 + \cdots + k^3 + (k+1)^3$$

$$= \frac{k^2(k+1)^2}{4} + (k+1)^3 = \frac{(k+1)^2}{4} \cdot [k^2 + 4(k+1)] = \frac{(k+1)^2(k+2)^2}{4}.$$

Thus, P_{k+1} is also true. By induction, P_n is true for each $n \in \mathbb{N}$.

Now for any positive integer n,

$$1^3 + 3^3 + \cdots + (2n-1)^3 = 1^3 + 2^3 + \cdots + (2n)^3 - (2^3 + 4^3 + \cdots + (2n)^3)$$

$$= \left[\frac{(2n)(2n+1)}{2}\right]^2 - 2^3(1^3 + 2^3 + \cdots + n^3) = n^2(2n+1)^2 - 8 \cdot \frac{n^2(n+1)^2}{4}$$

$$= n^2(2n+1)^2 - 2n^2(n+1)^2 = n^2[(2n+1)^2 - 2(n+1)^2] = n^2(2n^2 - 1),$$

as desired.

Example 2. (CWMO/2008) $\{a_n\}$ is a sequence of real numbers, satisfying $a_0 \neq 0, 1, a_1 = 1 - a_0, a_{n+1} = 1 - a_n(1 - a_n)$ for all $n \in \mathbb{N}$. Prove that for any positive integer n,

$$a_0 a_1 \cdots a_n \left(\frac{1}{a_0} + \frac{1}{a_1} + \cdots + \frac{1}{a_n} \right) = 1.$$

Solution   The given condition yields $1 - a_{n+1} = a_n(1 - a_n), n \in \mathbb{N}_0$, hence

$$1 - a_{n+1} = a_n a_{n-1}(1 - a_{n-1}) = \cdots = a_n \cdots a_1(1 - a_1) = a_n \cdots a_1 a_0,$$

namely

$$a_{n+1} = 1 - a_0 a_1 \cdots a_n, \qquad n = 1, 2, \ldots.$$

We now prove the original conclusion by induction.

For $n = 1, a_0 a_1 \left(\dfrac{1}{a_0} + \dfrac{1}{a_1} \right) = a_0 + a_1 = 1$, the conclusion is proven.

Assume that the proposition is true for $n = k$, then for $n = k + 1$,

$$a_0 a_1 \cdots a_{k+1} \left(\frac{1}{a_0} + \frac{1}{a_1} + \cdots + \frac{1}{a_{k+1}} \right)$$

$$= a_0 a_1 \cdots a_k \left(\frac{1}{a_0} + \frac{1}{a_1} + \cdots + \frac{1}{a_k} \right) a_{k+1} + a_0 a_1 \cdots a_k$$

$$= a_{k+1} + a_0 a_1 \cdots a_k = 1.$$

Thus, the proposition is true for $n = k + 1$ also. By induction, the conclusion is true for all $n \in \mathbb{N}$.

Example 3. (IMO/Shortlist/2006) The sequence of real numbers $a_0, a_1, a_2, \ldots$ is defined recursively by

$$a_0 = -1, \quad \sum_{k=0}^{n} \frac{a_{n-k}}{k+1} = 0 \ \text{ for } n \geq 1.$$

Show that $a_n > 0$ for $n \geq 1$.

Solution   The proof goes by induction on n. For $n = 1$ the formula yields $a_1 = 1/2$. Take $n \geq 1$ and assume $a_1, \ldots, a_n > 0$. Write the recurrence formula for n and $n + 1$ respectively as

$$\sum_{k=0}^{n} \frac{a_k}{n-k+1} = 0 \quad \text{and} \quad \sum_{k=0}^{n+1} \frac{a_k}{n-k+2} = 0.$$

Subtraction yields

$$0 = (n+2) \sum_{k=0}^{n+1} \frac{a_k}{n-k+2} - (n+1) \sum_{k=0}^{n} \frac{a_k}{n-k+1}$$

$$= (n+2)a_{n+1} + \sum_{k=0}^{n} \left(\frac{n+2}{n-k+2} - \frac{n+1}{n-k+1} \right) a_k.$$

The coefficient of a_0 vanishes, so

$$
\begin{aligned}
a_{n+1} &= -\frac{1}{n+2}\sum_{k=1}^{n}\left(\frac{n+2}{n-k+2}-\frac{n+1}{n-k+1}\right)a_k \\
&= \frac{1}{n+2}\sum_{k=1}^{n}\frac{k}{(n-k+1)(n-k+2)}a_k.
\end{aligned}
$$

The coefficients of $a_1,\ldots,a_n$ are all positive. Therefore, $a_1,\ldots,a_n > 0$ implies $a_{n+1} > 0$.

Example 4. (CHNMO/2010) (edited) If $0 < x_1 \le x_2 \le \ldots \le x_s, 0 < y_1 \le y_2 \le \ldots \le y_s$ satisfy the condition that

$$x_1^r + x_2^r + \cdots + x_s^r = y_1^r + y_2^r + \cdots + y_s^r, \quad \text{for all } r \in \mathbb{N},$$

prove that $x_i = y_i, i = 1, 2, \ldots, s$.

Solution We prove by induction on s. Let $P_s, s = 1, 2, \ldots$ be the statement to be proven.

For $s = 1$, taking $r = 1$ yields $x_1 = y_1$. Assume that P_s is true for $s = k$ ($k \ge 1$) i.e.,

$$x_1 = y_1, x_2 = y_2, \ldots, x_k = y_k.$$

then for $s = k + 1$, if $x_{k+1} \ne y_{k+1}$, without loss of generality we assume that $x_{k+1} < y_{k+1}$. Then

$$\left(\frac{x_1}{y_{k+1}}\right)^r + \cdots + \left(\frac{x_{k+1}}{y_{k+1}}\right)^r = \left(\frac{y_1}{y_{k+1}}\right)^r + \cdots + \left(\frac{y_k}{y_{k+1}}\right)^r + 1 \ge 1.$$

But on the other hand, since $0 < \dfrac{x_i}{y_{k+1}} < 1, i = 1, 2, \ldots, k+1$, letting $r \to +\infty$ yields $0 \ge 1$, a contradiction. Hence $x_{k+1} = y_{k+1}$, then the inductive assumption implies that $x_i = y_i, i = 1, 2, \ldots, k$. Thus, P_{k+1} is also true. By induction, P_s is true for each $s \in \mathbb{N}$.

Example 5. (AM-GM Inequality) For real numbers $a_1, a_2, \ldots, a_n > 0$, prove that

$$a_1 + a_2 + \cdots + a_n \ge n \cdot \sqrt[n]{a_1 a_2 \cdots a_n}, \qquad (*)$$

and that equality holds if and only if $a_1 = a_2 = \cdots = a_n$.

Solution We first prove $(*)$ for $n = 2^t, t \in \mathbb{N}$ by induction on t.
When $t = 1$, $(*) \Leftrightarrow (\sqrt{a_1} - \sqrt{a_2})^2 \ge 0$, so it's true.

Assume that $(*)$ is true for $t = k$, then for $t = k + 1$, by the induction assumption,

$$\sum_{j=1}^{2^{k+1}} a_j = \sum_{j=1}^{2^k} a_j + \sum_{j=1}^{2^k} a_{j+2^k} \geq 2^k \sqrt[2^k]{\prod_{j=1}^{2^k} a_j} + 2^k \cdot \sqrt[2^k]{\prod_{j=1}^{2^k} a_{j+2^k}}$$

$$\geq 2^k \cdot 2 \cdot \sqrt[2^{k+1}]{\prod_{j=1}^{2^{k+1}} a_j} = 2^{k+1} \cdot \sqrt[2^{k+1}]{\prod_{j=1}^{2^{k+1}} a_j}.$$

Thus, the proposition is proven for $t = k + 1$. By induction, $(*)$ is proven for all $n = 2^t, t = 1, 2, \ldots$.

Next we prove $(*)$ for each $n \in \mathbb{N}$. For any $n \in \mathbb{N}$ which is not of the form 2^t, there exist an $s \in \mathbb{N}$ such that $2^{s-1} < n < 2^s$. Since

$$a_1 + a_2 + \cdots + a_{2^s} \geq 2^s \cdot \sqrt[2^s]{a_1 a_2 \cdots a_{2^s}}, \qquad (**)$$

By letting $a_{n+1} = a_{n+2} = \cdots = a_{2^s} = \sqrt[n]{a_1 a_2 \cdots a_n}$ in $(**)$, $(**)$ yields

$$a_1 + a_2 + \cdots + a_n + (2^s - n) \sqrt[n]{a_1 a_2 \cdots a_n}$$

$$\geq 2^s \cdot \sqrt[2^s]{(a_1 a_2 \cdots a_n)(a_1 a_2 \cdots a_n)^{\frac{2^s-n}{n}}} = 2^s \cdot \sqrt[n]{a_1 a_2 \cdots a_n},$$

$$\therefore a_1 + a_2 + \cdots + a_n \geq n \cdot \sqrt[n]{a_1 a_2 \cdots a_n},$$

hence $(*)$ is proven for general $n \in \mathbb{N}$.

Example 6. (CMC/2007) For each positive integer $n > 1$, prove that

$$\frac{2n}{3n+1} < \frac{1}{n+1} + \frac{1}{n+2} + \cdots + \frac{1}{n+n} < \frac{25}{36}.$$

Solution To prove the left inequality, define

$$f(n) = \frac{1}{n+1} + \frac{1}{n+2} + \cdots + \frac{1}{n+n} - \frac{2n}{3n+1}, \quad n \in \mathbb{N}.$$

Then $f(n+1) = \frac{1}{n+2} + \frac{1}{n+3} + \cdots + \frac{1}{2n+2} - \frac{2n+2}{3n+4}$, so

$$f(n+1) - f(n) = \left(\frac{1}{2n+1} - \frac{1}{2n+2} \right) + \left(\frac{2n}{3n+1} - \frac{2n+2}{3n+4} \right)$$

$$= \frac{1}{(2n+1)(2n+2)} - \frac{2}{(3n+1)(3n+4)}$$

$$= \frac{n(n+3)}{(2n+1)(2n+2)(3n+1)(3n+4)} > 0,$$

i.e., $f(n)$ is an increasing function of n. Thus, for $n \geq 2$,

$$f(n) \geq f(2) = \frac{1}{3} + \frac{1}{4} - \frac{4}{7} = \frac{1}{84} > 0,$$

the left inequality is proven.

Below we prove the right inequality by proving a stronger one by induction. Let the proposition $P_n, n \geq 2$ be

$$\frac{1}{n+1} + \frac{1}{n+2} + \cdots + \frac{1}{n+n} < \frac{25}{36} - \frac{1}{4n+1}.$$

When $n = 2$, the left hand side of P_2 is $\frac{1}{3} + \frac{1}{4} = \frac{7}{12}$, and the right hand side is $\frac{25}{36} - \frac{1}{9} = \frac{7}{12}$, so P_2 is true. Assume P_n is true for $n = k$ ($k \geq 2$), i.e.,

$$\frac{1}{k+1} + \frac{1}{k+2} + \cdots + \frac{1}{k+k} < \frac{25}{36} - \frac{1}{4k+1}.$$

Then for $n = k + 1$, the left hand side of P_{k+1} is

$$\frac{1}{k+2} + \cdots + \frac{1}{2k+2} = \frac{1}{k+1} + \cdots + \frac{1}{2k+1} - \frac{1}{2k+2}$$

$$\leq \frac{25}{36} - \frac{1}{4k+1} + \frac{1}{2k+1} - \frac{1}{2k+2}$$

$$= \frac{25}{36} - \frac{1}{4k+5} + \frac{1}{4k+5} - \frac{1}{4k+1} + \frac{1}{2k+1} - \frac{1}{2k+2}.$$

Since

$$\frac{1}{4k+5} - \frac{1}{4k+1} + \frac{1}{2k+1} - \frac{1}{2k+2} = -\frac{3}{(2k+1)(2k+2)(4k+1)(4k+5)} < 0, \text{ therefore } \frac{1}{k+2} + \cdots + \frac{1}{2k+2} < \frac{25}{36} - \frac{1}{4k+5}.$$

Thus, P_{k+1} is true also. The inductive proof is completed.

Example 7. Suppose that $a_1, a_2, \cdots, a_n, \cdots$ are real numbers with $0 \leq a_i \leq 1$ for $i = 1, 2, \cdots$. Prove that for every $n \geq 2$,

$$\sum_{i=1}^{n} \frac{1}{1+a_i} \leq \frac{n}{1 + \sqrt[n]{a_1 a_2 \cdots a_n}}. \qquad (*)$$

Solution   As the first step, we prove the conclusion for all n with the form 2^m by induction on m. For $m = 1$, the inequality $(*)$ becomes

$$\frac{1}{1+a_1} + \frac{1}{1+a_2} \leq \frac{2}{1 + \sqrt{a_1 a_2}}$$

or equivalently:

$$2(1 + a_1)(1 + a_2) \geq (2 + a_1 + a_2)(\sqrt{a_1 a_2} + 1),$$
$$a_1 + a_2 + 2a_1 a_2 \geq (2 + a_1 + a_2)\sqrt{a_1 a_2},$$
$$(a_1 + a_2)(1 - \sqrt{a_1 a_2}) - 2\sqrt{a_1 a_2}(1 - \sqrt{a_1 a_2}) \geq 0,$$
$$(\sqrt{a_1} - \sqrt{a_2})^2 (1 - \sqrt{a_1 a_2}) \geq 0.$$

Therefore the conclusion is true. Assume that for $m = k$ $(k \geq 1)$ the conclusion is true, i.e.

$$\sum_{i=1}^{2^k} \frac{1}{1 + a_i} \leq \frac{2^k}{1 + \sqrt[2^k]{a_1 a_2 \cdots a_{2^k}}}.$$

Then, for $m = k + 1$

$$\sum_{i=1}^{2^{k+1}} \frac{1}{1 + a_i} = \sum_{i=1}^{2^k} \left(\frac{1}{1 + a_{2i-1}} + \frac{1}{1 + a_{2i}} \right) \leq 2 \sum_{i=1}^{2^k} \frac{1}{1 + \sqrt{a_{2i-1} a_{2i}}}$$

$$\leq 2 \cdot \frac{2^k}{1 + \sqrt[2^k]{\sqrt{a_1 a_2} \sqrt{a_3 a_4} \cdots \sqrt{a_{2^{k+1}-1} a_{2^{k+1}}}}}$$

$$= \frac{2^{k+1}}{1 + \sqrt[2^{k+1}]{a_1 a_2 \cdots a_{2^{k+1}}}}.$$

Therefore the conclusion is true for all $n = 2^m$, $m \in \mathbb{N}$.

Next, to show that $(*)$ is true for all $n \in \mathbb{N}$, it suffices to show that $(*)$ is true for $n = k - 1$ $(k \geq 2)$ if $(*)$ is true for $n = k$. In fact, by letting $a_k = \sqrt[k-1]{a_1 a_2 \cdots a_{k-1}}$, we have

$$\sum_{i=1}^{k-1} \frac{1}{1 + a_i} + \frac{1}{1 + \sqrt[k-1]{a_1 a_2 \cdots a_{k-1}}}$$

$$\leq \frac{k}{1 + \sqrt[k]{a_1 a_2 \cdots a_{k-1} \cdot \sqrt[k-1]{a_1 a_2 \cdots a_{k-1}}}} = \frac{k}{1 + \sqrt[k-1]{a_1 a_2 \cdots a_{k-1}}},$$

therefore we have

$$\sum_{i=1}^{k-1} \frac{1}{1 + a_i} \leq \frac{k - 1}{1 + \sqrt[k-1]{a_1 a_2 \cdots a_{k-1}}}.$$

By backward induction, $(*)$ is true for each $n \in \mathbb{N}$.

Example 8. Let $\{F_n\}$, $n \in \mathbb{N}$ be the Fibonacci sequence. Prove that $F_{n+1}^2 + F_n^2 = F_{2n+1}$ for each $n \in \mathbb{N}$.

Solution Let $P_n : F_{n+1}^2 + F_n^2 = F_{2n+1}$, $Q_n : 2F_{n+1}F_n + F_{n+1}^2 = F_{2n+2}$ for $n = 1, 2, \ldots$.

When $n = 1$, then P_1, Q_1 are obvious since $F_1 = F_2 = 1$.

When P_k, Q_k are true for $n = k$ ($k \geq 1$), i.e., $F_{k+1}^2 + F_k^2 = 2F_{2k+1}$ and $2F_{k+1}F_k + F_{k+1}^2 = F_{2k+2}$, then

$$
\begin{aligned}
F_{k+2}^2 + F_{k+1}^2 &= (F_{k+1} + F_k)^2 + F_{k+1}^2 \\
&= (F_{k+1}^2 + 2F_{k+1}F_k) + (F_k^2 + F_{k+1}^2) \\
&= F_{2k+2} + F_{2k+1} = F_{2k+3} = F_{2(k+1)+1},
\end{aligned}
$$

i.e., P_{k+1} is true. Further, P_{k+1} and Q_k are true yields

$$
\begin{aligned}
2F_{k+2}F_{k+1} + F_{k+2}^2 &= 2(F_{k+1} + F_k)F_{k+1} + F_{k+2}^2 \\
&= (F_{k+1}^2 + F_{k+2}^2) + (F_{k+1}^2 + 2F_{k+1}F_k) \\
&= F_{2k+3} + F_{2k+2} = F_{2k+4} = F_{2(k+1)+2},
\end{aligned}
$$

namely Q_{k+1} is true. By the cross-active induction, P_n, Q_n are true for each $n \in \mathbb{N}$.

Example 9. m boxes are arranged in a row. Prove that there are $\dfrac{(n+m-1)!}{n!(m-1)!}$ different ways to put n identical balls into these boxes.

Solution By $F(n, m)$ we denote the number of ways to put n identical balls into m boxes. The propositions to be proven are

$$
F(n, m) = \frac{(n+m-1)!}{n!(m-1)!}
$$

for $n, m \in \mathbb{N}$.

For $n = 1$, we have $F(1, m) = m = \dfrac{m!}{(m-1)!}$, hence the conclusion is true for $n = 1$ and all $m \in \mathbb{N}$. For $m = 1$ and any $n \in \mathbb{N}$, we have $F(n, 1) = 1 = \dfrac{n!}{n!}$, therefore the conclusion is true for $m = 1$ and all $n \in \mathbb{N}$.

Suppose that $F(n+1, m) = \dfrac{(n+m)!}{(n+1)!(m-1)!}$ and $F(n, m+1) = \dfrac{(n+m)!}{n!m!}$.

To evaluate $F(n+1, m+1)$, by considering whether the number of balls in the $m + 1$st box is 0 or greater than 0, we have

$$
\begin{aligned}
&F(n+1, m+1) \\
&= F(n+1, m) + F(n, m+1) = \frac{(n+m)!}{(n+1)!(m-1)!} + \frac{(n+m)!}{n!m!} \\
&= \frac{(n+m)!(m+n+1)}{(n+1)!m!} = \frac{(m+n+1)!}{(n+1)!(m!)} = \frac{[(n+1)+(m+1)-1]!}{(n+1)!(m+1-1)!}.
\end{aligned}
$$

The inductive proof is completed.

Testing Questions (A)

1. (USAMO/2002) Let S be a set with 2002 elements, and let N be an integer with $0 \le N \le 2^{2002}$. Prove that it is possible to color every subset of S either blue or red so that the following conditions hold:

   (a) the union of any two red subsets is red;

   (b) the union of any two blue subsets is blue;

   (c) there are exactly N red subsets.

2. (CHNMO/TST/2005) Let k be a positive integer. Prove that it is always possible to partition the set $\{0, 1, 2, 3, \ldots, 2^{k+1} - 1\}$ into two disjoint subsets $\{x_1, x_2, \ldots, x_{2^k}\}$ and $\{y_1, y_2, \ldots, y_{2^k}\}$ such that $\displaystyle\sum_{i=1}^{2^k} x_i^m = \sum_{i=1}^{2^k} y_i^m$ holds for each $m \in \{1, 2, \ldots, k\}$.

3. (CMC/2006) Let $f(x) = x^2 + a$. Define $f^1(x) = f(x)$, $f^n(x) = f(f^{n-1}(x))$, $n = 2, 3, \ldots$ and $M = \{a \in \mathbb{R} \mid |f^n(0)| \le 2 \text{ for all } n \in \mathbb{N}\}$. Prove that

$$M = \left[-2, \frac{1}{4}\right].$$

4. (BELARUS/2007) For positive real numbers $x_1, x_2, \ldots, x_{n+1}$, prove that

$$\frac{1}{x_1} + \frac{x_1}{x_2} + \frac{x_1 x_2}{x_3} + \cdots + \frac{x_1 x_2 \cdots x_n}{x_{n+1}} \ge 4(1 - x_1 x_2 \cdots x_{n+1}).$$

5. (CMC/2010) Given the sequence $\{a_n\}$, $a_1 = 2$ and $a_{n+1} = a_1 a_2 \cdots a_n + 1$ for $n = 1, 2, \ldots$. Prove that

$$\frac{1}{a_1} + \frac{1}{a_2} + \cdots + \frac{1}{a_n} \ge \frac{1}{2} + \frac{1}{4} + \frac{1}{8} + \cdots + \frac{1}{2^n} \quad n \in \mathbb{N}.$$

6. (AUSTRIA/2009) Prove that $3^{n^2} > (n!)^4$ for all positive integers n.

7. (CHNMO/TST/2008) Let a and b be positive integers with $(a, b) = 1$, and a, b having different parities. Let the set S have the following properties: (1) $a, b \in S$; (ii) If $x, y, z \in S$, then $x + y + z \in S$. Prove that all the integers greater than $2ab$ are in S.

8. (USAMO/2007) Let S be a set containing $n^2 + n - 1$ elements, for some positive integer n. Suppose that the n-element subsets of S are partitioned into two classes. Prove that there are at least n pairwise disjoint sets in the same class.

Testing Questions   (B)

1.   (CHNMO/TST/2010) It is given that the real numbers a_i, b_i $(i = 0, 1, \ldots, 2n)$ satisfy the following three conditions:

   (i) $a_i + a_{i+1} \geq 0$ for $i = 0, 1, \ldots, 2n - 1$;

   (ii) $a_{2j+1} \leq 0$ for $j = 0, 1, \ldots, n - 1$;

   (iii)   For any integers p, q with $0 \leq p \leq q \leq n$, $\displaystyle\sum_{k=2p}^{2q} b_k > 0$.

   Prove that $\displaystyle\sum_{i=0}^{2n}(-1)^i a_i b_i \geq 0$, and find the condition under which the equality holds.

2.   (USAMO/2007) Prove that for every nonnegative integer n, the number $7^{7^n} + 1$ is the product of at least $2n + 3$ (not necessarily distinct) primes.

3.   (BULGARIA/2007) Let $a_1 > \dfrac{1}{12}$, $a_{n+1} = \sqrt{(n + 2)a_n + 1}, n = 1, 2, \ldots.$. Prove that

   (i) $a_n > n - \dfrac{2}{n}, n \in \mathbb{N}$;   (ii)   The sequence $b_n = 2^n \left(\dfrac{a_n}{n} - 1\right), n \in \mathbb{N}$ is convergent.

4.   (USAMO/2010) There are n students standing in a circle, one behind the other. The students have heights $h_1 < h_2 < \ldots < h_n$. If a student with height h_k is standing directly behind a student with height h_{k-2} or less, the two students are permitted to switch places. Prove that it is not possible to make more than $\binom{n}{3}$ such switches before reaching a position in which no further switches are possible.

5.   (IMO/2010) In each of six boxes $B_1, B_2, B_3, B_4, B_5, B_6$ there is initially one coin. There are two types of operation allowed:

   Type 1 : Choose a nonempty box B_j with $1 \leq j \leq 5$. Remove one coin from B_j and add two coins to B_{j+1}.

   Type 2 : Choose a nonempty box B_k with $1 \leq k \leq 4$. Remove one coin from B_k and exchange the contents of (possibly empty) boxes B_{k+1} and B_{k+2}.

   Determine whether there is a finite sequence of such operations that results in boxes B_1, B_2, B_3, B_4, B_5 being empty and box B_6 containing exactly $2010^{2010^{2010}}$ coins. (Note that $a^{b^c} = a^{(b^c)}$.)

Lecture 17

Arithmetic Progressions and Geometric Progressions

Definition 17.1. A sequence $a_1, a_2, a_3, \ldots$ is said to be an **Arithmetic Progression**, which is abbreviated as **A.P.**, if it has a *common difference d* given by

$$d = a_2 - a_1 = a_3 - a_2 = a_{n+1} - a_n, \quad \text{for } n = 1, 2, 3, \ldots.$$

(I)   One of the most important properties for an A.P. is its formula for the general term a_n given by

$$a_n = a_1 + (n-1)d, \quad n = 1, 2, 3, \ldots.$$

It implies that an A.P. is completely determined by its *initial term* a_1 and its common difference d.

(II) Let S_n denote its *partial sum*, i.e., $S_n = a_1 + a_2 + \cdots + a_n, n \in \mathbb{N}$. Then

$$S_n = \frac{n}{2}(a_1 + a_n) = \frac{n}{2}[2a_1 + (n-1)d] = \frac{d}{2}n^2 + \frac{2a_1 - d}{2}n, n \in \mathbb{N}.$$

Note that, if $S_n = an^2 + bn, n \in \mathbb{N}$, then $\{a_n\}$ must be an A.P..

(III) An often used property for an A.P. is that if a, b, c are three consecutive terms in an A.P., then

$$a + c = 2b.$$

Definition 17.2. A sequence $a_1, a_2, a_3, \ldots$ is said to be a **Geometric Progression**, which is abbreviated as **G.P.**, if it has a *common ratio r* given by

$$r = \frac{a_2}{a_1} = \frac{a_3}{a_2} = \frac{a_{n+1}}{a_n}, \quad \text{for } n = 1, 2, 3, \ldots.$$

11

(I)   For a G.P. $\{a_n\}$, the formula for its general term a_n is given by

$$a_n = a_1 \cdot r^{n-1}, \quad n = 1, 2, 3, \ldots .$$

It implies that a G.P. is completely determined by its initial term a_1 and its common ratio r.

(II)   For a G.P. $\{a_n\}$, if $S_n = a_1 + a_2 + \cdots + a_n$, $n \in \mathbb{N}$ with $r \neq 1$, then

$$S_n = a_1 \cdot \frac{1 - r^n}{1 - r}.$$

and the finite limit $\lim\limits_{n \to \infty} S_n = \dfrac{a_1}{1 - r}$ exists (denoted by S_∞) when $|r| < 1$.

(III)   An often used property for an G.P. is that if a, b, c are three consecutive terms in an G.P., then

$$ac = b^2.$$

Examples

Example 1. (CMC/2009) Let S_n be the sum of the first n terms of an A.P. $\{a_n\}$. If $S_{15} > 0$ and $S_{16} < 0$, find the maximum ratio among $\dfrac{S_1}{a_1}, \dfrac{S_2}{a_2}, \ldots, \dfrac{S_{15}}{a_{15}}$.

Solution   Let d be the common difference of the A.P. $\{a_n\}$. Then $a_1 = a_8 - 7d$, $a_2 = a_8 - 6d, \ldots, a_{15} = a_8 + 7d$. Hence $S_{15} = 15a_8 > 0$, implying that $a_8 > 0$.
Since

$$a_1 + a_{16} = a_2 + a_{15} = \cdots = a_8 + a_9 = 2a_8 + d \Rightarrow S_{16} = 8(a_8 + a_9) \Rightarrow a_9 < 0,$$

so $d = a_9 - a_8 < 0$ and $a_1 > a_2 > \ldots > a_8 > 0 > a_9 > a_{10} > \cdots > a_{15}$.
Thus,

$$0 < S_1 < S_2 < \cdots < S_8 \text{ and } S_8 > S_9 > \cdots > S_{15} > 0,$$

which implies that

$$\frac{S_1}{a_1} < \frac{S_2}{a_2} < \cdots < \frac{S_8}{a_8} \text{ and } 0 > \frac{S_9}{a_9}, \frac{S_{10}}{a_{10}}, \ldots, \frac{S_{15}}{a_{15}}.$$

Hence, $\dfrac{S_8}{a_8}$ is the maximum ratio.

Example 2. (CMC/2009) It is given that the sequence $\{a_n\}$ satisfies $a_1 = 0$, $a_{n+1} = a_n + 1 + 2\sqrt{1 + a_n}$ for $n = 1, 2, \ldots .$ Then a_{2009} is
    (A) 4036080   (B) 4036078   (C) 4036082   (D) 4036099.

Solution The given condition gives

$$a_{n+1} + 1 = a_n + 1 + 2\sqrt{1 + a_n} + 1 = (\sqrt{a_n + 1} + 1)^2.$$

By induction it is easy to see that $a_{n+1} > 0$ for any $n \in \mathbb{N}$. Therefore

$$\sqrt{a_{n+1} + 1} = \sqrt{a_n + 1} + 1, \qquad n \in \mathbb{N}.$$

Hence $\{\sqrt{a_n + 1}\}$ is an A.P. with initial term 1 and common difference 1, so $\sqrt{a_n + 1} = n$, i.e., $a_n = n^2 - 1 = (n - 1)(n + 1), n \in \mathbb{N}$. Thus, $a_{2009} = 2008 \times 2010 = 4036080$, the answer is (A).

Example 3. (CMC/2009) $\{a_n\}$ consists of positive numbers, and for any positive integer n, $\dfrac{a_n + 2}{2} = \sqrt{2S_n}$. Find the formula for its general term a_n in terms of n.

Solution For each $n \in \mathbb{N}$, $\dfrac{a_n + 2}{2} = \sqrt{2S_n} \Rightarrow a_n^2 + 4a_n + 4 = 8S_n$, therefore $a_{n-1}^2 + 4a_{n-1} + 4 = 8S_{n-1}$. Taking the difference of previous two equations yields

$$a_n^2 - a_{n-1}^2 + 4(a_n - a_{n-1}) = 8a_n,$$
$$(a_n - a_{n-1})(a_n + a_{n-1}) = 4(a_n + a_{n-1}),$$
$$\therefore a_n - a_{n-1} = 4.$$

Thus, $\{a_n\}$ is an A.P. with common difference $d = 4$. Since $S_1 = a_1$, so

$$a_n^2 + 4a_n + 4 = 8S_n \Rightarrow a_1^2 - 4a_1 + 4 = 0 \Rightarrow (a_1 - 2)^2 = 0 \Rightarrow a_1 = 2.$$

Thus, $a_n = 2 + 4(n - 1)$, namely $a_n = 4n - 2$ for $n = 1, 2, \ldots$.

Example 4. (CHNMO/TST/2008) If the infinite A.P. $\{a_n\}$ of positive integers is not a constant and contains a cubic number as a term, prove that it contains also a term which is a cubic number but is not a perfect square number.

Solution Let d be the common difference of $\{a_n\}$, then d is a positive integer.

Without loss of generality we assume that $a_1 = a^3$ for some natural number a, then all the numbers of form $(a + md)^3, m \in \mathbb{N}$ are in the A.P..

Take $m = a^2 d$. The term

$$(a + a^2 d^2)^3 = a^3 + 3a^4 d^2 + 3a^5 d^4 + a^6 d^6 = a^3 + (3a^4 d + 3^5 a^5 d^3 + a^6 d^5)d$$

is a term in the A.P.. If it is a perfect square number, then it must be a power 6 of a positive integer. However

$$(ad)^2 < a + a^2 d^2 < (ad + 1)^2$$

implies that $a + a^2 d^2$ is not a perfect square. Hence $(a + a^2 d^2)^3$ cannot be a power 6 of a positive integer, and so it is not a perfect square.

Example 5. (CMC/2007) The common difference d of an A.P. $\{a_n\}$ is not zero, the common ratio q of a G.P. $\{b_n\}$ is a positive rational number less than 1. Further, if $a_1 = d, b_1 = d^2$, and $\dfrac{a_1^2 + a_2^2 + a_3^2}{b_1 + b_2 + b_3}$ is a positive integer, find q.

Solution From $a_1^2 + a_2^2 + a_3^2 = d^2 + 4d^2 + 9d^2 = 14d^2, b_1 + b_2 + b_3 = d^2 + d^2 q + d^2 q^2$,

$$\frac{a_1^2 + a_2^2 + a_3^2}{b_1 + b_2 + b_3} = \frac{14}{1 + q + q^2} \Rightarrow 1 + q + q^2 = \frac{14}{m}$$

for some positive integer m. Hence,

$$q = -\frac{1}{2} + \sqrt{\frac{1}{4} + \frac{14}{m} - 1} = -\frac{1}{2} + \sqrt{\frac{56 - 3m}{4m}}.$$

Now $0 < q < 1 \Rightarrow 5 \leq m \leq 13$. Note also that, $\dfrac{56 - 3m}{4m}$ is a square of a rational number, so $m = 8$ and $q = \dfrac{1}{2}$.

Example 6. (CMC/2008) Let $x_1, x_2, x_3, \ldots$ be distinct positive real numbers. Prove that $\{x_n\}$ is a G.P. if and only if for all integers $n \geq 2$

$$\frac{x_1}{x_2} \sum_{k=1}^{n-1} \frac{x_n^2}{x_k x_{k+1}} = \frac{x_n^2 - x_1^2}{x_2^2 - x_1^2}. \tag{$*$}$$

Solution *Necessity:* When $\{x_n\}$ is a G.P., let $x_k = ar^{k-1}, k \in \mathbb{N}$, then

$$\frac{x_1}{x_2} \sum_{k=1}^{n-1} \frac{x_n^2}{x_k x_{k+1}} = \frac{r^{2(n-1)}}{r} \sum_{k=1}^{n-1} \frac{1}{r^{2k-1}} = 1 + r^2 + \cdots + r^{2(n-2)}$$

$$= \frac{r^{2(n-1)} - 1}{r^2 - 1} = \frac{x_n^2 - x_1^2}{x_2^2 - x_1^2} \Rightarrow (*) \text{ holds.}$$

Sufficiency: When $(*)$ holds, we prove that the finite sequence $x_1, x_2, \ldots, x_n$ has a common ratio for each $n \geq 2$ by induction.

For $n = 2$, the both sides of $(*)$ are equal to 1. When $n = 3$, $(*)$ gives

$$\frac{x_1}{x_2} \left(\frac{x_3^2}{x_1 x_2} + \frac{x_3^2}{x_2 x_3} \right) = \frac{x_3^2 - x_1^2}{x_2^2 - x_1^2} \Rightarrow x_1 x_3 = x_2^2 \Rightarrow \frac{x_3}{x_2} = \frac{x_2}{x_1},$$

so x_1, x_2, x_3 is a G.P.. Assume that $x_1, x_2, \ldots, x_{n-1}$ $(n \geq 4)$ is a G.P.. Write $x_k = ar^{k-1}, k = 1, 2, \ldots, n-1$, and write $x_n = au_n$. Then $(*)$ yields

$$\frac{u_n^2}{r}\left(\frac{1}{r} + \frac{1}{r^3} + \cdots + \frac{1}{r^{2n-5}} + \frac{1}{r^{n-2}u_n}\right) = \frac{u_n^2 - 1}{r^2 - 1}.$$

Multiplying its two sides by $(r^2 - 1)r^{2n-4}$ yields

$$[u_n^2(1 + r^2 + r^4 + \cdots + r^{2n-6}) + r^{n-3}u_n](r^2 - 1) = (u_n^2 - 1)r^{2n-4},$$

hence $u_n^2 - (r^{n-1} - r^{n-3})u_n - r^{2n-4} = 0 \Rightarrow (u_n - r^{n-1})(u_n + r^{n-3}) = 0$. Since $u_n > 0$, we must have $u_n = r^{n-1}$, namely $x_n = ar^{n-1}$. Thus, $x_1, x_2, x_3, \ldots, x_n$ is a G.P. as well. The inductive proof is completed. By induction, $x_1, x_2, \ldots, x_n, \ldots$ is a G.P..

Example 7. (CMC/2009) Given that a_1, a_2, a_3 is an A.P. in that order, $a_1 + a_2 + a_3 = 15$; b_1, b_2, b_3 is a G.P. in that order, and $b_1 b_2 b_3 = 27$. If $a_1 + b_1, a_2 + b_2, a_3 + b_3$ are positive integers and form a G.P. in that order, find the maximum possible value of a_3.

Solution   Let $a_1 = 5 - d, a_2 = 5, a_3 = 5 + d, b_1 = \dfrac{3}{q}, b_2 = 3, b_3 = 3q$,

then the given conditions indicate that $5 - d + \dfrac{3}{q}$ and $5 + d + 3q$ are all positive integers and

$$\left(5 - d + \frac{3}{q}\right)(5 + d + 3q) = 64.$$

It is easy to check that for getting maximum positive d, there are only four possibilities for $\left(5 - d + \dfrac{3}{q}, 5 + d + 3q\right)$: $(1, 64)$, $(2, 32)$, $(4, 16)$ and $(8, 8)$.

(i)   By solving the system of equations corresponding to $(1, 64)$, it follows that

$$3q + \frac{3}{q} = 55 \Rightarrow q_{1,2} = \frac{55 \pm 7\sqrt{61}}{6}, d_{1,2} = 4 + \frac{3}{q_{1,2}},$$

$$\Rightarrow d_{\max} = 4 + \frac{3 \cdot 6}{55 - 7\sqrt{61}} = 4 + \frac{55 + 7\sqrt{61}}{2}.$$

(ii)   By solving the system of equations corresponding to $(2, 32)$, it follows that

$$q + \frac{1}{q} = 8 \Rightarrow q_{1,2} = \frac{8 \pm \sqrt{60}}{2} = 4 \pm \sqrt{15}, d_{1,2} = 3 + \frac{3}{q_{1,2}},$$

$$\Rightarrow d_{\max} = 3 + \frac{3}{4 - \sqrt{15}} = 15 + 3\sqrt{15}.$$

(iii)   By solving the system of equations corresponding to $(4, 16)$, it follows that

$$3q + \frac{3}{q} = 10 \Rightarrow q = 3 \text{ or } \frac{1}{3}, d_{1,2} = 1 + \frac{3}{q_{1,2}} = 2 \text{ or } 10 \Rightarrow d_{\max} = 10.$$

(iv)   By solving the system of equations corresponding to $(8, 8)$, it follows that

$$q + \frac{1}{q} = 2 \Rightarrow q = 1, d = 0.$$

In summary, $d_{\max} = 4 + \dfrac{55 + 7\sqrt{61}}{2}$, hence the maximum possible value of a_3 is $9 + \dfrac{55 + 7\sqrt{61}}{2}$.

Example 8. (CHINA/2008) $\{a_n, n = 1, 2, 3, \ldots\}$ is a sequence of real numbers. We arrange the terms in the following form

$$
\begin{array}{llll}
a_1 & & & \\
a_2 & a_3 & & \\
a_4 & a_5 & a_6 & \\
a_7 & a_8 & a_9 & a_{10} \\
\multicolumn{4}{c}{\cdots}
\end{array}
$$

Let $\{b_n\}$ denote the sequence formed by $a_1, a_2, a_4, a_7, \ldots$, where $b_1 = a_1 = 1$. Assume that its partial sum $S_n = \displaystyle\sum_{k=1}^{n} b_k$ satisfies $\dfrac{2b_n}{b_n S_n - S_n^2} = 1$ for $n \geq 2$.

(i)   Prove that the sequence $\{1/S_n\}$ is a A.P., and find the formula for b_n in terms of n.

(ii)   If in above table, starting from the third row, each row (from left to right) is a G.P., and their common ratio are all equal to thee same positive numbers, find the sum of all numbers in the kth row if $a_{81} = -\dfrac{4}{91}$.

Solution   (i)   For $n \geq 2$, $\dfrac{2b_n}{b_n S_n - S_n^2} = 1 \Rightarrow \dfrac{2(S_n - S_{n-1})}{(S_n - S_{n-1})S_n - S_n^2} = 1$

$\Rightarrow \dfrac{2(S_n - S_{n-1})}{-S_{n-1}S_n} = 1 \Rightarrow \dfrac{1}{S_n} - \dfrac{1}{S_{n-1}} = \dfrac{1}{2}$. Since $S_1 = b_1 = a_1 = 1$, so $\{1/S_n\}$ is an arithmetic progression with initial term 1 and common ratio $\frac{1}{2}$. Thus,

$$\frac{1}{S_n} = 1 + \frac{1}{2}(n-1) = \frac{1}{2}(n+1) \text{ or } S_n = \frac{2}{n+1} \text{ for } n = 1, 2, \ldots, \text{ hence}$$

$$b_n = S_n - S_{n-1} = \frac{2}{n+1} - \frac{2}{n} = -\frac{2}{n(n+1)} \text{ for } n \geq 2.$$

In summary, $b_n = 1$ for $n = 1$ and $b_n = -\dfrac{2}{n(n+1)}$ for $n \geq 2$.

(ii) Suppose that, starting from the third row of the table, each row has the common ratio $q > 0$. On the first 12 rows of the table, there are a total of $1 + 2 + \cdots + 12 = 78$ terms of $\{a_n\}$, so a_{81} is on the 13th row and third column, hence

$$a_{81} = b_{13}q^2 = -\frac{4}{91} \Rightarrow q^2 = \frac{4}{91} \cdot \frac{13 \cdot 14}{2} = 4 \Rightarrow q = 2.$$

Thus, the sum of all numbers on the k th row is given by

$$\frac{b_k(1 - q^k)}{1 - q} = -\frac{2}{k(k+1)} \cdot \frac{1 - 2^k}{1 - 2} = \frac{2(1 - 2^k)}{k(k+1)}, \quad \text{for } k \geq 3.$$

Testing Questions   (A)

1. (CMC/2010) Let S_n be sum of the first n terms of an A.P. $\{a_n\}$. If $S_6 = S_9$, find the ratio $a_3 : a_5$.

    (A) $9 : 5$ (B) $5 : 9$ (C) $3 : 5$ (D) $5 : 3$.

2. (CMC/2009) If in the arithmetic progression $\{a_n\}$, $a_1 > 0$, $3a_8 = 5a_{13}$, then among the following listed partial sums the largest one is

    (A) S_{10},   (B) S_{11},   (C) S_{20},   (D) S_{21}.

3. (CMC/2009) It is given that for the sequence $\{a_n\}$, its sum of first n terms $S_n = n^2 + 3n + 4, n = 1, 2, \ldots$. Find $a_1 + a_3 + a_5 + \cdots + a_{21}$.

4. (AUSTIA/2005) Given that each term of an A.P. $\{a_n\}$ with $a_n = a_0 + nd$ is a positive real number. If there are two terms which can expressed as distinct powers of some integer c $(c > 1)$, prove that there are infinitely many terms in the A.P. such that they form a G.P..

5. (CMC/2010) When the three nonzero real numbers $x(y - z)$, $y(z - x)$, $z(y - x)$ form a geometric progression in that order, find its common ratio q.

6. (SLOVENIA/2004) Given that $\{a_n\}$ is an infinite geometric progression of positive integers, where at least two terms are not divisible by 4. If one term of the sequence is 2004, find the expression of a_n in terms of n.

7. (ROMANIA/2010) Let $\{a_n\}$ $(n \geq 0)$ be a sequence of positive real numbers satisfying $\displaystyle\sum_{k=0}^{n} \binom{n}{k} a_k a_{n-k} = a_n^2$. Prove that $\{a_n\}$ is a geometric progression.

Testing Questions   (B)

1. (CMC/2010) Given that each term of the sequence $\{a_n\}$ is not zero, S_n is the sum of its first n terms, and $(1 - p)S_n = p - pa_n$ for any $n \in \mathbb{N}$, where p is a constant greater than 1. Let

$$f(n) = \frac{1 + \binom{n}{1}a_1 + \binom{n}{2}a_2 + \cdots + \binom{n}{n}a_n}{2^n S_n}, \quad n \in \mathbb{N}.$$

   (i) Compare the sizes of $f(n + 1)$ and $\dfrac{p + 1}{2p} f(n)$ for $n \in \mathbb{N}$.

   (ii) Prove   that $(2n - 1)f(n) \leq \sum\limits_{k=1}^{2n-1} f(k) \leq$

$$\frac{p + 1}{p - 1}\left[1 - \left(\frac{p + 1}{2p}\right)^{2n-1}\right] \text{ for } n \in \mathbb{N}.$$

2. (CHINA/2009) Let S_n be the sum of the first n terms of the sequence $\{a_n\}$. $a_n = 5S_n + 1$ holds for any $n \in \mathbb{N}$. Let $b_n = \dfrac{4 + a_n}{1 - a_n}, n \in \mathbb{N}$.

   (i) Find the formulas for a_n and b_n in terms of n.

   (ii) Let R_n denote the sum of first n terms of $\{b_n\}$. Determine if there exists a positive integer k such that $R_k \geq 4k$? Find a such k if so, otherwise show why k does not exist?

   (iii) Define $c_n = b_{2n} - b_{2n-1}, n \in \mathbb{N}$. Let T_n be the sum of first n terms of $\{c_n\}$. Prove that $T_n < 3/2$ for any $n \in \mathbb{N}$.

3. (APMO/2009) Prove that for any positive integer k, there exists an arithmetic sequence

$$\frac{a_1}{b_1}, \quad \frac{a_2}{b_2}, \quad \cdots, \quad \frac{a_k}{b_k}$$

   of rational numbers, where a_i, b_i are relatively prime positive integers for each $i = 1, 2, \ldots, k$, such that the positive integers $a_1, b_1, a_2, b_2, \ldots, a_k, b_k$ are all distinct.

4. (CHNMO/TST/2009) Prove that in an arithmetic progression consisting of distinct 40 positive integers, there is at least one term that cannot be expressed in the form $2^k + 3^l$, where k, l are nonnegative integers.

Lecture 18

Recursive Sequences

Definition 18.1. A sequence $\{a_n, n \in \mathbb{N}\}$ is said to be a **recursive sequence**, if starting from some term each term is determined by several of the previous terms, i.e., it can be expressed as a function of several previous terms:

$$a_{n+m} = f(a_n, a_{n+1} \ldots, a_{n+m-1}), \qquad \text{for } n \geq n_0.$$

f is said to be the **recursive formula**.

Recursive sequences are very widespread, and its investigation has important theoretical and applied meaning in various fields of science and applications. However, in this lecture we can only introduce some basic types of recursive sequences and their related recursive formulas.

Type I: $a_{n+1} = pa_n + q, n \geq 1, (p \neq 0, 1)$.

Letting $a = \dfrac{q}{1-p}$ yields $a_{n+1} - a = p(a_n - a)$, so the sequence $\{b_n\}$, where $b_n = a_n - a$ for $n = 1, 2, \ldots$, is a geometric progression with initial term $b_1 = a_1 - a$ and common ratio p. From $b_n = b_1 p^{n-1}$ we have

$$a_n = a + (a_1 - a)p^{n-1} = a_1 p^{n-1} + \frac{q}{1-p}(1 - p^{n-1}), \quad n = 1, 2, \ldots.$$

$$(18.1)$$

Type II: $a_{n+1} = pa_n + q_n, n \geq 1, (p \neq 0)$.

Letting $b_n = \dfrac{a_n}{p^n}$ for $n = 1, 2, \ldots$ yields $b_{n+1} = b_n + r_n$, where $r_n = \dfrac{q_n}{p^{n+1}}$, therefore $b_{n+1} = b_1 + \displaystyle\sum_{k=1}^{n} r_k$ which yields

19

$$a_{n+1} = a_1 \cdot p^n + p^{n+1} \sum_{k=1}^{n} r_k, \quad n = 1, 2, \ldots . \qquad (18.2)$$

Type III: $a_{n+1} = p_n a_n, n \geq 1, (p_n \neq 0).$

By using the recursive formula repeatedly n times, we obtain

$$a_{n+1} = p_n p_{n-1} \cdots p_1 a_1, \quad n = 1, 2, \ldots . \qquad (18.3)$$

All the models discussed above are first order recursive models. Below we discuss a second order recursive model.

Type IV: $a_{n+1} = p a_n + q a_{n-1}, n \geq 2, (q \neq 0).$

We write the given recursive formula into the following new form:

$$a_{n+1} - \alpha a_n = \beta(a_n - \alpha a_{n-1}), \qquad (18.4)$$

and then determine α and β. (4) gives $a_{n+1} = (\alpha + \beta)a_n - \alpha\beta a_{n-1}$, so

$$\begin{cases} \alpha + \beta &= p, \\ \alpha\beta &= -q. \end{cases}$$

Thus, α, β are the two roots of the quadratic equation $t^2 - pt - q = 0$, which is called the **Characteristic Equation** of the given recursive formula.

Theorem I. *For a sequence $\{a_n\}$ given by*

$$a_{n+1} = p a_n + q a_{n-1}, \quad n = 2, 3, 4, \cdots,$$

where a_1, a_2 are given as the initial values, if its characteristic equation $t^2 = pt + q$ has two real roots α and β, then for $n \geq 2$

(i) $a_n = A\alpha^n + B\beta^n$ *if $\alpha \neq \beta$;*

(ii) $a_n = (An + B)\alpha^n$ *if $\alpha = \beta$,*

where A, B are constants determined by the initial values a_1 and a_2.

Proof.   See the Appendix A (on page 257). $\qquad\qquad\qquad\qquad\qquad$ □

Type V: $\quad a_{n+1} = \dfrac{pa_n}{qa_n + r}, n \geq 1, (a_1 > 0, pqr \neq 0).$

The substitution $b_n = \dfrac{1}{a_n}$ yields the model $b_{n+1} = \dfrac{r}{p}b_n + \dfrac{q}{p}.$

Type VI: $\quad a_{n+1} = \dfrac{pa_n + q}{ra_n + s}, n \geq 1, (pqrs \neq 0).$

Let λ be a real root of the equation $x = \dfrac{px + q}{rx + s}$ and $b_n = a_n - \lambda$

for all $n \in \mathbb{N}$, then $b_{n+1} = \dfrac{(p - r\lambda)b_n}{rb_n + (r\lambda + s)}.$

Besides the above basic types, there are more complicated non-linear recursive sequences. Appropriate transformations are useful tools to convert them to simpler problems, and the other tools for dealing with sequences, like mathematical induction, telescopic sum etc. are often used. The readers can find examples in Testing question (A) and (B).

Examples

Example 1. (CMC/2010) Given that the initial term of $\{a_n\}$ is $a_1 = 4$, and the sum of first n terms is S_n satisfies $S_{n+1} - 3S_n - 2n - 4 = 0$ for $n \in \mathbb{N}$.
(1)   Find the expression of a_n in terms of n.
(2)   Let $f_n(x) = a_n + 2a_{n-1}x + \cdots + na_1x^{n-1}$ and $b_n = f_n(1)$. Find the expression of b_n in terms of n, and prove that $\{b_n\}$ is monotone sequence.

Solution (1) For $n \geq 2$, the difference of $S_{n+1} = 3S_n + 2n + 4$ and $S_n = 3S_{n-1} + 2(n-1) + 4$ gives

$$a_{n+1} = 3a_n + 2 \Rightarrow a_{n+1} + 1 = 3(a_n + 1), \quad n \geq 2.$$

Since $a_1 + 1 = 5$, the sequence $\{a_n + 1, n \geq 1\}$ is a G.P. of initial term 5 and common ratio $r = 3$ hence $a_n = 5 \cdot 3^{n-1} - 1, n \geq 1.$

(2) $b_n = f_n(1) = \displaystyle\sum_{k=1}^{n} ka_k = \sum_{k=1}^{n} k(5 \cdot 3^{n-k} - 1) = 5\sum_{k=1}^{n} k \cdot 3^{n-k},$

$-\dfrac{n(n+1)}{2} = 5S - \dfrac{n(n+1)}{2}$, where $S = \displaystyle\sum_{k=1}^{n} k \cdot 3^{n-k}$. Then

$3S \quad = \quad 3^n + 2 \cdot 3^{n-1} + \cdots + a_n \cdot 3 \Rightarrow 2S = 3^n + 3^{n-1} + \cdots + 3 - n$

$\Rightarrow \quad S = \dfrac{3^{n+1} - 3}{4} - \dfrac{n}{2} \Rightarrow b_n = \dfrac{5 \cdot 3^{n+1} - 15}{4} - \dfrac{n(n+6)}{2}.$

Hence

$$
\begin{aligned}
b_{n+1} - b_n &= \frac{5 \cdot 3^{n+2} - 15}{4} - \frac{(n+1)(n+7)}{2} - \frac{5 \cdot 3^{n+1} - 15}{4} + \frac{n(n+6)}{2} \\
&= \frac{15 \cdot 3^n}{2} - n - \frac{7}{2} > 0,
\end{aligned}
$$

i.e., the sequence $\{b_n, n \geq 1\}$ is monotonically increasing.

Example 2. (CHINA/2009) It is known that the sum of the first n terms of $\{a_n\}$ is $S_n = -a_n - \left(\frac{1}{2}\right)^{n-1} + 2, n \in \mathbb{N}$. Find the expression of a_n in terms of n.

Solution   Letting $n = 1$ in the given equality yields $a_1 = -a_1 - 1 + 2$, so $a_1 = \frac{1}{2}$. When $n \geq 2$,

$$
\begin{aligned}
S_n - S_{n-1} &= -a_n - \left(\frac{1}{2}\right)^{n-1} + 2 + a_{n-1} + \left(\frac{1}{2}\right)^{n-2} - 2 \\
&\Rightarrow a_n = -a_n + a_{n-1} + \left(\frac{1}{2}\right)^{n-1} \Rightarrow a_n = \frac{1}{2}a_{n-1} + \frac{1}{2^n}.
\end{aligned}
$$

Multiplying the both sides of the last equality by 2^n leads to

$$
2^n a_n = 2^{n-1} a_{n-1} + 1, \qquad n \geq 2,
$$

so $\{2^n a_n, n \geq 1\}$ is an A.P. with initial term and common difference both being 1, hence

$$
2^n a_n = n \quad \text{or} \quad a_n = \frac{n}{2^n}, \qquad n \in \mathbb{N}.
$$

Example 3. (SSSMO/2006) Suppose $x_0, x_1, x_2, \cdots$ is a sequence of numbers such that $x_0 = 1000$, and

$$
x_n = -\frac{1000}{n}(x_0 + x_1 + x_2 + \cdots + x_{n-1})
$$

for all $n \geq 1$. Find the value of

$$
\frac{1}{2^2}x_0 + \frac{1}{2}x_1 + x_2 + 2x_3 + 2^2 x_4 + \cdots + 2^{997} x_{999} + 2^{998} x_{1000}.
$$

Solution   For $n \geq 1$,

$$
x_n = -\frac{x_0}{n}\left[-\frac{(n-1)}{x_0}x_{n-1} + x_{n-1}\right] = -\frac{x_0 - (n-1)}{n}x_{n-1}.
$$

By using this reduction formula repeatedly, we obtain that for $1 \le n \le 1000$,

$$x_n = \left(-\frac{x_0 - (n-1)}{n}\right) x_{n-1} = \left(\frac{x_0 - (n-1)}{n}\right)\left(\frac{x_0 - (n-2)}{n-1}\right) x_{n-2}$$

$$= \cdots = (-1)^n \frac{(x_0 - (n-1))(x_0 - (n-2)) \cdots \cdot x_0}{n!} x_0 = (-1)^n \binom{1000}{n} x_0.$$

Now define the polynomial $f(x)$ for real x by

$$f(x) = x_0 + x_1 x + x_2 x^2 + \cdots + x_{1000} x^{1000}.$$

Then $f(x) = (1-x)^{1000} x_0$, hence

$$\frac{1}{2^2} x_0 + \frac{1}{2} x_1 + x_2 + 2x_3 + 2^2 x_4 + \cdots + 2^{997} x_{999} + 2^{998} x_{1000} = \frac{1}{4} f(2)$$

$$= \frac{1}{4} x_0 = 250.$$

Example 4. (CHINA/2009) Given that p, q are real numbers and the equation $x^2 - px + q = 0$ has two real roots α and β, and that the sequence $\{a_n\}$ satisfies $a_1 = p, a_2 = p^2 - q$ and $a_n = pa_{n-1} - qa_{n-2}, n = 3, 4, \ldots$.
(1)   Find the expression of a_n, in terms of α, β.
(2)   Find the sum of first n terms of $\{a_n\}$ when $p = 1, q = \frac{1}{4}$.

Solution   (1)   By Viete's Theorem, $\alpha + \beta = p, \alpha \cdot \beta = q$.
(i)   When $\alpha \ne \beta$, namely $p^2 - 4q > 0$, then

$$a_n = A\alpha^n + B\beta^n \Rightarrow p = \alpha A + \beta B, p^2 - q = \alpha^2 A + \beta^2 B.$$

By solving the system, it is obtained that $A = -\frac{\alpha}{\beta - \alpha}, B = \frac{\beta}{\beta - \alpha}$.
Therefore

$$a_n = \frac{1}{\beta - \alpha}(-\alpha^{n+1} + \beta^{n+1}).$$

(ii)   When $\alpha = \beta$, namely $p^2 = 4q$, then

$$a_n = (A + Bn)\alpha^n \Rightarrow p = (A + B)\alpha, p^2 - q = (A + 2B)\alpha^2$$
$$\Rightarrow A = B = 1 \Rightarrow a_n = (1 + n)\alpha^n, n \ge 1.$$

(2)   When $p = 1, q = \frac{1}{4}$, then $\alpha = \beta = \frac{1}{2}, a_n = \frac{1+n}{2^n}$ for $n \ge 1$, therefore

$$S_n = \sum_{k=1}^{n} \frac{1+k}{2^k} = \frac{2}{2^1} + \frac{3}{2^2} + \frac{4}{2^3} + \cdots + \frac{n+1}{2^n}.$$

Then

$$\frac{1}{2}S_n = S_n - \frac{1}{2}S_n = 1 + \sum_{k=2}^{n}\frac{1}{2^k} - \frac{n+1}{2^{n+1}} = 1 + \left(\frac{1}{2} - \frac{1}{2^n}\right) - \frac{n+1}{2^{n+1}} = \frac{3}{2} - \frac{n+3}{2^{n+1}},$$

hence $S_n = 3 - \dfrac{n+3}{2^n}, n \geq 1.$

Example 5. (CMC/2010) In the sequence $\{a_n\}$, $a_1 = 1, a_2 = \dfrac{1}{4}$ and $a_{n+1} = \dfrac{(n-1)a_n}{n-a_n}$ for $n = 2, 3, \ldots$.

    (1)   Find the expression of a_n in terms of n.

    (2)   Prove that $\displaystyle\sum_{k=1}^{n} a_k^2 < \frac{7}{6}$ for $n \geq 1$.

    Solution   (1)   The given recursive relation gives

$$\frac{1}{a_{n+1}} = \frac{n - a_n}{(n-1)a_n} = \frac{n}{(n-1)a_n} - \frac{1}{n-1}.$$

Dividing both sides by n and simplifying it, then

$$\frac{1}{na_{n+1}} - \frac{1}{(n-1)a_n} = -\left(\frac{1}{n-1} - \frac{1}{n}\right).$$

Hence

$$\frac{1}{(n-1)a_n} - \frac{1}{a_2} = \sum_{k=2}^{n-1}\left[\frac{1}{ka_{k+1}} - \frac{1}{(k-1)a_k}\right] = -\sum_{k=2}^{n-1}\left(\frac{1}{k-1} - \frac{1}{k}\right)$$

$$= -\left(1 - \frac{1}{n-1}\right), \quad \text{for all } n \geq 2,$$

therefore

$$\frac{1}{(n-1)a_n} = 3 + \frac{1}{n-1} = \frac{3n-2}{n-1} \Rightarrow a_n = \frac{1}{3n-2}, \quad n \geq 2.$$

Considering $a_1 = 1 = \dfrac{1}{3-2}$, it is obtained that $a_n = \dfrac{1}{3n-2}$ for $n \geq 1$.

(2) For $k \geq 2$, $a_k^2 = \dfrac{1}{(3k-2)^2} < \dfrac{1}{(3k-4)(3k-1)}$, therefore for $n \geq 2$

$$\sum_{k=1}^{n} a_k^2 < 1 + \sum_{k=2}^{n} \frac{1}{(3k-4)(3k-1)} = 1 + \frac{1}{3}\sum_{k=2}^{n}\left(\frac{1}{3k-4} - \frac{1}{3k-1}\right)$$

$$= 1 + \frac{1}{3}\left[\left(\frac{1}{2} - \frac{1}{5}\right) + \left(\frac{1}{5} - \frac{1}{8}\right) + \cdots + \left(\frac{1}{3n-4} - \frac{1}{3n-1}\right)\right]$$

$$= 1 + \frac{1}{3}\left(\frac{1}{2} - \frac{1}{3n-1}\right) < 1 + \frac{1}{6} = \frac{7}{6}.$$

Since $a_1^2 = 1 < \dfrac{7}{6}$, so $\displaystyle\sum_{k=1}^{n} a_k^2 < \frac{7}{6}$ for $n \geq 1$.

Example 6. (CHINA/2007) In the sequence $\{b_n\}$, $b_1 = 2, b_{n+1} = \dfrac{3b_n + 4}{2b_n + 3}, n \in$
$\mathbb{N}$, find an expression of b_n in terms of n.

Solution First of all solve the equation $\lambda = \dfrac{3\lambda + 4}{2\lambda + 3}$.

$$\lambda = \frac{3\lambda + 4}{2\lambda + 3} \Rightarrow 2\lambda^2 + 3\lambda = 3\lambda + 4 \Rightarrow \lambda = \pm\sqrt{2}.$$

Then $\dfrac{b_{n+1} - \sqrt{2}}{b_{n+1} + \sqrt{2}} = \dfrac{3b_n + 4 - \sqrt{2}(2b_n + 3)}{3b_n + 4 + \sqrt{2}(2b_n + 3)} = \dfrac{(\sqrt{2}-1)^2}{(\sqrt{2}+1)^2} \cdot \dfrac{b_n - \sqrt{2}}{b_n + \sqrt{2}}$. Let
$\dfrac{b_n - \sqrt{2}}{b_n + \sqrt{2}} = c_n, n \in \mathbb{N}$, then $c_{n+1} = (\sqrt{2}-1)^4 c_n$ and $c_1 = \dfrac{2 - \sqrt{2}}{2 + \sqrt{2}} =$
$(\sqrt{2}-1)^2$. Hence $c_n = (\sqrt{2}-1)^2(\sqrt{2}-1)^{4(n-1)} = (\sqrt{2}-1)^{4n-2}$. Then

$$\frac{b_n - \sqrt{2}}{b_n + \sqrt{2}} = c_n \Rightarrow \frac{b_n}{\sqrt{2}} = \frac{1 + c_n}{1 - c_n} \Rightarrow b_n = \sqrt{2}\cdot\frac{1 + (\sqrt{2}-1)^{4n-2}}{1 - (\sqrt{2}-1)^{4n-2}}.$$

Example 7. In the sequence $\{a_n\}$, $a_0 = x, a_1 = y, a_{n+1} = \dfrac{a_n a_{n-1} + 1}{a_n + a_{n-1}}$.
(1)   Find x, y such that there exists positive integer n_0 satisfying the condition that a_n is a constant for all $n \geq n_0$.
(2)   Find an expression for a_n in terms of n.

Solution The equality

$$a_{n+1} - a_n = \frac{a_n a_{n-1} + 1}{a_n + a_{n-1}} - a_n = \frac{1 - a_n^2}{a_n + a_{n-1}}, n \in \mathbb{N} \qquad (18.1)$$

implies that $a_{n+1} = a_n$ for some n if and only if $a_n^2 = 1$ and $a_n + a_{n-1} \neq 0$. Applying it to $n = 1$, then

$$|y| = 1 \quad \text{and} \quad x \neq -y. \tag{18.2}$$

For $n > 1$,

$$a_n - 1 = \frac{a_{n-1}a_{n-2} + 1}{a_{n-1} + a_{n-2}} - 1 = \frac{(a_{n-1} - 1)(a_{n-2} - 1)}{a_{n-1} + a_{n-2}}, \tag{18.3}$$

$$a_n + 1 = \frac{a_{n-1}a_{n-2} + 1}{a_{n-1} + a_{n-2}} + 1 = \frac{(a_{n-1} + 1)(a_{n-2} + 1)}{a_{n-1} + a_{n-2}}. \tag{18.4}$$

By multiplying the previous two equations, we obtain

$$a_n^2 - 1 = \frac{a_{n-1}^2 - 1}{a_{n-1} + a_{n-2}} \cdot \frac{a_{n-2}^2 - 1}{a_{n-1} + a_{n-2}}. \tag{18.5}$$

By repeatedly using (18.5), it is obtained that (18.2) holds or

$$|x| = 1 \qquad \text{and} \qquad y \neq -x. \tag{18.6}$$

Conversely, if (18.2) or (18.6) holds, then $a_n = $ constant for $n \geq 2$, and the constant is 1 or -1.

(2)   Combining (18.3) and (18.4) yields

$$\frac{a_n - 1}{a_n + 1} = \frac{a_{n-1} - 1}{a_{n-1} + 1} \cdot \frac{a_{n-2} - 1}{a_{n-2} + 1} \quad \text{for } n \geq 2. \tag{18.7}$$

Let $b_n = \dfrac{a_n - 1}{a_n + 1}, n \in \mathbb{N}$, then for $n \geq 2$

$$b_n = b_{n-1}b_{n-2} = b_{n-2}^2 b_{n-3} = (b_{n-3})^3 (b_{n-4})^2 = \cdots$$

$$= \left(\frac{y - 1}{y + 1}\right)^{F_{n-1}} \cdot \left(\frac{x - 1}{x + 1}\right)^{F_{n-2}}, \tag{18.8}$$

here $F_0 = F_1 = 1$ and $F_n = F_{n-1} + F_{n-2}$ for $n \geq 2$. In fact, it is easy to get

$$F_n = \frac{1}{\sqrt{5}} \left[\left(\frac{1 + \sqrt{5}}{2}\right)^{n+1} - \left(\frac{1 - \sqrt{5}}{2}\right)^{n+1} \right]. \tag{18.9}$$

By defining $F_{-1} = 0, F_0 = 1$, then (18.8) holds for $n \geq 1$. From (18.8) a_n can be solved in the form

$$a_n = \frac{(x + 1)^{F_{n-2}}(y + 1)^{F_{n-1}} + (x - 1)^{F_{n-2}}(y - 1)^{F_{n-1}}}{(x + 1)^{F_{n-2}}(y + 1)^{F_{n-1}} - (x - 1)^{F_{n-2}}(y - 1)^{F_{n-1}}}. \tag{18.10}$$

Testing Questions (A)

1. (CMC/2009) In the sequence $\{a_n\}$, $a_1 = 1, a_{n+1} = 1 + a_n + \sqrt{1 + 4a_n}$ $(n \in \mathbb{N})$. Find a_n in terms of n.

2. (CMC/2010) It is given that the sum S_n of the first n terms of the sequence $\{a_n\}$ satisfies $(1 - b)S_n = -ba_n + 4^n$, $n \in \mathbb{N}$, where $b > 0$.

   (i) Find a_n in terms of n;

   (ii) Let $c_n = \dfrac{a_n}{4^n}$ $(n \in \mathbb{N})$. If $|c_n| \le 2$, find the range of b.

3. (CMC/2009) If the sequence $\{a_n\}$ satisfies $a_1 = 1, 2^n(a_{n+1} - a_n) = a_n$, prove that $2 - \dfrac{1}{2^{n-1}} < a_n < \left(\dfrac{3}{2}\right)^{n-1}$ for $n \ge 3$.

4. (CMC/2009) $\{a_n\}$ is a sequence of positive numbers, satisfying $a_1 = 1, a_2 = 2, a_n = \dfrac{a_{n-2}}{a_{n-1}}, n \ge 3$. Express a_n in terms of n.

5. (CMC/2009) In $\{a_n\}$, $a_1 = 2, a_{n+1} = 1 - \dfrac{1}{a_n}$ for $n \ge 1$. Let P_n be the product of its first n terms, then the value of P_{2009} is

   (A) $-\dfrac{1}{2}$ (B) -1 (C) $\dfrac{1}{2}$ (D) 1.

6. (CMC/2009) Given that the sequence $\{a_n\}$ satisfies $a_1 = 1, a_2 = 2, \dfrac{a_{n+2}}{a_n} = \dfrac{a_{n+1}^2 + 1}{a_n^2 + 1}$ for $n \ge 1$.

   (i) Find the recursive formula f such that $a_{n+1} = f(a_n)$;

   (ii) Prove that $63 < a_{2008} < 78$.

7. (CMC/2010) The sequence $\{a_n\}$ is given by $a_1 = 3, a_{n+1} = a_n^2 + a_n - 1$ for $n \ge 1$. Prove that

   (i) $a_n \equiv 3 \pmod 4$ for all $n \in \mathbb{N}$.

   (ii) $(a_m, a_n) = 1$ if $m \ne n$.

8. (CMC/2010) The sequence $\{a_n\}$ is given by $a_1 = 1, a_2 = 6, 4a_{n-1} + a_{n+1} = 4a_n, n \ge 2$.

   (i) Find a_n in terms of n.

   (ii) Find the sum of the first terms S_n of $\{a_n\}$.

Testing Questions   (B)

1. (CMC/2010) $\{a_n\}$ is a sequence satisfying $a_1 = 0, a_n = \dfrac{2}{1 + a_{n-1}}, n \geq 2$. Express a_n in terms of n.

2. (CMC/2010) $\{a_n\}$ is a sequence satisfying $a_1 = \dfrac{12}{11}, a_{n+1} = \dfrac{4a_n}{2a_n + 1}$.

    (1)   Express a_n in terms of n, and prove that

$$a_n \geq \frac{3}{2 + x} - \frac{3}{(2 + x)^2}\left(\frac{3}{4^n} - x\right), \quad \text{for any } x > 0, n \geq 1.$$

    (2)   Prove that $a_1 + a_2 + \cdots + a_n > \dfrac{3n^2}{2n + 1}$.

3. (VIETNAM/2009) Let a sequence $\{x_n\}$ be defined by

$$x_1 = \frac{1}{2}, x_n = \frac{\sqrt{x_{n-1}^2 + 4x_{n-1}} + x_{n-1}}{2}, n \geq 2.$$

    Prove that the sequence defined by $y_n = \displaystyle\sum_{i=1}^{n} \frac{1}{x_i^2}$ converges and find its limit.

4. (TURKEY/TST/2008) The sequence $\{x_n\}$ is defined by $x_1 = a, x_2 = b, x_{n+2} = 2008x_{n+1} - x_n$. Prove that there exist positive integers a, b such that $1 + 2006x_{n+1}x_n$ is a perfect square for all $n \in \mathbb{N}$.

5. (CROATIA/TST/2007) The sequence $\{a_n\}$ is given by $a_0 = 3, a_n = 2 + a_0a_1 \cdots a_{n-1}$ for $n \geq 1$.

    (1)   Prove that any two terms of $\{a_n\}$ are relatively prime;

    (2)   Find a_{2007}.

Lecture 19

Summation of Various Sequences

Summation of sequences is a common tool used in each of branches of mathematical olympiad competitions, so there are various methods for evaluating the summation for sequences.

Besides summations of a partial sum of an A.P., a G.P. and a recursive sequence that we have considered before, in this lecture we will introduce other sequences which appeared in Algebra, Trigonometry, Polynomials, Number Theory and Combinatorics.

However, thus lecture is not comprehensive. We are not going to mention inequalities of summations and complex numbers and complicated combinatorial identities so that the readers in the senior section level can understand more easily.

Examples

Example 1. The sequences $\{a_n\}$ and $\{b_n\}$ satisfy $a_k b_k = 1, k = 1, 2, \ldots$. If the sum of first n terms of $\{a_n\}$ is $A_n = \dfrac{n(n+1)(n+2)}{3}$, find the sum B_n of the first n terms of $\{b_n\}$.

Solution $a_1 = A_1 = \dfrac{1 \cdot 2 \cdot 3}{3} = 2$, and for $n \geq 2$

$$a_n = A_n - A_{n-1} = \frac{n(n+1)(n+2)}{3} - \frac{(n-1)n(n+1)}{3}$$

$$= \frac{n(n+1)[(n+2) - (n-1)]}{3} = n(n+1).$$

Therefore $b_n = \dfrac{1}{a_n} = \dfrac{1}{n(n+1)}$ for $n \geq 1$. Then

$$B_n = \sum_{k=1}^{n} \frac{1}{k(k+1)} = \sum_{k=1}^{n} \left(\frac{1}{k} - \frac{1}{k+1} \right) = 1 - \frac{1}{n+1} = \frac{n}{n+1}$$

for $n = 1, 2, \ldots$.

Example 2. (CMC/2010) Let n be any natural number. Prove the identity

$$\sum_{k=1}^{n} \frac{k}{k^4 + k^2 + 1} = \frac{1}{n^2 + n + 1} \sum_{k=1}^{n} k.$$

Solution   For any natural number n,

$$\sum_{k=1}^{n} \frac{k}{k^4 + k^2 + 1} = \sum_{k=1}^{n} \frac{k}{(k^4 + 2k^2 + 1) - k^2} = \sum_{k=1}^{n} \frac{k}{(k^2 + 1)^2 - k^2}$$

$$= \sum_{k=1}^{n} \frac{k}{(k^2 + 1 - k)(k^2 + 1 + k)} = \frac{1}{2} \sum_{k=1}^{n} \left(\frac{1}{k^2 + 1 - k} - \frac{1}{k^2 + 1 + k} \right)$$

$$= \frac{1}{2} \left(1 - \frac{1}{n^2 + 1 + n} \right) = \frac{1}{2} \cdot \frac{n^2 + n}{n^2 + 1 + n}$$

$$= \frac{1}{n^2 + n + 1} \cdot \frac{n(n + 1)}{2} = \frac{1}{n^2 + n + 1} \sum_{k=1}^{n} k.$$

Example 3. (JAPAN/2009) Evaluate $\dfrac{\displaystyle\sum_{n=1}^{99} \sqrt{10 + \sqrt{n}}}{\displaystyle\sum_{n=1}^{99} \sqrt{10 - \sqrt{n}}}$.

Solution   Let $S = \displaystyle\sum_{n=1}^{99} \sqrt{10 + \sqrt{n}}$, $T = \displaystyle\sum_{n=1}^{99} \sqrt{10 - \sqrt{n}}$. Note that $\sqrt{a + b + 2\sqrt{ab}} = \sqrt{a} + \sqrt{b}$ for positive real numbers a and b. In particular, for each fixed positive integer n with $1 \le n \le 99$, choose $a > b > 0$ satisfying $a + b = 20, ab = n$, then a, b are roots of the equation $x^2 - 20x + n = 0$, i.e.,

$$a = 10 + \sqrt{100 - n}, \qquad b = 10 - \sqrt{100 - n}.$$

So $\sqrt{2}S = \displaystyle\sum_{n=1}^{99} \sqrt{20 + 2\sqrt{n}} = \sum_{n=1}^{99} \left(\sqrt{10 + \sqrt{100 - n}} + \sqrt{10 - \sqrt{100 - n}} \right)$

$$= \sum_{n=1}^{99} \left(\sqrt{10 + \sqrt{n}} + \sqrt{10 - \sqrt{n}} \right) = S + T.$$

Thus, $(\sqrt{2} - 1)S = T \Rightarrow \dfrac{S}{T} = \dfrac{1}{\sqrt{2} - 1} = 1 + \sqrt{2}$.

Example 4. (CMC/2008) In the sequence $\{x_n\}$, $x_1 = 1$ and $x_{n+1} = \dfrac{\sqrt{3}x_n + 1}{\sqrt{3} - x_n}$

for $n \in \mathbb{N}$. Find $\displaystyle\sum_{k=1}^{2008} x_k$.

Solution Note that if we let $x_n = \tan \theta_n$, $n \in \mathbb{N}$, where $\theta_1 = \dfrac{\pi}{4}$, then

$$x_{n+1} = \frac{\sqrt{3}x_n + 1}{\sqrt{3} - x_n} = \frac{x_n + \frac{1}{\sqrt{3}}}{1 - \frac{1}{\sqrt{3}}x_n} = \tan\left(\theta_n + \frac{\pi}{6}\right).$$

Therefore $\theta_{n+1} \equiv \theta_n + \frac{\pi}{6} \pmod{\pi}$ for $n \in \mathbb{N}$, and so $\{x_n\}$ is periodic with a period of length 6.

It is easy to find that

$$x_1 = 1, \ x_2 = 2+\sqrt{3}, \ x_3 = -2-\sqrt{3}, \ x_4 = -1, \ x_5 = -2+\sqrt{3}, \ x_6 = 2-\sqrt{3}.$$

Then $x_1 + x_2 + \cdots + x_5 + x_6 = 0$ and $2008 = 6 \times 334 + 4$ implies that

$$\sum_{k=1}^{2008} x_k = x_1 + x_2 + x_3 + x_4 = 0.$$

Example 5. (CROATIA/2008) Find the simplest expression of

$$S_n = \lfloor\sqrt{1}\rfloor + \lfloor\sqrt{2}\rfloor + \cdots + \lfloor\sqrt{n^2 - 1}\rfloor,$$

in terms of n.

Solution The inequalities $(n-1)^2 < n^2 - 1 < n^2$ implies that $\lfloor\sqrt{n^2 - 1}\rfloor = n - 1$, so the last term of S_n is $n - 1$.

Below we compute the number of terms in S_n that are equal to k for $1 \le k \le n - 1$. Since these terms are

$$\lfloor\sqrt{k^2}\rfloor, \ \lfloor\sqrt{k^2 + 1}\rfloor, \ \cdots, \ \lfloor\sqrt{(k + 1)^2 - 1}\rfloor,$$

so the number is $[(k + 1)^2 - 1] - [k^2 - 1] = (k + 1)^2 - k^2$. Hence,

$$\begin{aligned} S_n &= \sum_{k=1}^{n-1} k[(k + 1)^2 - k^2] = \sum_{k=1}^{n-1}(2k^2 + k) \\ &= \frac{(n - 1)n(2n - 1)}{3} + \frac{n(n - 1)}{2} = \frac{n(n - 1)(4n + 1)}{6}. \end{aligned}$$

Example 6. (AUSTRALIA/2008) Find the remainder of $S = \displaystyle\sum_{i=1}^{2009} i \cdot \binom{2008}{i - 1}$

when it is divided by 2008.

Solution   Note that

$$2S = \sum_{i=1}^{2009}\binom{2008}{i-1} + \sum_{i=1}^{2009}(2010-i)\binom{2008}{i-1} = 2010\sum_{i=1}^{2009}\binom{2008}{i-1}$$

$$= 2010\sum_{i=0}^{2008}\binom{2008}{i} = 2010\times 2^{2008} \Rightarrow S = 2010\cdot 2^{2007}.$$

Thus, $S \equiv 2\cdot 2^{2007} \equiv 2^{2008}$ (mod 2008). By Fermat's Little Theorem, $2^{250} \equiv 1$ (mod 251), so

$$2^{2008} = (2^{250})^8 \cdot 2^8 = 251m + 2^8 \Rightarrow m = 8k \Rightarrow 2^{2008} = 2008k + 256,$$

i.e., the remainder is 256.

Example 7. (APMO/2009) Let a_1, a_2, a_3, a_4, a_5 be real numbers satisfying the following equations:

$$\frac{a_1}{k^2+1} + \frac{a_2}{k^2+2} + \frac{a_3}{k^2+3} + \frac{a_4}{k^2+4} + \frac{a_5}{k^2+5} = \frac{1}{k^2} \quad \text{for } k = 1, 2, 3, 4, 5.$$

Find the value of $\dfrac{a_1}{37} + \dfrac{a_2}{38} + \dfrac{a_3}{39} + \dfrac{a_4}{40} + \dfrac{a_5}{41}$. (Express the value as a single fraction.)

Solution   Let $R(x) := \dfrac{a_1}{x^2+1} + \dfrac{a_2}{x^2+2} + \dfrac{a_3}{x^2+3} + \dfrac{a_4}{x^2+4} + \dfrac{a_5}{x^2+5}.$

Then

$$R(\pm 1) = 1, R(\pm 2) = \frac{1}{4}, R(\pm 3) = \frac{1}{9}, R(\pm 4) = \frac{1}{16}, R(\pm 5) = \frac{1}{25}$$

and $R(6)$ is the value to be found.

Define $P(x) := (x^2+1)(x^2+2)(x^2+3)(x^2+4)(x^2+5)$ and $Q(x) := R(x)P(x)$. Then for $k = \pm 1, \pm 2, \pm 3, \pm 4, \pm 5$, we get $Q(k) = R(k)P(k) = \dfrac{P(k)}{k^2}$, that is, $P(k) - k^2Q(k) = 0$. Since $P(x) - x^2Q(x)$ is a polynomial of degree 10 with roots $\pm 1, \pm 2, \pm 3, \pm 4, \pm 5$, we get

$$P(x) - x^2Q(x) = A(x^2-1)(x^2-4)(x^2-9)(x^2-16)(x^2-25). \qquad (*)$$

Putting $x = 0$, we get $A = \dfrac{P(0)}{(-1)(-4)(-9)(-16)(-25)} = -\dfrac{1}{120}.$

Finally, dividing both sides of $(*)$ by $P(x)$ yields

$$1 - x^2R(x) = 1 - x^2\frac{Q(x)}{P(x)} = -\frac{1}{120}\cdot\frac{(x^2-1)(x^2-4)(x^2-9)(x^2-16)(x^2-25)}{(x^2+1)(x^2+2)(x^2+3)(x^2+4)(x^2+5)}$$

and hence

$$1 - 36R(6) = -\frac{35 \times 32 \times 27 \times 20 \times 11}{120 \times 37 \times 38 \times 39 \times 40 \times 41} = -\frac{3 \times 7 \times 11}{13 \times 19 \times 37 \times 41}$$

$$= -\frac{231}{374699},$$

which implies $R(6) = \frac{1}{6^2}\left[1 + \frac{1}{120}\prod_{k=1}^{5}\frac{(6^2 - k^2)}{(6^2 + k)}\right] = \frac{187465}{6744582}.$

Remark. We can get $a_1 = \dfrac{1105}{72}, a_2 = -\dfrac{2673}{40}, a_3 = \dfrac{1862}{15}, a_4 = -\dfrac{1885}{18}, a_5 = \dfrac{1323}{40}$ by solving the given system of linear equations, which is extremely messy and takes a lot of time.

Example 8. (TURKEY/2008) $f : \mathbb{N} \times \mathbb{Z} \mapsto \mathbb{Z}$ satisfies the given conditions
(a) $f(0, 0) = 1, f(0, 1) = 1;$
(b) $\forall k \notin \{0, 1\}, f(0, k) = 0;$ and
(c) $\forall n \geq 1$ and k, $f(n, k) = f(n - 1, k) + f(n - 1, k - 2n).$

Find the sum $\displaystyle\sum_{k=0}^{\binom{2009}{2}} f(2008, k).$

Solution   First of all by induction on n we prove that $f(n, k) = 0$ for $k < 0$ or $n^2 + n + 1 < k$.

For $n = 0$, the condition (b) gives $f(0, k) = 0$ if $k < 0$ or $1 < k$. Assume that $f(n, k) = 0$ if $k < 0$ or $n^2 + n + 1 < k$, then for $n + 1$, by the condition (c),

$$f(n + 1, k) = f(n, k) + f(n, k - 2(n + 1)) = 0 \ \ \text{if } k < 0 \text{ and}$$

if $(n + 1)^2 + (n + 1) + 1 < k$, then $n^2 + n + 1 < k - 2(n + 1)$, so

$$f(n + 1, k) = f(n, k) + f(n, k - 2(n + 1)) = 0.$$

The inductive proof is completed.

Below by induction on n we show that for $k \in \mathbb{Z}$

$$f(n, n^2 + n + 1 - k) = f(n, k); \tag{19.1}$$

$$\sum_{k=0}^{n^2+n+1} f(n, k) = 2^{n+1}. \tag{19.2}$$

For $n = 0$ the conclusions are clear. Assume that (19.1) and (19.2) are true, then verify that they are true still for $n + 1$.

$$\begin{aligned}
&f(n + 1, (n + 1)^2 + (n + 1) + 1 - k) \\
&= f(n, (n + 1)^2 + (n + 1) + 1 - k) + \\
&\quad + f(n, (n + 1)^2 + (n + 1) + 1 - k - 2(n + 1)) \\
&= f(n, n^2 + n + 1 - (k - 2(n + 1))) + f(n, n^2 + n + 1 - k) \\
&= f(n, k - 2(n + 1)) + f(n, k) = f(n + 1, k).
\end{aligned}$$

So (19.1) is true for $n + 1$. For (19.2),

$$\sum_{k=0}^{(n+1)^2+(n+1)+1} f(n+1, k) = \sum_{k=0}^{(n+1)^2+(n+1)+1} [f(n, k - 2(n+1)) + f(n, k)]$$

$$= \sum_{k=0}^{(n+1)^2+(n+1)+1} f(n, k - 2(n+1)) + \sum_{k=0}^{(n+1)^2+(n+1)+1} f(n, k)$$

$$= \sum_{k=2(n+1)}^{(n+1)^2+(n+1)+1} f(n, k - 2(n+1)) + \sum_{k=0}^{n^2+n+1} f(n, k)$$

$$= \sum_{k=0}^{n^2+n+1} f(n, k) + \sum_{k=0}^{n^2+n+1} f(n, k) = 2 \cdot 2^{n+1} = 2^{n+2}.$$

Thus, (19.2) is true also for $n + 1$. Hence

$$\sum_{k=0}^{\binom{2009}{2}} f(2008, k) = \frac{1}{2} \sum_{k=0}^{2008^2+2008+1} f(2008, k) = \frac{1}{2} \cdot 2^{2009} = 2^{2008}.$$

Example 9. (THAILAND/2006) Evaluate $\displaystyle\sum_{k=84}^{8000} \binom{k}{84} \cdot \binom{8084 - k}{84}$.

Solution Let $S = \{0, 1, 2, \ldots, 8084\}$. Then there are a total of $\binom{8085}{169}$ ways to select $84 + 1 + 84 = 169$ distinct elements from S.

Let k be the median of the chosen 169 numbers. Then the range of the possible values of k is the set $T = \{84, 85, \ldots, 8000\}$.

For each value of k in T, there are $\binom{k}{84}$ ways choose 84 numbers which are less than k, and $\binom{8084-k}{84}$ ways choose 84 numbers more which are greater than k, so that k is the median. Therefore there are a total of $\binom{k}{84} \cdot \binom{8084 - k}{84}$ ways to get the median k.

Thus, $\displaystyle\sum_{k=84}^{8000} \binom{k}{84} \cdot \binom{8084 - k}{84} = \binom{8085}{169}$.

Testing Questions   (A)

1. (SSSMO/2002) Let $s_n = \sum_{k=1}^{n} \dfrac{k}{(k+1)!}$. Find the value of $\dfrac{1 - s_{2001}}{1 - s_{2002}}$.

2. (CMC/2009) In $\{a_n\}$, $a_1 = 4$ and $a_{n+1} = \sqrt{\dfrac{3 + a_n}{2}}$. Let $b_n = |a_{n+1} - a_n|$, $n \in \mathbb{N}$ and $S_n = \sum_{k=1}^{n} b_k$. Prove that $S_n < \dfrac{5}{2}$.

3. (CMC/2009) Given that the two consecutive terms a_{2k-1}, a_{2k} of the sequence $\{a_n\}$ are the roots of the equation $x^2 - (3k + 2^k)x + 3k \cdot 2^k = 0$.
    (i)   Find the sum S_{2n} of the first $2n$ terms of $\{a_n\}$;
    (ii) Let $f(n) = \dfrac{1}{2}\left(\dfrac{|\sin n|}{\sin n} + 3\right)$ and $T_n = \sum_{k=1}^{n} \dfrac{(-1)^{f(k+1)}}{a_{2k-1}a_{2k}}$. Prove that $\dfrac{1}{6} \le T_n \le \dfrac{5}{24}$, where $n \in \mathbb{N}$.

4. (SSSMO/2001) Given a sequence of numbers $a_1, a_2, a_3, \cdots$, which satisfies $a_{n+2} = a_{n+1} - a_n$ for all $n = 1, 2, 3, \cdots$. If $a_2 = 1001$, find the value of $a_1 + a_2 + \cdots + a_{2001}$.

5. (GREECE/TST/2009) If a is a positive even integer and
    $$A = a^n + a^{n-1} + \cdots + a + 1 \quad (\text{where } n \in \mathbb{N})$$
    is a perfect square number, prove that a is a multiple of 8.

6. (CMC/2008) In $\{a_n\}$, $a_n \ge 0$, $a_1 = 1$, $S_n = \dfrac{1}{2}\left(a_n + \dfrac{1}{a_n}\right)$. Find $\left\lfloor \sum_{k=1}^{100} \dfrac{1}{S_k} \right\rfloor$.

7. (SMO/2005/Q7) Let $f(x) = a_0 + a_1 x + a_2 x^2 + \cdots + a_n x^n$, where a_i are nonnegative integers for $i = 0, 1, 2, \ldots, n$. If $f(1) = 21$ and $f(25) = 78357$, find the value of $f(10)$.

8. (CMC/2007) Let $a_n = \sum_{k=1}^{n} \dfrac{1}{k(n+1-k)}$. Prove that $a_{n+1} < a_n$ when $n \ge 2$.

9. Prove the identity $f(n) = \sum_{k=0}^{n} \binom{n+k}{k} \dfrac{1}{2^k} = 2^n$ for $n \in \mathbb{N}$.

10   (SMO/2005) Let $a_1 = 21, a_2 = 90$, and for $n \geq 3$, let a_n be the last two
     digits of $a_{n-1} + a_{n-2}$. What is the remainder of $a_1^2 + a_2^2 + \cdots + a_{2005}^2$ when
     it is divided by 8?

Testing Questions   (B)

1.   (BALKAN/2008) Is there a sequence $a_1, a_2, \cdots$ of positive reals which si-
     multaneously satisfies the following inequalities for all positive integers :

     (a) $\displaystyle\sum_{i=1}^{n} a_i \leq n^2;$ (b) $\displaystyle\sum_{i=1}^{n} \frac{1}{a_i} \leq 2008?$

2.   (JAPAN/2009) Find the sum of all real numbers x that satisfies the equation

     $$\sum_{k=1}^{9} \lfloor kx \rfloor = 44x.$$

3.   (AUSTRIA/2009) (i) a, b are given positive integers with $a < b$. Let
     $M(a, b) = \dfrac{\sum_{k=a}^{b} \sqrt{k^2 + 3k + 3}}{b - a + 1}$. Evaluate $K(a, b) = \lfloor M(a, b) \rfloor$;

     (ii) Evaluate $N(a, b) = \dfrac{\sum_{k=a}^{b} \lfloor \sqrt{k^2 + 3k + 3} \rfloor}{b - a + 1}$.

4.   (CMC/2008) Let $a_k > 0, k = 1, 2, \ldots, 2008$. Prove that there exists a se-
     quence $\{x_n\}$ satisfying the following conditions

     (i $0 = x_0 < x_n < x_{n+1}, n = 1, 2, 3, \ldots;$

     (ii) $\lim\limits_{n \to \infty} x_n$ exists; and

     (iii) $x_n - x_{n-1} = \displaystyle\sum_{k=1}^{2008} a_k x_{n+k} - \sum_{k=0}^{2007} a_{k+1} x_{n+k}, n = 1, 2, 3, \ldots$

     if and only if $\displaystyle\sum_{k=1}^{2008} a_k > 1.$

5.   (CSMO/2009) Let $X = (x_1, x_2, \cdots, x_9)$ be a permutation of the set $\{1, 2, \cdots,$
     $9\}$ and let A be the set of all such X. For any $X \in A$, denote $f(X) =$
     $x_1 + 2x_2 + 3x_3 + \cdots + 9x_9$ and $M = \{f(X) \mid X \in A\}$. Find $|M|$. ($|M|$
     denotes number of members of the set M.)

Lecture 20

Some Fundamental Theorems on Congruence

Definition 20.1. A set S of integers is said to be a **complete residue system modulo** m, if every integer is congruent to exactly one integer in S modulo m.

For example, for any natural number m, the set $\{0, 1, 2, \cdots, m-1\}$ is a complete residue system, and is said to be the **least nonnegative residue system modulo** m.

If a natural number n is relatively prime to a natural number m, then all numbers in the same congruence class modulo m as n are also relatively prime to m. Such a class is said to be a **reduced class** modulo m, and the number of reduced classes is denoted by $\varphi(m)$, which is call the **Euler's phi-function**.

For a given natural number $m > 1$, each reduced class contains one unique number less than m, therefore the value of $\varphi(m)$ is actually equal to the number of natural numbers n which are relatively prime to m and not greater than m. For example, $\varphi(1) = 1, \varphi(p) = p - 1$ for each prime number p and $\varphi(m) < m - 1$ for each composite m.

The following formula is useful for calculating the value of $\varphi(m)$:

Theorem I. *When m has the prime factorization $m = p_1^{\alpha_1} \cdots p_k^{\alpha_k}$, then*

$$\varphi(m) = m(1 - \frac{1}{p_1})(1 - \frac{1}{p_2}) \cdots (1 - \frac{1}{p_k}).$$

In particular, when $(m, n) = 1$, then $\varphi(mn) = \varphi(m)\varphi(n)$.

Proof. We use the inclusion and exclusion principle to show the equality.

Let $S = \{1, 2, 3, \cdots, m\}$, $S_i = \{x \in S : p_i \mid x\}$ for $i = 1, 2, \ldots, k$. Then $\varphi(m) = |\overline{S_1 \cup S_2 \cup \cdots \cup S_k}|$. It is obvious that

$$|S| = m, \quad |S_i| = \frac{m}{p_i}, \quad |S_i \cap S_j| = \frac{m}{p_i p_j}, \quad \cdots.$$

37

Hence, by the inclusion and exclusion principle,

$$
\begin{aligned}
\varphi(m) &= m - \left[\sum_{i=1}^{k} \frac{m}{p_i} - \sum_{1 \le i < j \le k} \frac{m}{p_i p_j} + \cdots + (-1)^{k-1} \frac{m}{p_1 p_2 \cdots p_k} \right] \\
&= m \left(1 - \frac{1}{p_1} \right) \left(1 - \frac{1}{p_2} \right) \cdots \left(1 - \frac{1}{p_k} \right).
\end{aligned}
$$

In particular, if $(m, n) = 1$, then m and n have no common prime factors, and so by grouping the factors of the last product as two groups, the formula $\varphi(mn) = \varphi(m) \cdot \varphi(n)$ is obtained. □

Definition 20.2. A set S of $\varphi(m)$ distinct integers is called a **reduced residue system modulo m**, if these integers are all relatively prime to m and any two are not congruent modulo m. For example, the set of $\varphi(m)$ natural numbers which are not greater than m and relatively prime to m is a reduced residue system, and is called the **least reduced residue system modulo m**.

Theorem II. *For any natural number $m > 1$ and an integer a, the congruence equation $ax \equiv b$ has always solution for any integer b if and only if $(a, m) = 1$.*

Proof. If $(a, m) = 1$, then the system $S = \{ac_1 - b, ac_2 - b, \ldots, ac_m - b\}$ is a complete residue system modulo m provided $\{c_1, c_2, \ldots, c_m\}$ is a complete residue system modulo m, since any two numbers in S have different remainders modulo m. Therefore there must be one element in S which is divisible by m, say $m \mid (ac_1 - b)$, so $ac_1 - b \equiv 0 \pmod{m}$, and which is equivalent to that c_1 is a solution of the equation $ac_1 \equiv b \pmod{m}$.

Conversely, If $ax \equiv b \pmod{m}$ always has a solution for any integer b, it is equivalent to the statement that $ax \equiv 1 \pmod{m}$ has a solution. Let x_0 be a solution of the equation $ax \equiv 1 \pmod{m}$, then $ax_0 - 1 = km$ for some integer k. Suppose that $(a, m) = d > 1$, then $d \mid 1$, a contradiction. Thus, $(a, m) = 1$. □

Note: Any solution of the equation $ax \equiv 1 \pmod{m}$ is said to be an **inverse of a modulo m**, denoted by $a^{-1} \pmod{m}$ or $\dfrac{1}{a} \pmod{m}$. In particular, there is a unique a^{-1} satisfying $1 \le a^{-1} < m$.

Theorem III. (Euler's Theorem)   Let $a, m \in \mathbb{Z}$ with $m \ge 1$. If $(a, m) = 1$, then

$$
a^{\varphi(m)} \equiv 1 \pmod{m}.
$$

Proof.   Suppose that $\{r_1, r_2, \cdots, r_{\varphi(m)}\}$ is the least reduced residue system modulo m. Then $\{ar_1, ar_2, \cdots, ar_{\varphi(m)}\}$ is also a reduced system modulo m since the $\varphi(m)$ elements of the latter system have distinct remainders modulo m. As such,

$$
(ar_1)(ar_2) \cdots (ar_{\varphi(m)}) \equiv r_1 r_2 \cdots r_{\varphi(m)} \pmod{m}
$$

Since $(r_i, m) = 1$ for all $i = 1, 2, \cdots, \varphi(m)$, we have $(r_1 r_2 \cdots r_{\varphi(m)}, m) = 1$, hence, by cancelling $r_1 r_2 \cdots r_{\varphi(m)}$ from both sides of the last congruence equality, the conclusion is obtained at once. □

Theorem IV. *(Fermat's Little Theorem)* *For any prime number p, if $a \in \mathbb{Z}$ and $p \nmid a$, then*

$$a^{p-1} \equiv 1 \pmod{p}.$$

Proof. The application of Euler's Theorem yields the conclusion at once when we let $m = p$. □

Note: Fermat's Little theorem sometimes is restated in the following form: For any prime number p and any integer a, $a^p \equiv a \pmod{p}$. This statement is also known as *Fermat's Great Theorem.*

Theorem V. (Wilson's Theorem) *For any prime number p,*

$$(p-1)! \equiv -1 \pmod{p}.$$

Proof. The conclusion is obvious for $p = 2$. If $p \geq 3$ and $1 \leq a \leq p - 1$, then there is a unique a^{-1} with $1 \leq a^{-1} \leq p - 1$, such that $aa^{-1} \equiv 1 \pmod{p}$. Since

$$a \equiv a^{-1} \pmod{p} \iff a^2 \equiv 1 \pmod{p} \iff a \equiv \pm 1 \pmod{p},$$

so $a = 1$ or $a = p - 1$ only. The rest of the $p - 3$ numbers $2, 3, \cdots, p - 2$ can be grouped into $(p - 3)/2$ pairs such that each pair is of the form of (a, a^{-1}) modulo p. Thus,

$$(p-1)! \equiv 1 \cdot (p-1) \equiv -1 \pmod{p}. \qquad \square$$

Examples

Example 1. (AUSTRIA/2009) The sequence $\{a_n\}$ is defined by $a_0 = a$ ($a \in \mathbb{N}$), $a_n = a_{n-1} + 40^{n!}$ ($n > 0$). Prove that the sequence $\{a_n\}$ contains infinitely many terms divisible by 2009.

Solution $2009 = 41 \cdot 49$ implies that $(40, 2009) = 1$. Therefore, by the Euler's Theorem,

$$40^{k \cdot \varphi(2009)} \equiv 1 \pmod{2009} \qquad \text{for all } k \in \mathbb{N}_0.$$

When $n > \varphi(2009)$, then $\varphi(2009) \mid n!$, so $a_{n+1} \equiv a_n + 1 \pmod{2009}$, hence $\{a_n\}$ is a periodic sequence starting from some term with 2009 as a period in modulo 2009, and this period forms a complete residue system modulo 2009. In conclusion, $\{a_n\}$ contains infinitely many terms divisible by 2009.

Example 2. Find all the positive integers n such that $\varphi(2n) = \varphi(3n)$.

Solution   Write $n = 2^\alpha 3^\beta M$, where α, β are non-negative integers and M is a positive integer with $(6, M) = 1$.

Suppose that $\beta > 0$, then $\varphi(2^{\alpha+1}) = 2^\alpha, \varphi(3^\beta) = 3^{\beta-1} \cdot 2$, so

$$\varphi(2n) = \varphi(2^{\alpha+1}) \cdot \varphi(3^\beta) \cdot \varphi(M) = 2^\alpha \cdot 2(3^{\beta-1}) \cdot \varphi(M),$$
$$\varphi(3n) = \varphi(2^\alpha) \cdot \varphi(3^{\beta+1}) \cdot \varphi(M) = \varphi(2^\alpha) \cdot 2(3^\beta) \cdot \varphi(M),$$

which implies $\varphi(2n) \neq \varphi(3n)$. Thus, $\beta = 0$, namely $n = 2^\alpha M$. For such n,

$$\varphi(2n) = \varphi(2^{\alpha+1}) \cdot \varphi(M) = 2^\alpha \cdot \varphi(M) \text{ and } \varphi(3n) = \varphi(2^\alpha) \cdot 2 \cdot \varphi(M)$$
$$\Rightarrow \varphi(2^\alpha) = 2^{\alpha-1} \Rightarrow \alpha \geq 1.$$

Thus, $n = 2^\alpha M$, where α is a positive integer and M is a positive integer with $(6, M) = 1$.

Example 3. Find the remainder of the sum $9^{10} + 9^{10^2} + \cdots + 9^{10^{100}}$ when it is divided by 7.

Solution   Since $(9, 7) = 1$, by the Euler's Theorem, $9^6 \equiv 1 \pmod 7$, so

$$9^{10} \equiv 9^4 \equiv 3^8 \equiv 3^2 \equiv 2 \pmod 7.$$

Similarly,

$$9^{10^2} \equiv (9^{10})^{10} \equiv 2^{10} \equiv 2 \pmod 7.$$

Assume that $9^{10^n} \equiv 2 \pmod 7$ for $n = k$, then for $n = k + 1$,

$$9^{10^{k+1}} \equiv (9^{10^k})^{10} \equiv 2^{10} \equiv 2 \pmod 7.$$

By induction, $9^{10^n} \equiv 2 \pmod 7$ for all $n \in \mathbb{N}$. Thus,

$$9^{10} + 9^{10^2} + \cdots + 9^{10^{100}} \equiv 2 \cdot 100 \equiv 200 \equiv 4 \pmod 7.$$

Example 4. (MOSCOW MO) For positive integers a, b and n, prove that

$$n! \mid b^{n-1} a(a + b)(a + 2b) \cdots [a + (n - 1)b].$$

Solution   It suffices to show that for any prime $p \leq n$, if $p^\alpha \mid n!$, where α is the index of p in the prime factorization of $n!$, then $p^\alpha \mid b^{n-1} a(a + b)(a + 2b) \cdots (a + (n - 1)b)$.

(i)   When $p \mid b$, then $\alpha = \sum_{k=1}^{\infty} \left[\frac{n}{p^k}\right] < \sum_{k=1}^{\infty} \frac{n}{p^k} = \frac{n}{p - 1} \leq n$ implies $\alpha \leq n - 1$, and so the conclusion is clear.

(ii) When $p \nmid b$, then $(b, p) = 1 \Rightarrow (b, p^\alpha) = 1$, so there is a b_1 such that $bb_1 \equiv 1 \pmod{p^\alpha}$. Hence

$$b_1^n a(a + b)(a + 2b) \cdots (a + (n - 1)b)$$
$$\equiv ab_1(ab_1 + 1)(ab_1 + 2) \cdots (ab_1 + (n - 1)) \pmod{p^\alpha}.$$

The right hand side is a product of n consecutive integers, hence is divisible by $n!$, and so it is divisible by p^α. Since $(b, p) = 1 \Rightarrow (b_1, p) = 1 \Rightarrow (b_1^n, p^\alpha) = 1$, we have

$$p^\alpha \mid a(a + b)(a + 2b) \cdots (a + (n - 1)b),$$

thus, $p^\alpha \mid b^{n-1}(a + b)(a + 2b) \cdots (a + (n - 1)b)$.

Example 5. (AUSTRALIA/2008) Find all the functions $f : \mathbb{N}_0 \mapsto \mathbb{N}_0$ such that
(i) $f(mn) = f(m) + f(n)$;   (ii) $f(2008) = 0$;
(iii) $f(n) = 0$ for all n with $n \equiv 39 \pmod{2008}$.

Solution Note that $2008 = 2^3 \times 251$. If f is a solution, (i) and (ii) yields $3f(2) + f(251) = 0$. The range of f is the set of non-negative integers, so $f(2) = f(251) = 0$. Therefore for any $n = 2^a \cdot 251^b \cdot m$ with $(m, 2008) = 1$

$$f(n) = af(2) + bf(251) + f(m) = f(m). \tag{20.1}$$

Since $(m, 2008) = 1$, the equation $mx \equiv 1 \pmod{2008}$ must have a solution k such that
$$km \equiv 1 \pmod{2008}.$$

(iii) then yields

$$f(39km) = f(39) + f(k) + f(m) = 0 \Rightarrow f(39) = f(k) = f(m) = 0.$$

Thus, $f(m) = 0$ for any positive integer m with $(m, 2008) = 1$, and hence by (20.1)
$$f(n) = 0, \qquad \text{for all } n \in \mathbb{N}.$$

Example 6. (CHNMO/2009) Find all the ordered pairs (p, q) of two primes such that $pq \mid 5^p + 5^q$.

Solution When $2 \mid pq$ we assume that $p = 2$, then $2q \mid 5^2 + 5^q \Rightarrow q \mid 25 + 5^q$.

From the Fermat's Little Theorem, $q \mid (5^q - 5)$, so $q \mid 30$, namely q may be $2, 3, 5$. By checking, $(2, 3), (2, 5)$ are two solutions for (p, q) but $\{2, 2\}$ is not acceptable.

When p, q are both odd and $5 \mid pq$, we assume that $p = 5$, then $5q \mid (5^5 + 5^q)$, so $q \mid 5^{q-1} + 625$. When $q = 5$, the pair $(5, 5)$ satisfies the requirement. When $q \neq 5$, by Fermat's Little Theorem, $q \mid (5^{q-1} - 1)$, so $q \mid 626$. Since q is an odd prime and $626 = 2 \cdot 313$, we must have $q = 313$. The pair $(5, 313)$ is qualified by checking.

When p, q are both not 2 and 5, then $pq \mid 5^{p-1} + 5^{q-1}$, and so

$$5^{p-1} + 5^{q-1} \equiv 0 \pmod{p}. \tag{20.2}$$

By Fermat's Little Theorem,

$$5^{p-1} \equiv 1 \pmod{p}. \tag{20.3}$$

Combining (20.2) and (20.3) yields

$$5^{q-1} \equiv -1 \pmod{p}. \tag{20.4}$$

Write $p - 1 = 2^k(2r - 1), q - 1 = 2^l(2s - 1)$, where k, l, r, s are all positive integers.

If $k \leq l$, from (20.3) and (20.4),

$$
\begin{aligned}
1 = 1^{2^{l-k}(2s-1)} &\equiv (5^{p-1})^{2^{l-k}(2s-1)} \equiv 5^{2^l(2r-1)(2s-1)} \\
&= (5^{q-1})^{2r-1} \equiv (-1)^{2r-1} \equiv -1 \pmod{p},
\end{aligned}
$$

which contradicts the assumption that $p \neq 2$! Thus, $k > l$.

However, after exchanging p and q and doing reasoning similarly, it is obtained that $k < l$. Thus, there are no solutions for (p, q) in this case.

Finally, considering the symmetry of p and q in the question, the desired pairs for (p, q) are

$$(2, 3), \ (3, 2), \ (2, 5), \ (5, 2), \ (5, 5), \ (5, 313), \ (313, 5).$$

Example 7. (CHNMO/TST/2010) Prove that there exists an unbounded increasing sequence $\{a_n\}, n = 1, 2, \ldots$ of positive integers, such that there exists a positive integer M with the following property: for all positive integers $n \geq M$, all the prime factors of $n! + 1$ must be greater than $n + a_n$ provided $n + 1$ is not prime.

Solution   Let p be a prime factor of $n! + 1$ and $n + 1$ is not prime, then $p > n + 1$. Let $t = p - n - 1$, then t is a positive integer. By Wilson's Theorem,

$$(p - 1)! \equiv -1 \pmod{p}.$$

Since $(p - t - 1)!t! \equiv (-1)^t(p - 1)! \equiv (-1)^{t+1} \pmod{p}$, so $n!t! \equiv (-1)^{t+1} \pmod{p}$, therefore $p \mid t! + (-1)^{t+1}$. Thus, $t! \geq p - 1$. Since $p - 1 = n + t$, so $t! - t \geq n$ which implies that $t! > n$.

Now define a_n be the minimum positive integer m satisfying $m! > n$, then $\{a_n\}$ is increasing sequence, and $a_n \to +\infty$ as $n \to +\infty$. Otherwise, there is a constant C such that $C! > n$ for all $n \in \mathbb{N}$.) Thus, $t \geq a_n$ implies that

$$p = n + t + 1 > n + a_n.$$

Testing Questions   (A)

1. Prove that
    $$\varphi(n) \cdot \tau(n) \geq n \qquad \text{for all } n \in \mathbb{N},$$
    where $\tau(n)$ denotes the number of positive divisors of n.

2. Prove that if $(m, n) = 2$ then $\varphi(mn) = 2\varphi(m)\varphi(n)$.

3. (GERMANY/2005) If $Q(n)$ denotes the sum of digits of the positive integer n, prove that
    $$Q(Q(Q(2005^{2005}))) = 7.$$

4. (CHNMO/TST/2005) Find all polynomials with integer coefficients, f, such that $f(n) \mid (2^n - 1)$ for all positive integers n.

5. (CWMO/2008) It is given that $m, a_1, a_2, \ldots, a_m \in \mathbb{N}$ and $m \geq 2$. Prove that there exist infinitely many positive integers n such that the numbers given by
    $$a_1 \cdot 1^n + a_2 \cdot 2^n + \cdots + a_m \cdot m^n$$
    are all composite numbers.

6. (CHNMO/TST/2008) It is given that the integer $n > 1$ divides $2^{\varphi(n)} + 3^{\varphi(n)} + \cdots + n^{\varphi(n)}$. Let $p_1, p_2, \ldots, p_k$ be the total of distinct prime factors of n. Prove that the number given by
    $$\frac{1}{p_1} + \frac{1}{p_2} + \cdots + \frac{1}{p_k} + \frac{1}{p_1 p_2 \cdots p_k}$$
    is an integer. (Here $\varphi(n)$ denotes the number of positive integers $\leq n$ which are relatively prime to n.)

7. (TURKEY/2007) It is given that $p = 6k + 1$ (where $k \in \mathbb{N}, k > 1$) is a prime number and $m = 2^p - 1$. Prove that the number $\dfrac{2^{m-1} - 1}{127m}$ is an integer.

8. (SMO/2009/R2) A palindromic number is a number which is unchanged when the order of its digits is reversed. Prove that the arithmetic progression $18, 37, \ldots$ contains infinitely many palindromic numbers.

9. (MACEDONIA/2009) In the range of integers find the solutions of equation

$$x^{2010} - 2006 = 4y^{2009} + 4y^{2008} + 2007y.$$

Testing Questions   (B)

1. (SMO/TST/2008) Find all odd primes p, if any, so that p divides $\sum\limits_{n=1}^{103} n^{p-1}$.

2. (ESTONIA/2007) Let $n \geq 2$ be a natural number. Prove that if there exists a positive integer b such that $\dfrac{b^n - 1}{b - 1}$ is a power of some prime number, then n must be prime.

3. (JAPAN/2010) Let k be a positive integer and m be an odd number. Prove that there exists a positive integer n such that $n^n - m$ is divisible by 2^k.

4. (SERBIA/2009) Find the minimum positive integer m, such that $2009 \mid m$, and the sum of digits of the decimal representation of m is equal to 2009.

5. (CMO/2008) Determine all functions f defined on the natural numbers that take values among the natural numbers for which

$$(f(n))^p \equiv n \pmod{f(p)}$$

for all $n \in \mathbb{N}$ and all prime numbers p.

6. (CHNMO/TST/2008) For any positive integer n, prove that $n^7 + 7$ is not a perfect square number.

Lecture 21

Chinese Remainder Theorem and Order of Integer

Theorem I. (Chinese Remainder Theorem*)* *If $m_1, m_2, \cdots, m_k$ are positive integers with $(m_i, m_j) = 1$ for any $1 \le i < j \le k$, then the system*

$$x \equiv b_1 \pmod{m_1}, \quad x \equiv b_2 \pmod{m_2}, \quad \cdots, \quad x \equiv b_k \pmod{m_k} \quad (21.1)$$

must have solutions for any integers $b_1, b_2, \cdots, b_k$, and the solution set is a congruence class modulo $m_1 m_2 \cdots m_k$, that is, the solutions are formed by the numbers

$$x \equiv M_1 M_1^{-1} b_1 + M_2 M_2^{-1} b_2 + \cdots + M_k M_m^{-1} b_k \pmod{m_1 m_2 \cdots m_k}, \quad (21.2)$$

where $M_i = m_1 m_2 \cdots m_k / m_i$ and M_i^{-1} is the inverse of M_i modulo m_i for $i = 1, 2, \cdots, k$.

Proof. By S_1 and S_2 we denote the solution set of the system (21.1) and the congruence class given by (21.2) respectively. Below we show that $S_1 = S_2$.

From the conditions $(m_i, M_i) = 1$ for $i = 1, 2, \cdots, k$, the inverse of M_i denoted by M_i^{-1} exists for $i = 1, 2, \cdots, k$. Also, the fact that M_j has the factor m_i for any $j \ne i$ implies that $m_i \mid M_j M_j^{-1} b_j$ for all $j \ne i$. Then, for any $x \in S_2$ and each $i \in \{1, 2, \ldots, k\}$,

$$x \equiv \sum_{j=1}^{k} M_j M_j^{-1} b_j \equiv M_i M_i^{-1} b_i \equiv b_i \pmod{m_i},$$

i.e. $x \in S_2 \Rightarrow x \in S_1$, so $S_2 \subseteq S_1$.

Conversely, to show that $S_1 \subseteq S_2$, it is enough to show that all the $x \in S_1$ are in a same congruence class modulo $m_1 m_2 \ldots m_k$. Let x and x' be any two different solutions of (21.1), then $x \equiv x' \pmod{m_i}$ for $i = 1, 2, \cdots, k$, hence $x \equiv x' \pmod{m_1 m_2 \cdots m_k}$, so x and x' are in a same congruence class. $\qquad \square$

Note: The Chinese Remainder Theorem indicates the existence of solution for the very general system (21.1), but it does not mean that the formula (21.2) is the unique method for obtaining the solutions, sometimes there are more direct methods instead of finding $M_1^{-1}, \ldots, M_k^{-1}$, as calculating them may take a long time. For example, to solve the system

$$x \equiv 1 \quad (\text{mod } 3), \quad x \equiv 3 \quad (\text{mod } 5), \quad x \equiv 9 \quad (\text{mod } 11),$$

it is convenient to use the fact that $x + 2$ is divisible by 3, 5 and 11.

Definition 21.1. For integers a and m with $m > 0$ and $(a, m) = 1$, the least positive integer n satisfying $a^n \equiv 1 \pmod{m}$ is called the **order of a modulo m,** denoted by $\mathbf{ord}_m a$.

Some basic properties of the order of an integer a

(i)   If $(a, m) = 1$ and $\text{ord}_m a = k$, then for $u, v \in \mathbb{N}$

$$a^u \equiv a^v \pmod{m} \Leftrightarrow u \equiv v \pmod{k}.$$

In particular, $a^u \equiv 1 \pmod{m}$ if and only if $k \mid u$.

Proof: *Sufficiency:* If $u \equiv v \pmod{k}$, letting $u \geq v$, then $u - v = kn$ for some $n \in \mathbb{N}_0$ and

$$a^u - a^v = a^v(a^{u-v} - 1) = a^v((a^k)^n - 1) \equiv 0 \pmod{m}.$$

Necessity: Suppose that $u > v$ and $a^u - a^v \equiv 0 \pmod{m}$. Since $(a, m) = 1, a^v(a^{u-v}-1) \equiv 0 \Rightarrow a^{u-v} \equiv 1 \pmod{m}$. Let $u-v = kq+r$, where $0 \leq r < k$, then $a^r \equiv 1 \pmod{m}$, and $k = \text{ord}_m a$ implies that $r = 0$, namely $u - v \equiv 0 \pmod{k}$.

(ii)   If $(a, m) = 1$ and $\text{ord}_m a = k$, then the remainders of the integers $a, a^2, a^3, \cdots$ modulo m are periodic with the minimum period k. Therefore the remainders of $a, a^2, \cdots, a^k$ modulo m are distinct.
Proof It's a direct consequence of (i).

(iii)   If $(a, m) = 1$ and $\text{ord}_m a = k$, then $k \mid \varphi(m)$. In particular, $\text{ord}_m a \leq \varphi(m)$, and $k \mid (p - 1)$ if $m = p$ is a prime.

Proof: $a^{\varphi(m)} \equiv 1 \pmod{m} \Rightarrow a^{\varphi(m)} \equiv a^k \pmod{m}$, therefore $\varphi(m) \equiv k \pmod{k}$ from the property (i), hence $\varphi(m) \equiv 0 \pmod{k}$. Thus, $k \mid \varphi(m)$. In particular, for $m = p$, we have $\varphi(p) = p - 1$, hence $k \mid (p - 1)$.

Definition 21.2. For integers r and m with $m > 0$ and $(r, m) = 1$, r is said to be a **primitive root modulo** m if $\text{ord}_m r = \varphi(m)$.

Note It has been proven in number theory that primitive roots exist only for $m = 1, 2, 4, p^n, 2p^n$, where p is an odd prime number.

Examples

Example 1. (IMO/Shortlist/2005) Let a and b be positive integers such that $a^n + n$ divides $b^n + n$ for every positive integer n. Show that $a = b$.

Solution Assume that $b \neq a$. Taking $n = 1$ shows that $a + 1$ divides $b + 1$, so that $b > a$. Let $p > b$ be a prime and let n be a positive integer such that

$$n \equiv 1 \quad (\text{mod } p - 1) \qquad \text{and} \qquad n \equiv -a \quad (\text{mod } p).$$

Such an n exists by the Chinese remainder theorem. (Without the Chinese remainder theorem, one could notice that $n = (a + 1)(p - 1) + 1$ has this property.)

By Fermat's little theorem, $a^n = a(a^{p-1})^k \equiv a \pmod{p}$, and therefore $a^n + n \equiv 0 \pmod{p}$. p divides $a^n + n$, and hence p also divides $b^n + n$. However, by Fermat's little theorem again, we have analogously $b^n + n \equiv b - a \pmod{p}$. We are therefore led to the conclusion $p \mid (b - a)$, which is a contradiction.

Example 2. (CZECH-SLOVAK-POLAND/2008) Prove that there is a positive integer n such that $k^2 + k + n$ has no prime factor less than 2008 for any integer k.

Solution Let p be a specified prime less than 2008. Then there exists an integer $r = r(p)$ (i.e., the value of r depends on p), such that $k^2 + k \not\equiv r \pmod{p}$ for any integer k.

In fact, when $k \equiv 0$ or $p - 1 \pmod{p}$, then $k^2 + k \equiv 0 \pmod{p}$. Therefore the system of p numbers

$$A = \{(k^2 + k) \pmod{p} \mid k \equiv 0, 1, 2, \ldots, (p - 1) \pmod{p}\}$$

is not a complete residue system modulo p, and so there exists integer $r \notin A$, such that $k^2 + k \not\equiv r \pmod{p}$. Let the set $\{p_1, p_2, \ldots, p_m\}$ consist of all primes less than 2008. Take n be a positive integer satisfying the system

$$n \equiv p_j - r(p_j) \pmod{p_j}, \quad j = 1, 2, \ldots, m.$$

The Chinese Remainder Theorem indicates that there must be a positive solution n. Then $k^2 + k + n \equiv p_j + (k^2 + k - r(p_j)) \not\equiv 0 \pmod{p_j}$ for $j = 1, 2, \ldots, m$, i.e., n is not divisible by each of $p_j, j = 1, 2, \ldots, m$. In conclusion, n satisfies the requirement in the question.

Example 3. (RUSMO/2008) For which integers $n > 1$ do there exist natural numbers $b_1, b_2, \ldots, b_n$ not all equal such that $(b_1 + k)(b_2 + k) \cdots (b_n + k) = a^b$ with $a, b \in \mathbb{N}$ and $a, b > 1$ for each natural number k? (The exponents a, b may depend on k.)

Solution   When n is composite, then $n = rs$ where $r > 1$ and $s > 1$. Let $b_1 = b_2 = \cdots = b_r = 1, b_{r+1} = \cdots = b_n = 2$, then

$$(b_1 + k)(b_2 + k) \cdots (b_n + k) = (1 + k)^r (2 + k)^{r(s-1)} = [(1 + k)(2 + k)^{s-1}]^r$$

which is the rth power of the positive integer $(1 + k)(2 + k)^{s-1}$.

When n is a prime number, suppose that there exist positive integers $b_1, b_2, \ldots,$ b_n not all equal such that $(b_1 + k)(b_2 + k) \cdots (b_n + k) = a^b$ with $a, b \in \mathbb{N}$ and $a, b > 1$ for each natural number k (where a, b may depend on k). Without loss of generality we assume that $b_1, b_2, \ldots, b_l$ ($l > 1$) are distinct but each of $b_{l+1}, \ldots, b_n$ is equal to one of $b_1, b_2, \ldots, b_l$, and in $b_1, b_2, \ldots, b_n$, the number of b_i is s_i for $i = 1, 2, \ldots, l$, so $s_1 + s_2 + \cdots + s_l = n$.

Let $p_1, p_2, \ldots, p_l > \max\{b_1, b_2, \ldots, b_l\}$ be l distinct prime numbers. By the Chinese Remainder Theorem, the system in m

$$b_1 + m \equiv p_1 \pmod{p_1^2}, \cdots, b_l + m \equiv p_l \pmod{p_l^2}$$

has positive integer solutions. This means that for any positive integer solution m the number $m + b_i$ is a multiple of p_i but not of p_i^2 for $i = 1, 2, \ldots, l$. Since $|b_i - b_j| < p_i$ for all $j \neq i, 1 \leq i, j \leq l, b_j + m$ cannot be a multiple of p_i, hence

$$a^b = (b_1 + m)(b_2 + m) \cdots (b_n + m)$$

is a multiple of $p_i^{s_i}$ but not of $p_i^{s_i+1}$, thus, $b \mid s_i$ for $i = 1, 2, \cdots, l$, therefore $b \mid n$. However, $b < n$ since $l > 1$, so $b = 1$, a contradiction.

Example 4. The numbers of the form $2^p - 1$ where p is prime are called the **Mersenne numbers** in number theory. Prove that if p is an odd prime and q is a prime factor of $2^p - 1$, then (i) $q > p$; (ii) q has the form $2np + 1$, where $n \in \mathbb{N}$.

Solution q must be odd. Let $k = \text{ord}_q 2$. Then $2^p - 1 \equiv 0 \pmod{q} \Rightarrow$ $2^p \equiv 1 \pmod{q} \Rightarrow k \mid p$, so $k = 1$ or $k = p$. $q > 1$ implies that $2 \not\equiv 1$ $\pmod{q}$, therefore $k = p$.

By Fermat's Little theorem, $2^{q-1} \equiv 1 \pmod{q}$, therefore $p \mid (q - 1)$ and the quotient must be even, hence (i) $p \leq q - 1 \Rightarrow q > p$; and (ii) $q - 1 = 2np$, namely $q = 2pn + 1$.

Example 5. The numbers of the form $2^{2^n} + 1$ (denoted by F_n), $n \in \mathbb{N}$, are called **Fermat numbers** in number theory. Prove that
(i)   Each prime factor of F_n must have the form $2^{n+1}k + 1$, where $k \in \mathbb{N}$.
(ii)   Prove that there are infinitely many primes of the form $2^s m + 1$, $s, m \in \mathbb{N}$.

Solution   (i)   Let p be a prime factor of $F_n = 2^{2^n} + 1$. Let $k = \mathrm{ord}_p 2$. Then
$$2^{2^n} \equiv -1 \pmod{p} \Rightarrow 2^{2^{n+1}} \equiv 1 \pmod{p} \Rightarrow k \mid 2^{n+1},$$
i.e., k is a power of 2. Let $k = 2^l$, where $0 \le l \le n + 1$. If $l \le n$, then
$$2^{2^n} = (2^{2^l})^{2^{n-l}} \equiv 1 \pmod{p}$$
which contradicts $2^{2^n} \equiv -1 \pmod{p}$, therefore $l = n + 1$ i.e. $k = 2^{n+1}$. Then Fermat's Little Theorem yields $2^{n+1} \mid (p - 1)$, hence $p = 2^{n+1}m + 1$.
   (ii)   The result of (i) implies that any prime factor p of $F_n = 2^{2^n} + 1$ satisfies $p \equiv 1 \pmod{2^{n+1}}$, hence these primes are of the form $p = 2^{n+1}k + 1 = 2^s m + 1$ for all $n \ge s$.
   Below we show that $(F_n, F_m) = 1$ for $n > m$. Since $2^m \mid 2^n$, so $F_n - 2 = 2^{2^n} - 1$ is divisible by F_m. Suppose that $(F_n, F_m) = d$, then $d \mid F_m$ implies $d \mid F_n - 2$, so $d \mid 2$. Since F_m, F_n are both odd, so $d \ne 2$, hence $d = 1$.
   Thus, all the prime factors of F_n and those of F_m are distinct for $n \ne m$, so there are infinitely many primes of the form $2^s m + 1$.

Example 6. (SMO/2009/R2) A palindromic number is a number which is unchanged when the order of its digits is reversed. Prove that the arithmetic progression $18, 37, \ldots$ contains infinitely many palindromic numbers.

Solution   In the Question A8 of Lecture 20, we proved that the terms $\dfrac{10^k - 1}{9}$ are in the given A.P. if $k = 6 + 18t$, $t \in \mathbb{N}_0$. Here we prove that the condition $10^k \equiv 11 \pmod{19}$ is a sufficient for the term $\dfrac{10^k - 1}{9}$ being in the A.P..

   Let $k_0 = \mathrm{ord}_{19} 10$. Since $10 = 2 \cdot 5$ is a primitive root, so $k_0 = \varphi(19) = 18$.
   Or, since $10^{18} \equiv 1 \pmod{19}$, so $k_0 \mid 18$, i.e., k_0 is one of $1, 2, 3, 6, 9, 12, 18$. The same result can be obtained from the following table of remainders of 10^k modulo 19 at once:

k	1	2	3	4	5	6	9	12	18
Remainder	10	5	12	6	3	11	18	7	1

If $10^k \equiv 11 \pmod{19}$ for some $k > 6$, then
$$10^{k-6} \equiv 1 \pmod{19} \Leftrightarrow k - 6 = 18t \Leftrightarrow k = 6 + 18t,$$

so $10^k \equiv 11 \pmod{19}$ if and only if $k = 6 + 18t, t \in \mathbb{N}_0$.

Example 7. (KOREA/2003) m is a positive integer. If $2^{m+1} + 1$ divides $3^{2^m} + 1$, prove that $2^{m+1} + 1$ is a prime number.

Solution   Let $q = 2^{m+1} + 1$. Then $q \mid (3^{2^m} + 1)$, so

$$3^{2^m} \equiv -1 \pmod{q}. \tag{21.3}$$

Hence $(3, q) = 1$. Taking squares to both sides of (21.3):

$$3^{2^{m+1}} \equiv 1 \pmod{q}.$$

Let $k = \mathrm{ord}_q 3$, then k is a factor of $2^{m+1} = q - 1$. Hence k has the form of 2^r, where r is a positive integer with $r \leq m + 1$. If $r \leq m$ then $3^{2^m} \equiv 1 \pmod{q}$, which contradicts (21.3). Thus, $r = m + 1$.

On the other hand, from Euler's Theorem, k is a factor of $\varphi(q)$, so $2^{m+1} = q - 1$ divides $\varphi(q)$, i.e. $q - 1 \leq \varphi(q)$. Since $\varphi(q) \leq q - 1$, so $\varphi(q) = q - 1$, i.e. q is prime.

Example 8. (IMO/2003) Let p be a prime number. Prove that there exists a prime number q such that for every integer n, the number $n^p - p$ is not divisible by q.

Solution.   Since $(p^p - 1)/(p - 1) = 1 + p + p^2 + \cdots + p^{p-1} \equiv p + 1 \pmod{p^2}$, we can get at least one prime divisor of $(p^p - 1)/(p - 1)$ which is not congruent to 1 modulo p^2.

In fact, write $M = \dfrac{p^p - 1}{p - 1} = p_1^{\alpha_1} \cdots p_s^{\alpha_s}$. If $p_i \equiv 1 \pmod{p^2}$ for $i = 1, 2, \ldots, s$, then

$$M \equiv \prod_{i=1}^{s} (k_i p^2 + 1)^{\alpha_i} \equiv 1 \pmod{p^2},$$

a contradiction.

Denote such a prime divisor by q. This q satisfies the conditions in question. The proof is as follows.

It's clear that $q \neq p$. Assume that there exists an integer n such that $n^p \equiv p \pmod{q}$, then $q \nmid n$, and we have $n^{p^2} \equiv p^p \equiv 1 \pmod{q}$. On the other hand, from Fermat's little theorem, $n^{q-1} \equiv 1 \pmod{q}$.

Since p^2 does not divide $q - 1$, let $\mathrm{ord}_q n = k$, then $k \mid p^2$ and $k \mid q - 1$, so $k = 1$ or p. In each of these two cases we have $n^p \equiv 1 \pmod{q}$. Hence we have $p \equiv 1 \pmod{q}$. However, this implies $1 + p + .. + p^{p-1} \equiv p \pmod{q}$. From the definition of q, this leads to $p \equiv 0 \pmod{q}$, a contradiction.

Testing Questions   (A)

1. (CHINA/2000) Prove that for any $n \in \mathbb{N}$, there are n consecutive positive integers such that each is divisible by a perfect square number greater than 1.

2. Solve the system of congruence equations

    $$x \equiv 1 \pmod{4}, \quad x \equiv 2 \pmod{5}, \quad x \equiv 4 \pmod{7}.$$

3. (KOREA/2007) Find all ordered pairs (p, q) of two integers such that

    $$pq \mid (p^p + q^q + 1).$$

4. For any positive integer n, prove that there are n consecutive positive integers such that they are all not powers of prime numbers.

5. Given that $m, k \in \mathbb{N}$. Prove that there is a reduced residue system modulo m, such that each prime factor of every number in the residue system is greater than k.

6. For the Fermat number $F_n = 2^{2^n} + 1, n > 1$, prove that if q is a prime factor of F_n, then $q = 2^{n+2}k + 1$, where $k \in \mathbb{N}$.

7. Find all ordered pairs (p, q) of prime numbers such that $pq \mid 2^p + 2^q$.

8. Prove that $2^n - 1$ is not divisible by n for any positive integer $n > 1$.

Testing Questions   (B)

1. (CHNMO/2008) Let $n > 1$ be a given integer and A be an infinite set of positive integers that has the following property: For any prime $p \nmid n$, there exist infinitely many elements in A not divisible by p.

    Prove that for any integer with $m > 1$ and $(m, n) = 1$, there exists a finite subset of distinct elements in A, such that the sum S of its elements satisfies $S \equiv 1 \pmod{m}$ and $S \equiv 0 \pmod{n}$.

2. (THAILAND/2006) When use p_k to denote the kth prime number for $k = 1, 2, 3, \ldots$, find the remainder of the sum

$$\sum_{k=2}^{2550} p_k^{p_k^4 - 1}$$

    when it is divided by 2550.

3. (CHNMO/TST/2004) Let u be a given positive integer. Prove that the equation $n! = u^a - u^b$ has at most finitely many positive integer solutions (n, a, b).

4. (CHNMO/TST/2005) Let $F_n = 2^{2^n} + 1, n \in \mathbb{N}$. Prove that F_n must have a prime factor greater than $2^{n+2}(n + 1)$.

5. (COLOMBIA/2009) Find all triples (a, b, n) of positive integers satisfying $a^b = 1 + b + \cdots + b^n$.

Lecture 22

Diophantine Equations (III)

In the lecture notes for Junior section we discussed linear and quadratic Diophantine equations. In this lecture we will discuss the general non-linear Diophantine equations.

As shown in dealing with quadratic Diophantine equations, we have the following usual methods for solving non-linear Diophantine equations.

(I) Using algebraic manipulations: it includes using completing squares, factorizations, substitution of variables or expressions.

(II) Estimation by using inequalities: its purpose is to determine a narrow range for the variables.

(III) Using various tools in number theory: it includes using divisibility, parity analysis, congruence, various theorems on congruence, quadratic residue.

(IV) Infinite descent technique.

(V) Construction method: According to the given conditions in question, construct a qualified special solution or a recursive formula for a solution. Here, mathematical induction is sometimes a useful tool.

There are usually three types of questions involving Diophantine equations: (i) To find its solutions; (ii) to determine if the equation has a required solution; and (iii) to determine if the given equation has finitely many or infinitely many solutions.

Examples

Example 1. (IRE/2008) Find all integers x such that $x(x + 1)(x + 7)(x + 8)$ is a perfect square.

Solution We wish to find all the ordered pairs (x, y) of integers satisfying

$$x(x + 1)(x + 7)(x + 8) = y^2 \quad (x \in \mathbb{Z}, y \in \mathbb{N}_0).$$

53

Let $z = x + 4$, then

$$x(x + 1)(x + 7)(x + 8) = y^2 \Rightarrow (z - 4)(z - 3)(z + 3)(z + 4) = y^2$$
$$\Rightarrow (z^2 - 16)(z^2 - 9) = y^2 \Rightarrow z^4 - 25z^2 + 12^2 = y^2$$
$$\Rightarrow (2z^2)^2 - 50(2z^2) + 25^2 - (2y)^2 = 49$$
$$\Rightarrow (2z^2 - 25)^2 - (2y)^2 = 49$$
$$\Rightarrow (2z^2 - 25 - 2y)(2z^2 - 25 + 2y) = 49.$$

From $49 = (-49)(-1) = (-7)(-7) = 1 \cdot 49 = 7 \cdot 7$, four systems of two simultaneous equations can be obtained:

$$\begin{cases} 2z^2 - 25 - 2y = -49, \\ 2z^2 - 25 + 2y = -1; \end{cases} \text{or} \begin{cases} = -7, \\ = -7; \end{cases} \text{or} \begin{cases} = 1, \\ = 49; \end{cases} \text{or} \begin{cases} = 7, \\ = 7; \end{cases}$$

Let $A = 2z^2 - 25 - 2y$, $B = 2z^2 - 25 + 2y$, the solutions for (y, z) and then for x are obtained easily as shown in the following table:

A	B	y	z	x
-49	-1	12	0	-4
-7	-7	0	± 3	$-1, -7$
1	49	12	± 5	$1, -9$
7	7	0	± 4	$0, -8$

where the values of y, z, x are obtained by $B - A = 4y$, $2z^2 = 25 + A + 2y$, $x = z - 4$. By checking, all the values obtained for x are solutions.

Example 2. (CROATIA/2009) Find all ordered pairs (m, n) $(m, n > 1)$ of positive integers such that $n^3 - 1$ is divisible by $mn - 1$.

Solution Let $m, n > 1$ be positive integers satisfying $(mn - 1) \mid (n^3 - 1)$. Since $(n^3 - 1)m - n^2(mn - 1) = n^2 - m$, so $(mn - 1) \mid (n^2 - m)$. Further,

$$m(n^2 - m) - (mn - 1)n = n - m^2 \Rightarrow (mn - 1) \mid n - m^2.$$

(i)   If $n > m^2$, then $mn - 1 \leq n - m^2 \leq n - 1$, no solution.

(ii) If $n = m^2$, then $(m^3 - 1) \mid (m^6 - 1)$ implies that all pairs of integers with the form (m, m^2), $m > 1$ satisfy the requirement.

(iii) If $n < m^2$, then $mn - 1 \leq n^3 - 1 \Rightarrow \sqrt{n} < m \leq n^2$. If $m < n^2$, then $mn - 1 \leq n^2 - m < n^2 - 1 \Rightarrow m < n$, but then $mn - 1 \leq m^2 - n < m^2 - 1 \Rightarrow n < m$, a contradiction. Thus, $m = n^2$.
Since $(n^3 - 1) \mid (n^3 - 1)$ for any positive integer $n > 1$, so all the pairs (n^2, n), $n > 1$ are also solutions.
   In summary, all the pairs (k, k^2) and (k^2, k) where $k > 1$ and $k \in \mathbb{N}$ are the solutions.

Example 3. (HUNGARY/2008-2009) Find positive integers $k_1, k_2, \ldots, k_n$ and n, such that

$$k_1 + k_2 + \cdots + k_n = 5n - 4 \text{ and } \frac{1}{k_1} + \cdots + \frac{1}{k_n} = 1.$$

Solution By the Cauchy-Schwartz inequality,

$$\left(\frac{1}{k_1} + \frac{1}{k_2} + \cdots + \frac{1}{k_n} \right) (k_1 + k_2 + \cdots + k_n) \geq n^2. \qquad (22.1)$$

Therefore $5n - 4 \geq n^2$, and it follows that $1 \leq n \leq 4$. The equality holds if and only if $k_1 = k_2 = \cdots = k_n$ and $n = 1$ or 4. In fact,

When $n = 1$, then $k = 1$.

When $n = 4$, then $k_1 = k_2 = k_3 = k_4 = 4$.

When $n = 2$, then $k_1 + k_2 = 6$, $\frac{1}{k_1} + \frac{1}{k_2} = 1 \Rightarrow k_1 k_2 = 6$, k_1, k_2 are the roots of the equation $t^2 - 6t + 6 = 0$, so they are irrational numbers, no required solution in this case.

When $n = 3$, then $k_1 + k_2 + k_3 = 11$, $\frac{1}{k_1} + \frac{1}{k_2} + \frac{1}{k_3} = 1$. Letting $k_1 \leq k_2 \leq k_3$, then

$$1 = \frac{1}{k_1} + \frac{1}{k_2} + \frac{1}{k_3} \leq \frac{3}{k_1} \Rightarrow 1 < k_1 \leq 3.$$

If $k_1 = 2$, then $k_2 + k_3 = 9$, $\frac{1}{k_2} + \frac{1}{k_3} = \frac{1}{2} \Rightarrow k_2 = 3, k_3 = 6$.

If $k_1 = 3$, then $\frac{1}{k_1} + \frac{1}{k_2} + \frac{1}{k_3} = 1$ and $k_1 \leq k_2 \leq k_3$ implies that $k_1 = k_2 = k_3 = 3$, which contradicts $k_1 + k_2 + k_3 = 11$, so there is no solution in this case.

Thus, when $n = 3$, the solutions for ordered triple (k_1, k_2, k_3) are $(2, 3, 6)$ and its permutations.

Example 4. (KOREA/2007) Find all triples (x, y, z) of positive integers such that $1 + 4^x + 4^y = z^2$.

Solution By symmetry, we can assume that $x \leq y$.

(i) When $2x < y + 1$, then $(2^y)^2 < 1 + 4^x + 4^y < (1 + 2^y)^2$, which indicates that $1 + 4^x + 4^y$ is not a perfect square, there is no solution in this case.

(ii) When $2x = y + 1$, then $1 + 4^x + 4^y = 1 + 2^{y+1} + 2^{2y} = (1 + 2^y)^2$, so $(n, 2n - 1, 1 + 2^{2n-1}), n \in \mathbb{N}$ are solutions.

(iii) When $2x > y + 1$, then $x > 1$, so $y \geq x \geq 2$ and $4^x + 4^y = 4^x(1 + 4^{y-x}) = (z - 1)(z + 1)$. Since $\gcd(z - 1, z + 1) = 2$, one of $z - 1$ and $z + 1$ is divisible by 2^{2x-1}. However,

$$2(1 + 4^{y-x}) \leq 2(1 + 4^{x-2}) < 2^{2x-1} - 2,$$

a contradiction. Thus, there is no solution in this case. In conclusion,

$$(n, 2n - 1, 1 + 2^{2n-1}), n \in \mathbb{N} \text{ or } (2n - 1, n, 1 + 2^{2n-1}), n \in \mathbb{N}.$$

Example 5. (CWMO/2007) Find all positive integer n, such that there exist non-zero integers $x_1, x_2, \ldots, x_n, y$ satisfying the system of equations

$$\begin{cases} x_1 + x_2 + \cdots + x_n & = & 0, \\ x_1^2 + x_2^2 + \cdots + x_n^2 & = & ny^2. \end{cases}$$

Solution   It is clear that $n \neq 1$.
When $n = 2k, k \in \mathbb{N}$, let $x_{2i-1} = 1, x_{2i} = -1, i = 1, 2, \ldots, k$ and $y = 1$, the equations are satisfied.
When $n = 3 + 2k, k \in \mathbb{N}$, let $y = 2, x_1 = 4, x_2 = x_3 = x_4 = x_5 = -1$, and let

$$x_{2i} = 2, x_{2i+1} = -2, \quad \text{for } i = 3, 4, \ldots, k + 1,$$

then the equations are satisfied.
When $n = 3$, suppose that there exist non-zero integers x_1, x_2, x_3 and y such that

$$\begin{cases} x_1 + x_2 + x_3 & = & 0, \\ x_1^2 + x_2^2 + x_3^2 & = & 3y^2, \end{cases}$$

then $2(x_1^2 + x_2^2 + x_1 x_2) = 3y^2$. Since it can be assumed that $\gcd(x_1, x_2, x_3) = 1$, so it can be assumed that $(x_1, x_2) = 1$. Then $x_1^2 + x_2^2 + x_1 x_2$ must be odd when x_1 and x_2 are both odd (since x_1 and x_2 cannot be both even) or one odd and one even. However, since y is even, so $3y^2 \equiv 0 \pmod 4$, whereas $2(x_1^2 + x_2^2 + x_1 x_2) \not\equiv 0 \pmod 4$, a contradiction.
Thus, all the natural numbers but except for 1 and 3 satisfy the requirement of the question.

Example 6. (BALKAN/2009) Find all positive integer solutions of the equation $3^x - 5^y = z^2$.

Solution   By taking modulo 4 to both sides of the given equation,

$$z^2 \equiv (-1)^x - 1^y \equiv (-1)^x - 1 \pmod 4 \Rightarrow z, x \text{ are both even.}$$

Let $x = 2t$ ($t \in \mathbb{N}$), then the given equation becomes

$$(3^t - z)(3^t + z) = 5^y.$$

Therefore there exists $k \in \mathbb{N}_0$ such that $3^t - z = 5^k, 3^t + z = 5^{y-k}$, hence

$$5^k + 5^{y-k} = 2(3^t) \Rightarrow k = 0 \Rightarrow 2(3^t) = 5^y + 1.$$

When $t \geq 2$, then $5^y + 1 \equiv 0 \pmod{9} \Leftrightarrow y \equiv 3 \pmod{6}$ since $5^6 \equiv 1 \pmod{9}$ and $5^3 \equiv 8 \pmod{9}$, and in this case $5^y + 1 \equiv 5^3 + 1 \equiv 0 \pmod{7}$ since $5^6 \equiv 1 \pmod{7}$. Thus, $7 \mid (5^y + 1)$, whereas $7 \nmid 2(3^t)$, there are no solutions in this case.

When $t \leq 1$, then $t = 1$, which implies that $y = 1, x = 2, z = 2$. Thus, $(2, 1, 2)$ is the unique solution.

Example 7. (KOREA/2009) Find all ordered pairs (m, n) of positive integers, such that $3^m - 7^n = 2$.

Solution Since $3^6 \equiv 1 \pmod{7}$, the given equation yields $3^m - 2 \equiv 0 \pmod{7}$ implies that $m = 6k + 2, k \in \mathbb{N}_0$.

When $m = 2$, then $n = 1$, and it is easy to see that $(2, 1)$ is a solution.

When $m = 2s \geq 4, s \in \mathbb{N}$, then $27 \mid (2 + 7^n)$. Since $7^9 \equiv 1 \pmod{27}$ and $7^4 \equiv -2 \pmod{27}$, it is obtained that $n = 9t + 4$. The given equation becomes

$$9^s = 2 + 7^{9t+4}.$$

However, if take modulo 37 to both sides, on the right hand side $7^9 \equiv 1 \pmod{37}$ yields

$$2 + 7^{9t+4} \equiv 2 + 7^4 \equiv 35 \pmod{37},$$

and on the left hand side, the remainders $r(s)$ of 9^s modulo 37 cannot take the value 35, since $9^9 \equiv 1 \pmod{37}$, and for $1 \leq s \leq 9$, we have the table

s	1	2	3	4	5	6	7	8	9
$r(s)$	9	7	26	12	34	10	16	33	1

This implies that there is no solution for $m \geq 4$. Thus, the unique solution is $(2, 1)$.

Example 8. (BMO/2007) Show that there are infinitely many pairs of positive integers (m, n) such that

$$\frac{m+1}{n} + \frac{n+1}{m}$$

is a positive integer.

Solution It's clear that if $n = m$, then only $(2, 2)$ is a solution. For $n \neq m$, we can assume that $m < n$. By observation, $m = 1, n = 2$ is a solution. Suppose that (m, n) is a solution and

$$\frac{m+1}{n} + \frac{n+1}{m} = k \in \mathbb{N}.$$

Then

$$k = \frac{1}{n}\left[\frac{n(n+1)}{m} + m + 1\right] \Rightarrow kn = \frac{n(n+1)}{m} + m + 1 \in \mathbb{N} \Rightarrow \frac{n(n+1)}{m} \in \mathbb{N}.$$

Let $r = \dfrac{n(n+1)}{m}$, then

$$k = \frac{1}{n}(r+m+1) = \frac{1}{n}\left[r + \frac{n(n+1)}{r} + 1\right] = \frac{r+1}{n} + \frac{n+1}{r}.$$

Therefore (m, n) is a solution implies that (n, r) is also a new solution, since $r = \dfrac{n(n+1)}{m} > n+1 > n$. Thus, if we have a solution (m, n) with $m < n$, then a new solution (n, r) with $n < r$ must exist. By continuing this process infinitely many times, infinitely many distinct solutions are obtainable.

Example 9. (CHNMO/2008) Find all triples (p, q, n), where p, q are odd primes and n is an integer greater than 1, satisfying

$$q^{n+2} \equiv 3^{n+2} \pmod{p^n} \quad \text{and} \quad p^{n+2} \equiv 3^{n+2} \pmod{q^n}.$$

Solution　It is obvious that $(3, 3, n), n = 2, 3, 4, \ldots$ are solutions.

Suppose that (p, q, n) be another solution with $p \neq q, p \neq 3, q \neq 3$. We can assume that $q > p \geq 5$.

If $n = 2$, then $q^2 \mid (p^4 - 3^4)$, namely $q^2 \mid (p^2 - 3^2)(p^2 + 3^2)$. Since q cannot divide both $(p^2 - 3^2)$ and $(p^2 + 3^2)$ at the same time, q^2 divides only one of $(p^2 - 3^2)$ and $(p^2 + 3^2)$. However, $0 < p^2 - 3^2 < q^2$, $p^2 + 3^2 \neq q^2$ and $p^2 + 3^2 < 2p^2 < 2q^2$, a contradiction.

If $n \geq 3$, then $p^n \mid (q^{n+2} - 3^{n+2})$ and $q^n \mid (p^{n+2} - 3^{n+2})$ yield

$$p^n \mid (p^{n+2} + q^{n+2} - 3^{n+2}) \quad \text{and} \quad q^n \mid (p^{n+2} + q^{n+2} - 3^{n+2}).$$

$(p^n, q^n) = 1$ then yields

$$p^n q^n \mid (p^{n+2} + q^{n+2} - 3^{n+2}), \tag{22.2}$$

hence $p^n q^n \leq p^{n+2} + q^{n+2} - 3^{n+2} < 2q^{n+2}$, so $p^n < 2q^2$.

On the other hand, $q^n \mid p^{n+2} - 3^{n+2}$ and $p > 3$ implies that $q^n \leq p^{n+2} - 3^{n+2} < p^{n+2}$, so $q < p^{1+\frac{2}{n}}$. Combining it with $p^n < 2q^2$ then yields $p^n < 2p^{2+\frac{4}{n}} < p^{3+\frac{4}{n}}$, therefore $n < 3 + \frac{4}{n}$, namely $n = 3$. Thus, $p^3 \mid q^5 - 3^5$ and $q^3 \mid p^5 - 3^5$.

Since $5^5 - 3^5 = 2 \times 11 \times 131$ which does not contain any factors of the form q^3 for some prime q, so $p > 5$. $p^3 \mid (q^5 - 3^5)$ implies that $p \mid (q^5 - 3^5)$, and Fermat's Little Theorem yields $p \mid (q^{p-1} - 3^{p-1})$. When $(5, p-1) = 1$ and $p \nmid (q-3)$, then $(q^5 - 3^5)/(q-3), (q^{p-1} - 3^{p-1})/(q-3) \geq p > 1$. However, for example, if $p - 1 = 5k + 4$, then

$$\left(\frac{q^5 - 3^5}{q-3}, \frac{q^{p-1} - 3^{p-1}}{q-3}\right)$$
$$= (q^4 + q^3 \cdot 3 + \cdots + q \cdot 3^3 + 3^4, q^{p-2} + q^{p-3} \cdot 3 + \cdots + q \cdot 3^{p-3} + 3^{p-2})$$
$$= (q^4 + q^3 \cdot 3 + \cdots + 3^4, q^{p-2} + q^{p-3} \cdot 3 + q^{p-4} \cdot 3^2 + q^{p-5} \cdot 3^3)$$
$$= (q^4 + q^3 \cdot 3 + \cdots + 3^4, q^{p-6} \cdot 3^4) = (q^4 + q^3 \cdot 3 + \cdots + 3^4, q^{p-7} \cdot 3^4)$$
$$= \cdots = (q^4 + q^3 \cdot 3 + \cdots + 3^4, 3^4) = 1,$$

and the same result can also be obtained for $p - 1 = 5k + r, r = 1, 2, 3$.

Thus, $p \mid (q - 3)$ if $(5, p - 1) = 1$, i.e. $q \equiv 3 \pmod{p}$. Then

$$\frac{q^5 - 3^5}{q - 3} = q^4 + 3q^3 + \cdot + 3^4 \equiv 5 \times 3^4 \not\equiv 0 \pmod{p},$$

hence $p^3 \mid q - 3$. However, $q^3 \mid (p^5 - 3^5)$ gives

$$q^3 \le p^5 - 3^5 < p^5 = (p^3)^{\frac{5}{3}} < q^{\frac{5}{3}},$$

a contradiction. Similar contradiction is obtained if $(5, q - 1) = 1$.

Thus, we assume that $(5, p - 1) = (5, q - 1) = 5$. $(q, p - 3) = 1$ (since $q > p \ge 7$) and $q^3 \mid (p^5 - 3^5)$ implies that $q^3 \mid \left|\frac{p^5 - 3^5}{p - 3}\right|$, hence

$$q^3 \le \frac{p^5 - 3^5}{p - 3} = p^4 + p^3 \cdot 3 + p^2 \cdot 3^2 + p \cdot 3^3 + 3^4.$$

$5 \mid (p - 1)$ and $5 \mid (q - 1)$ implies that $p \ge 11, q \ge 31$. Therefore

$$q^3 \le p^4 \left(1 + \frac{3}{p} + \left(\frac{3}{p}\right)^2 + \left(\frac{3}{p}\right)^3 + \left(\frac{3}{p}\right)^4\right) < p^4 \cdot \left(1 - \frac{3}{p}\right)^{-1} \le \frac{11}{8} p^4,$$

hence $\left(\frac{q}{p}\right)^3 \le \frac{11}{8} p$ and $\frac{q^2}{p^3} = \left[\left(\frac{q}{p}\right)^3\right]^{\frac{2}{3}} \cdot \frac{1}{p} \le \left(\frac{11}{8}\right)^{\frac{2}{3}} \cdot p^{-\frac{1}{3}} < \frac{5}{8}$. Thus,

$$\frac{p^5 + q^5 - 3^5}{p^3 q^3} < \frac{p^2}{q^3} + \frac{q^2}{p^3} < \frac{1}{31} + \frac{5}{8} < 1,$$

which contradicts (22.2). Thus, the solutions are $(3, 3, n), n = 2, 3, 4, \ldots$.

Testing Questions   (A)

1. (KOREA/2007) Find all ordered pairs (p, q) of two primes, such that $pq \mid (p^p + q^q + 1)$.

2. (CGMO/2009) Prove that the equation $abc = 2009(a + b + c)$ has only finitely many positive integer solutions for (a, b, c).

3. (CSMO/2009) Find all the pairs $\{x, y\}$ of two integers, such that $x^2 - 2xy + 126y^2 = 2009$.

4. (BELARUS/2009) Find all pairs (m, n) of positive integers such that $m! + n! = m^n$.

5. (GERMANY/2010) Find all positive integer solutions (x, y, z) of the equation $(3x + 1)(3y + 1)(3z + 1) = 34xyz$.

6. (CZECH-POLISH-SLOVAK/2005) Find all pairs (x, y) of two integers that satisfy the equation $y(x + y) = x^3 - 7x^2 + 11x - 3$.

7. (SSSMO/2009) Find all pairs of positive integers n, m that satisfy the equation $3 \cdot 2^m + 1 = n^2$.

8. (IRAN/2010) Given that m, n are two relatively prime integers. Prove that the equation in x, t, y, s, v, r

$$x^m t^n + y^m s^n = v^m r^n$$

has infinitely many positive integer solutions.

Testing Questions   (B)

1. (GERMANY/2008) Find all real x such that $4x^5 - 7$ and $4x^{13} - 7$ are both perfect squares.

2. (BULGARIA/2009) It is known that positive integers $a > b > 1$ and the equation $\dfrac{a^x - 1}{a - 1} = \dfrac{b^y - 1}{b - 1}$ $(x > 1, y > 1)$ has at least two distinct positive integer solutions (x, y). Prove that a and b are relatively prime.

3. (SLOVENIA/TST/2008-2009) $k > 1$ is a positive integer. Prove that there always exist k positive integers $n_1, n_2, \ldots, n_k$ such that $n_1^2 + n_2^2 + \cdots + n_k^2 = 5^{m+k}$ for each nonnegative integer m.

4. (CANADA/2009) Find all ordered pairs (a, b) such that a and b are integers and $3^a + 7^b$ is a perfect square.

5. (ITALY/TST/2009) Find all ordered pairs (x, y) of integers such that $y^3 = 8x^6 + 2x^3 y - y^2$.

Lecture 23

Pythagorean Triples and Pell's Equations

Although the Diophantine equations $x^2 + y^2 = z^2$ and $x^2 - Dy^2 = 1$ both are quadratic, they are fundamental in number theory and are commonly used for dealing with many problems in number theory. Hence it is worth making them as the topic of an entire lecture.

Definition 23.1. A *Pythagorean triple* is a set of three integers x, y, z such that $x^2 + y^2 = z^2$; the triple is said to be *primitive* if $\gcd(x, y, z) = 1$.

Lemma I If x, y, z is a primitive Pythagorean triple, then x and y must have different parities.

Proof of Lemma I. If x, y are both even, then 2 divides each of x, y, z, so $\gcd(x, y, z) \geq 2$. If x, y are both odd, then z is even implies that $z^2 \equiv 0 \pmod 4$, whereas $x^2 + y^2 \equiv 2$, a contradiction.

Lemma II If positive integers a, b, c satisfy $ab = c^n$ for some $n \in \mathbb{N}$ and $(a, b) = 1$, then a and b are both nth powers; that is, there exist positive integers a_1, b_1 such that $a = a_1^n$, $b = b_1^n$.

Proof of Lemma II. The conclusion is obviously true if one of a, b is 1. Now assume that $a > 1$ and $b > 1$. If we write down a, b in their prime factorization forms

$$a = p_1^{k_1} p_2^{k_2} \cdots p_r^{k_r} \quad \text{and} \quad b = q_1^{j_1} q_2^{j_2} \cdots q_s^{j_s}$$

respectively, then $\{p_1, p_2, \cdots, p_r\} \cap \{q_1, q_2, \cdots, q_s\} = \emptyset$. Write $c = u_1^{l_1} u_2^{l_2} \cdots u_t^{l_t}$, then

$$p_1^{k_1} p_2^{k_2} \cdots p_r^{k_r} q_1^{j_1} q_2^{j_2} \cdots q_s^{j_s} = u_1^{nl_1} u_2^{nl_2} \cdots u_t^{nl_t}$$

which implies that each index on the left hand side is divisible by n. Define

$$a_1 = p_1^{k_1/n} p_2^{k_2/n} \cdots p_r^{k_r/n} \quad \text{and} \quad b_1 = q_1^{j_1/n} q_2^{j_2/n} \cdots q_s^{j_s/n},$$

then $a = a_1^n$, $b = b_1^n$, as desired.

Theorem I. *The positive integer solutions* (x, y, z) *of the Pythagorean equation*

$$x^2 + y^2 = z^2 \tag{23.1}$$

satisfying $\gcd(x, y, z) = 1$ *and* $2 \mid x$ *are given by the formulas*

$$x = 2st, \qquad y = s^2 - t^2, \qquad z = s^2 + t^2,$$

where s, t *are two positive integers satisfying*

$$s > t > 0, \quad (s, t) = 1, \quad and \quad s \not\equiv t \pmod{2}.$$

Proof.   Suppose that the triple (x, y, z) is a primitive root of the equation (23.1) and x is even. Then y, z are both odd. Let $z - y = 2u$, $z + y = 2v$, then

$$x^2 = (z - y)(z + y),$$

$$\left(\frac{x}{2}\right)^2 = \left(\frac{z - y}{2}\right)\left(\frac{z + y}{2}\right) = uv.$$

Since $(u, v) = ((z-y)/2, (y+z)/2) = (z, (y+z)/2) = (z, y+z) = (z, y) = 1$. By Lemma II, u, v are both perfect squares. Let

$$u = t^2, \qquad v = s^2$$

for some positive integers $s > t > 0$, then

$$z = v + u = s^2 + t^2, \qquad y = v - u = s^2 - t^2, \qquad x = \sqrt{4uv} = 2st.$$

Let $d = (s, t)$, then d divides each of x, y, z, therefore $d = 1$, i.e. $(s, t) = 1$, hence s, t cannot be both even. If s, t are both odd, then y is even which contradicts the fact that y is odd. Thus $s \not\equiv t \pmod{2}$.

Conversely, let s and t be two positive integers satisfying all the conditions described above, then the numbers given by $x = 2st$, $y = s^2 - t^2$, $z = s^2 + t^2$ form a primitive triple since

$$x^2 + y^2 = 4s^2t^2 + s^4 - 2s^2t^2 + t^4 = (s^2 + t^2)^2 = z^2,$$

and, since y, z are both odd, $(y, z) = (s^2 - t^2, s^2 + t^2) = (2s^2, s^2 + t^2) = (s^2, s^2 + t^2) = (s^2, t^2) = 1$, we have $(x, y, z) = 1$.   □

Examples

Example 1. Prove that the equation $x^4 + y^4 = z^2$ has no positive integer solution.

Solution We prove the conclusion by contradiction. Suppose that (x_0, y_0, z_0) is a positive integer solution such that z_0 is the minimum value among all the positive integer solutions.

If $(x_0, y_0) = d > 1$, let p be a prime factor of d, then $p^4 \mid z_0^2$ implies that $p^2 \mid z_0$ and $\left(\dfrac{x_0}{p}, \dfrac{y_0}{p}, \dfrac{z_0}{p^2} \right)$ is also a positive integer solution, contradicting that z_0 is the minimum value among all positive integer solutions. Therefore $(x_0, y_0) = 1$.

Since (x_0^2, y_0^2, z_0) is a primitive Pythagorean triple, there are $a, b \in \mathbb{N}$ with $a > b, a \not\equiv b \pmod 2$ such that $x_0^2 = 2ab$, $y_0^2 = a^2 - b^2$, $z_0 = a^2 + b^2$.

If a is even, then b is odd so $y_0^2 = a^2 - b^2 \equiv 3 \pmod 4$ which is impossible, so a is odd and b is even. From $\left(\dfrac{x_0}{2} \right)^2 = a \left(\dfrac{b}{2} \right)$ and $(a, b) = 1$, there are $u, v \in \mathbb{N}$ such that $a = u^2, b = 2v^2, (u, v) = 1$. Then $y_0^2 = u^4 - 4v^4 \Rightarrow 4v^4 + y_0^2 = u^4$, so $(2v^2, y_0, u^2)$ is also a Pythagorean triple with $(2v^2, u^2) = 1$, therefore $(y_0, 2v^2) = 1$, so $(2v^2, y_0, u^2)$ is a primitive Pythagorean triple. Thus, by using the Theorem I once again, there are integers $\rho > \sigma > 0$ with $(\rho, \sigma) = 1$ such that

$$2v^2 = 2\rho\sigma, \qquad y_0 = \rho^2 - \sigma^2, \qquad u^2 = \rho^2 + \sigma^2.$$

Since $v^2 = \rho\sigma$, let $\rho = r^2, \sigma = s^2$, then $r, s \in \mathbb{N}, r > s > 0$ and $(r, s) = 1$. Thus, $r^4 + s^4 = u^2$. However,

$$z_0 = a^2 + b^2 = u^4 + 4v^4 > u > 0,$$

which contradicts the fact that z_0 is the minimum value.

Thus, the conclusion is proven.

Example 2. Prove that the equation $x^4 - y^4 = z^2$ has no positive integer solution.

Solution Suppose that (x, y, z) with $x, y, z \in \mathbb{N}$ is a solution to the equation $x^4 - y^4 = z^2$, and x is the minimum among all such solutions. Then $(x, y) = 1$, and x is odd.

If y is odd, then z is even. Hence, from $(y^2)^2 + z^2 = (x^2)^2$, there exist $a, b \in \mathbb{N}$ with $(a, b) = 1$ and $a > b > 0$ such that

$$x^2 = a^2 + b^2, \qquad y^2 = a^2 - b^2, \qquad z = 2ab.$$

Therefore $x^2 y^2 = a^4 - b^4$. Thus, (a, b, xy) is also a positive integer solution with $a < x$, which contradicts the fact that x is minimum in all such solution.

If y is even, then there exist $a, b \in \mathbb{N}$ with $a > b, (a, b) = 1$ and a, b have opposite parities, such that

$$x^2 = a^2 + b^2, \qquad y^2 = 2ab, \qquad z = a^2 - b^2.$$

Without loss of generality we assume that a is even and b is odd. Since $y^2 = 2ab$, there exist $p, q \in \mathbb{N}$ with $(p, q) = 1$ such that $a = 2p^2, b = q^2$, with q is odd. Thus, $x^2 = 4p^4 + q^4, y = 2pq$, and hence there exist $r, s \in \mathbb{N}$ with $(r, s) = 1$ and $r > s$ such that

$$p^2 = rs, \qquad q^2 = r^2 - s^2.$$

Again, we can write $r = u^2, s = v^2$, where $r, s \in \mathbb{N}$ with $(u, v) = 1$. Therefore $u^4 - v^4 = q^2$, i.e., (u, v, q) is a solution of the original equation with $u < r < p^2 < a < x$, a contradiction. Thus, the conclusion is proven.

Example 3. Prove that the length of the inradius of any Pythagorean triangle must be an integer.

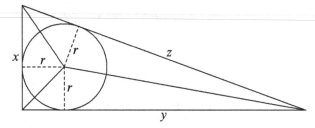

Solution   Let $x, y, z \in \mathbb{N}$ be the lengths of the three sides of the Pythagorean triangle with $x^2 + y^2 = z^2$, and r the length of inradius of the triangle. Then

$$x = 2kst, \qquad y = k(s^2 - t^2), \qquad z = k(s^2 + t^2)$$

for some positive integers k, s, t. Since $\frac{1}{2}xy = \frac{1}{2}r(x + y + z)$,

$$
\begin{aligned}
r &= \frac{xy}{x + y + z} = \frac{2k^2 st(s^2 - t^2)}{k(2st + s^2 - t^2 + s^2 + t^2)} = \frac{kt(s^2 - t^2)}{s + t}, \\
&= kt(s - t)
\end{aligned}
$$

which is an integer.

Example 4. Prove that for each natural number n, there are n right triangles with integer sides, such that they have equal perimeter and any two are not congruent.

Solution   Based on Theorem I, for any $n \in \mathbb{N}$ it is possible to get n primitive Pythagoreans such any two corresponding right triangles are not similar. Write these Pythagorean triples by $\{(a_k, b_k, c_k)$ with $0 < a_k < b_k < c_k$ for $k = 1, 2, \cdots, n$. Let $s_k = a_k + b_k + c_k, s = s_1 s_2 \cdots s_n$, and take

$$x_k = \frac{a_k s}{s_k}, \qquad y_k = \frac{b_k s}{s_k}, \qquad z_k = \frac{c_k s}{s_k}, \qquad k = 1, 2, \cdots, n,$$

then each (x_k, y_k, z_k) is a Pythagorean triple with $x_k + y_k + z_k = s$. Since the two triples (x_k, y_k, z_k) and (a_k, b_k, c_k) form two similar triangles, hence the two triangles formed by (x_k, y_k, z_k) and (x_m, y_m, z_m) cannot be similar if $k \neq m$.

Definition 23.2. An equation of the form

$$x^2 - dy^2 = 1. \tag{23.2}$$

is said to be a **Pell's Equation**, where d is a positive integer but is not a perfect square.

Definition 23.3. If (a, b) is a positive integer solution of (23.2) such that $a + \sqrt{d}b$ has a minimum value, then (a, b) is called the **minimum solution** or **fundamental solution** of (23.2).

Theorem II. *The equation (23.2) has at least one positive integer solution.*
Consequence *The Pell's equation (23.2) must have infinitely many positive integer solutions (x, y). If (a, b) is the minimum solution of (23.2), then all the positive integer solutions (x_n, y_n) are given by*

$$\begin{cases} x_n &= \dfrac{1}{2}[(a + \sqrt{d}b)^n + (a - \sqrt{d}b)^n], \\ y_n &= \dfrac{1}{2\sqrt{d}}[(a + \sqrt{d}b)^n - (a - \sqrt{d}b)^n]. \end{cases} \tag{23.3}$$

Theorem III. *If (a, b) is the minimum solution of (23.2), then (x, y) is a positive integer solution of (23.2) if and only if there exists an $n \in \mathbb{N}$ such that $x + \sqrt{d}y = (a + \sqrt{d}b)^n$.*

Theorem IV. *When d is a non-square positive integer, if the equation*

$$x^2 - dy^2 = -1 \tag{23.4}$$

has a positive integer solution, then equation (23.4) has infinitely many positive integer solutions. If (a, b) is the positive integer solution with minimum value of $x + \sqrt{d}y$ among all positive integer solutions (x, y), then all the positive integer solutions (x, y) of (23.4) can be expressed as

$$x + \sqrt{d}y = (a + \sqrt{d}b)^{2n+1},$$

and if (x_0, y_0) is the minimum solution of (23.2), then

$$x_0 + \sqrt{d}y_0 = (a + \sqrt{d}b)^2.$$

Example 5. Find all positive integer solutions of the equation $x^2 - 7y^2 = 1$.

Solution   First of all, we find the minimum solution (a, b). Since $x > y$, (a, b) can be found by letting $y = 1, 2, \cdots$ and checking the corresponding value of x.

When $y = 1$, no solution for x; when $y = 2$, no solution for x; when $y = 3$, we have $1 + 7(3^2) = 64 = 8^2$, therefore $x = 8$, i.e. $a = 8, b = 3$. By the Consequence of Theorem II, the positive integer solutions $\{(x_n, y_n)\}$ are given by

$$x_n = \frac{1}{2}[(8 + 3\sqrt{7})^n + (8 - 3\sqrt{7})^n],$$

$$y_n = \frac{1}{2\sqrt{7}}[(8 + 3\sqrt{7})^n - (8 - 3\sqrt{7})^n]$$

for $n = 1, 2, \cdots$.

Example 6. $k \geq 2$ is a natural number. Prove that there are infinitely many integers n such that $kn + 1$ and $(k + 1)n + 1$ are both perfect squares.

Solution   When $kn + 1$ and $(k + 1)n + 1$ are both perfect squares for some $n \in \mathbb{N}$, let

$$kn + 1 = u^2, \qquad (k + 1)n + 1 = v^2.$$

By eliminating n from the two equalities, we have

$$(k + 1)u^2 - kv^2 = 1. \tag{23.5}$$

Conversely, for any pairs (u, v) of two positive integers satisfying (23.5), letting $n = v^2 - u^2$, we find that $kn + 1 = u^2$ and $(k + 1)n + 1 = v^2$ which are both perfect squares. Thus, it suffices to show that equation (23.5) has infinitely many positive integer solutions (u, v). Under the substitutions

$$x = (k + 1)u - kv \qquad \text{and} \qquad y = v - u, \tag{23.6}$$

the equation (23.5) yields

$$x^2 - k(k + 1)y^2 = 1,$$

which is a Pell's equation, hence it has infinitely positive integer solutions (x, y). From (23.6) we obtain $u = x + ky$ and $v = x + (k + 1)y$. Thus, (23.5) has infinitely many positive integer solutions (u, v).

Example 7. (CHINA/2007) If x is an integer with $3 < x < 200$ such that $x^2 + (x + 1)^2$ is a perfect square number, find the value of x.

Solution   Let $x^2 + (x + 1)^2 = v^2$, then $(2x + 1)^2 + 1 = 2v^2$. Let $u = 2x + 1$, then

$$u^2 - 2v^2 = -1. \tag{23.7}$$

It is obvious that $(u_0, v_0) = (1, 1)$ is a solution of (23.7), and (u_0, v_0) is the positive integer solution with minimum value of $x + \sqrt{2}y$ among all positive integer solutions (x, y) of (23.7). Therefore, by Theorem IV, for any solution (u_n, v_n) of (23.7),

$$u_n + \sqrt{2}v_n = (1 + \sqrt{2})^{2n+1}, \quad n = 1, 2, \ldots.$$

Thus,

$$
\begin{aligned}
u_1 + \sqrt{2}v_1 &= (1 + \sqrt{2})^3 = 7 + 5\sqrt{2} \Rightarrow x_1 = 3, \\
u_2 + \sqrt{2}v_2 &= (1 + \sqrt{2})^5 = (7 + 5\sqrt{2})(3 + 2\sqrt{2}) \\
&= 41 + 29\sqrt{2} \Rightarrow x_2 = 20, \\
u_3 + \sqrt{2}v_3 &= (1 + \sqrt{2})^7 = (41 + 29\sqrt{2})(3 + 2\sqrt{2}) \\
&= 239 + 169\sqrt{2} \Rightarrow x_3 = 119, \\
u_4 + \sqrt{2}v_4 &= (1 + \sqrt{2})^9 = (239 + 169\sqrt{2})(3 + 2\sqrt{2}) \\
&= 1393 + 985\sqrt{2} \Rightarrow x_4 = 696.
\end{aligned}
$$

In conclusion, the desired solutions for x are 20 and 119.

Example 8. (HUNGARY/2009-2010) A positive integer is said to be a "multi-powered" number if in its prime factorization each prime factor has a power greater than or equal to 2. Prove that there are infinitely many pairs of adjacent positive integers such that they are both multi-powered.

Solution Note that each perfect square number is multi-powered, and the numbers of the form $8y^2 = 2^3 \cdot y^2$ are also multi-powered. Now consider the Pell's equation

$$x^2 - 8y^2 = 1.$$

Since $(3, 1)$ is its fundamental solution, therefore there are infinitely many integer solutions $(x_n, y_n), n = 1, 2, \ldots$. Since $x_n^2 - 8y_n^2 = 1$, each pair $(8y_n^2, x_n^2)$ is a desired pair.

Testing Questions   (A)

1. Find all right triangles with sides of integral lengths such that the lengths of the hypotenuse and one other side are two consecutive integers.

2. Given that (x, y, z) is a positive integer solution to the equation $x^2 + 2y^2 = z^2$, and $(x, y) = 1$. Prove that y is even, and there exist $m, n \in \mathbb{N}$ with $(m, n) = 1$ and m odd, such that $x = |m^2 - 2n^2|, y = 2mn, z = m^2 + 2n^2$.

3. Prove that the equation $x^4 + y^4 = z^4$ has no solutions with $x, y, z \in \mathbb{N}$.

4.   Prove that for any natural number $n > 1$, there exists a primitive Pythagorean triple such that its sum is an nth power of an integer.

5.   (AMC/2007) How many non-congruent right triangles with positive integer leg lengths have areas that are numerically equal to 3 times their perimeters?

(A) 6   (B) 7   (C) 8   (D) 10   (E) 12

6.   Find all positive integer solutions to the equation $x^2 - 2y^2 = 1$.

7.   Prove that the equation $x^2 - dy^2 = -1$ has no integer solution if $d \equiv 3 \pmod{4}$.

Testing Questions   (B)

1.   (IMO/Shortlist/2001) Consider the system

$$
\begin{aligned}
x + y &= z + u \\
2xy &= zu.
\end{aligned}
$$

Find the greatest value of the real constant m such that $m \leq x/y$ for any positive integer solution (x, y, z, u) of the system, with $x \geq y$.

2.   (IRAN/2002) Let $\{a_1, a_2, \cdots, a_n\}$ be a permutation of $\{1, 2, \cdots, n\}$. It is said to be a *quadratic permutation* if among the numbers

$$a_1, a_1 + a_2, a_1 + a_2 + a_3, \cdots, a_1 + a_2 + \cdots + a_n$$

at least one is a perfect square. Find all positive integers n such that each permutation of $\{1, 2, \cdots, n\}$ is a quadratic permutation.

3.   (SILK-ROAD MC/2009) Prove that for each prime number p, there exist infinitely many quartuples (x, y, z, t) formed by four distinct integers such that

$$(x^2 + pt^2)(y^2 + pt^2)(z^2 + pt^2)$$

is a perfect square number.

4.   (VIETNAM/TST/2009) Let a, b be positive integers. a, b and $a \cdot b$ are not perfect squares. Prove that at most one of following equations

$$ax^2 - by^2 = 1 \quad \text{and} \quad ax^2 - by^2 = -1$$

has solutions in positive integers.

5.   (INDIA/TST/2009) Given that p is a prime and $p \equiv 3 \pmod{8}$. Find all the integer solutions (x, y) of the equation $y^2 = x^3 - p^2 x$.

Lecture 24

Quadratic Residues

The general form of a quadratic congruence equation is

$$ax^2 + bx + c \equiv 0 \quad (\text{mod } p), \tag{24.1}$$

where $a \not\equiv 0 \,(\text{mod } p)$ and $p > 1$. When p is an odd prime number and $(a, p) = 1$, (24.1) is equivalent to

$$(2ax + b)^2 \equiv b^2 - 4ac \quad (\text{mod } p).$$

Let $y = 2ax + b$, $D = b^2 - 4ac$, then the standard form is obtained:

$$y^2 \equiv D \quad (\text{mod } p). \tag{24.2}$$

When (24.1) has a solution $x = x_0 \,(\text{mod } p)$, then (24.2) has a solution $y = 2ax_0 + b \,(\text{mod } p)$. Conversely, if $y = y_0 \,(\text{mod } p)$ satisfies (24.2), then the equation $2ax \equiv y_0 - b \,(\text{mod } p)$ can be solved to get a solution of (24.1). Thus, the problem of finding a solution of (24.1) for x is equivalent to finding a solution of a quadratic congruence equation of the form (24.2) and then a solution of a linear congruence equation.

Definition 24.1. For an odd prime p and an integer a with $(a, p) = 1$, when the quadratic congruence equation $x^2 \equiv a \,(\text{mod } p)$ has a solution for x, then a is said to be a **quadratic residue modulo p**. Otherwise, a is said to be a **quadratic non-residue modulo p**.

The theorems to be introduced below compose the basic theory of the quadratic residues, readers can find their proofs in Appendix B.

Theorem I. *For any odd prime p, the number of non-zero quadratic residues modulo p and quadratic non-residues modulo p are both $\dfrac{p-1}{2}$.*

69

Definition 24.2. For an odd prime p and any integer a with $(a, p) = 1$, define the **Legendre symbol** $\left(\dfrac{a}{p}\right)$ by

$$\left(\frac{a}{p}\right) = \begin{cases} 1 & \text{if } a \text{ is a quadratic residue modulo } p \\ -1 & \text{if } a \text{ is a quadratic non-residue modulo } p. \end{cases}$$

Theorem II. (Euler's Criterion) *Let p be an odd prime, a an integer with $(a, p) = 1$. Then a is a quadratic residue of p if and only if*

$$a^{\frac{p-1}{2}} \equiv 1 \pmod{p}.$$

Consequence I: $\left(\dfrac{a}{p}\right) \equiv a^{\frac{p-1}{2}} \pmod{p}$. *Consequence II:* $\left(\dfrac{a}{p}\right)\left(\dfrac{b}{p}\right) = \left(\dfrac{ab}{p}\right)$.

By considering the definition of the Legendre symbol and using the theorems II and I respectively, the following properties of the Legendre symbol are obtained:
For odd prime p and integers a and b satisfying $(a, p) = (b, p) = 1$,

(i) If $a \equiv b \pmod{p}$, then $\left(\dfrac{a}{p}\right) = \left(\dfrac{b}{p}\right)$;   (ii) $\left(\dfrac{a^2 b}{p}\right) = \left(\dfrac{b}{p}\right)$;

(iii) $\left(\dfrac{1}{p}\right) = 1$, $\left(\dfrac{-1}{p}\right) = (-1)^{\frac{p-1}{2}}$;   (iv) $\displaystyle\sum_{k=1}^{p-1} \left(\dfrac{k}{p}\right) = 0$.

Theorem III. (Gauss' Lemma) *For an odd prime number p and an integer a with $(a, p) = 1$, define the set S by $S = \left\{ a, 2a, 3a, \cdots, \left(\dfrac{p-1}{2}\right)a \right\}$.*

Among the remainders of the numbers in S mod p if n numbers are greater than $p/2$, then $\left(\dfrac{a}{p}\right) = (-1)^n$.

Theorem IV. (Quadratic Reciprocity Law) *For distinct odd primes p and q,*

$$\left(\frac{p}{q}\right)\left(\frac{q}{p}\right) = (-1)^{\frac{p-1}{2} \cdot \frac{q-1}{2}}.$$

Regarding the quadratic congruences with composite moduli, the following theorems give the basic results

Theorem V. *If p is an odd prime with $(a, p) = 1$, then $x^2 \equiv a \pmod{p^n}, n \geq 1$ has a solution if and only if $\left(\dfrac{a}{p}\right) = 1$.*

Theorem VI. *Let a be an integer and n a positive composite integer. When $n = p_1^{\alpha_1} p_2^{\alpha_2} \cdots p_r^{\alpha_r}$ is the prime factorization of n, then a is a quadratic residue modulo n if and only if a is a quadratic residue modulo p_i for all $i = 1, 2, \ldots, r$.*

Examples

Example 1. (SMO/2003) For any given prime p, determine whether the equation $x^2 + y^2 + pz = 2003$ always has integer solutions in x, y, z. Justify your answer.

Solution 1   In this solution we use the following lemma:

Lemma.   Any prime p with $p \equiv 1 \pmod{4}$ cab be expressed as a sum of two perfect square numbers.

Proof of Lemma.   Since $p = 4k + 1$, so $\left(\dfrac{-1}{p}\right) = (-1)^{\frac{p-1}{2}} = (-1)^{2k} = 1$, therefore there exists integer u such that $u^2 + 1 \equiv 0 \pmod{p}$, i.e., there exists $k \in \mathbb{N}$ such that $u^2 + 1 = kp$. Thus, there exists $k \in \mathbb{N}$ such that

$$kp = x^2 + y^2, \qquad \text{where } x, y \in \mathbb{Z}.$$

If $k = 1$, then the conclusion is proven. If $k > 1$, let $r \equiv x \pmod{k}$, $s \equiv y \pmod{k}$ such that $-\dfrac{k}{2} < r, s \le \dfrac{k}{2}$, then $r^2 + s^2 \equiv x^2 + y^2 \equiv 0 \pmod{k}$. Hence there exists $k_1 \in \mathbb{N}$ such that $r^2 + s^2 = k_1 k$, and so

$$(r^2 + s^2)(x^2 + y^2) = k_1 k \cdot kp = k_1 k^2 p.$$

Since $(r^2 + s^2)(x^2 + y^2) = (rx + sy)^2 + (ry - sx)^2$, so $\left(\dfrac{rx + sy}{k}\right)^2 + \left(\dfrac{ry - sx}{k}\right)^2 = k_1 p$. Since $rx + sy \equiv r^2 + y^2 \equiv 0 \pmod{k}$ and $ry - sx \equiv xy - yx \equiv 0 \pmod{k}$, so $\dfrac{rx + sy}{k}$ and $\dfrac{ry - sx}{k}$ are both integers, i.e. $k_1 p$ can be expressed as a sum of two perfect squares.

Since $k_1 k = r^2 + s^2 \le \left(\dfrac{k}{2}\right)^2 + \left(\dfrac{k}{2}\right)^2 = \dfrac{k^2}{2} \Rightarrow k_1 \le \dfrac{k}{2} < k$, so this process can be continued until $k_1 = 1$. Thus, p can be expressed as a sum of two perfect squares.

Below we return to the original problem.

(i)   If $p = 2$, then $(0, 1, 1001)$ is a solution;

(ii)   If $p = 2003$, then $(0, 0, 1)$ is a solution;

(iii)   If $p \ne 2$ and $p \ne 2003$, then

$$(2003 + 2p, 4p) = (2003, +2p, p) = (2003, p) = 1,$$

so, by Dirichlet's Theorem, the sequence $\{2003 + 2p + 4p \cdot n\}$ contains infinitely many primes. Assume that $q = 2003 + 4pn_0 + 2p$ is a prime, then $q \equiv 1$ (mod 4). Hence, by the lemma, there exist integers x, y such that

$$x^2 + y^2 + p(-4n_0 - 2) = 2003.$$

Solution 2　The answer is "Yes". When $p = 2$, $(2, 1, 999)$ is a solution.

If $p > 2$, then it is odd. Let $A = \left\{ n^2 : n \in \mathbb{N}_0, 0 \le n \le \dfrac{p-1}{2} \right\}$.

All the numbers in A have different residues modulo p: if $a^2, b^2 \in A$ and $a^2 \equiv b^2$ (mod p), then $a - b \equiv 0$ (mod p) or $a + b \equiv 0$ (mod p), but $0 < a + b < p$ if $a \ne b$, therefore $a = b$. Similarly, if

$$B = \left\{ -n^2 + 2003 : n \in \mathbb{N}_0, 0 \le n \le \frac{p-1}{2} \right\},$$

then all the numbers is B have different residues modulo p. Since $|A| + |B| = 2 \cdot \frac{p+1}{2} = p + 1$, by the pigeonhole principle, there exist a number $a^2 \in A$ and a number $-b^2 + 2003 \in B$ such that $a^2 \equiv -b^2 + 2003$ modulo p, i.e. there exists integer k such that

$$a^2 + b^2 - kp = 2003.$$

Thus, the given equation must have at least one integer solution $(a, b, -k)$.

Example 2. (IMO/2008) Prove that there exist infinitely many positive integers n such that $n^2 + 1$ has a prime divisor which is greater than $2n + \sqrt{2n}$.

Solution　For any prime number p with $p \equiv 1$ (mod 4), $\left(\dfrac{-1}{p} \right) = 1$ implies that there is $n \in \{1, 2, \ldots, p-1\}$ such that $n^2 \equiv -1$ (mod p). Therefore p is a prime factor of $n^2 + 1$. By using $p - n$ to replace n when $n > \dfrac{p}{2}$, it is always possible to let n be an integer such that $0 < n < \dfrac{p}{2}$.

Below we prove that $p > 2n + \sqrt{2n}$ if $p > 20$.

$$(p - 2n)^2 \equiv 4n^2 \equiv -4 \quad (\text{mod } p) \Rightarrow (p - 2n)^2 \ge p - 4$$

$$\Rightarrow p \ge 2n + \sqrt{p - 4} \ge 2n + \sqrt{2n + \sqrt{p - 4} - 4} > 2n + \sqrt{2n}.$$

There are infinitely many primes p with $p \equiv 1$ (mod 4), so we have infinitely many corresponding $n \in \mathbb{N}$. If (n_0, p_0) is such a pair, since $n_0^2 + 1 = k_0 p_0$ is a bounded value, n_0 cannot appear in infinitely many pairs. Hence, infinitely many distinct n must appear in the pairs.

Example 3. (KOREA/2003) If $2^{m+1} + 1$ is a prime number, where $m \in \mathbb{N}$, prove that $2^{m+1} + 1$ divides $3^{2^m} + 1$.

Solution Write $q = 2^{m+1} + 1$. Then $\varphi(q) = q - 1 = 2^{m+1}$. $(3, q) = 1$ since $q \geq 5$, and so by Fermat's Little Theorem $3^{q-1} = 3^{2^{m+1}} \equiv 1 \pmod{q}$. Hence

$$3^{2^m} = 3^{\frac{q-1}{2}} \equiv \pm 1 \pmod{q} \tag{24.3}$$

From $q \equiv 1 \pmod 4$ and $q \equiv 2 \pmod 3$, by the Quadratic Reciprocity Law,

$$\left(\frac{3}{q}\right) = (-1)^{\frac{(3-1)(q-1)}{4}} \left(\frac{q}{3}\right) = \left(\frac{2}{3}\right) = 2^{\frac{3-1}{2}} \equiv 2 \equiv -1 \pmod 3.$$

On the other hand, $\left(\dfrac{3}{q}\right) \equiv 3^{\frac{q-1}{2}} = 3^{2^m} \pmod{q}$, so $3^{2^m} \equiv -1 \pmod{q}$, i.e.,

$$q \mid (3^{2^m} + 1).$$

Example 4. (AUSTRIA/2007) For which non-negative integers $a < 2007$ the congruence $x^2 + a \equiv 0 \pmod{2007}$ has got exactly two different non-negative integer solutions?

That means, that there exist exactly two different non-negative integers u and v less than 2007, such that $u^2 + a$ and $v^2 + a$ are both divisible by 2007.

Solution Since $2007 = 3^2 \cdot 223$, so the number of solutions for x modulo 2007 is equal to the product of the number of solutions for x modulo 9 and that for modulo 223.

Since $x^2 \equiv 0, 1, 4, 7 \pmod 9$, so if $a \in \{2, 5, 8\}$, then $x^2 + a \equiv 0 \pmod 9$ has two solutions; if $a \in \{1, 3, 4, 6, 7\}$ then $x^2 + a \equiv 0 \pmod 9$ has no solution; if $a = 0$, then $x^2 + a \equiv 0 \pmod 9$ has three solutions.

For the equation $x^2 + a \equiv 0 \pmod{223}$, it has two solutions when $-a \in S = \{b_i \mid b_i \equiv i^2 \pmod{223}, 0 < b_i < 223, i = 1, 2, \ldots, 111\}$; if $-a \notin S \cup \{0\}$, then no solution and if $a \equiv 0 \pmod{223}$ then only one solution.

Thus, the original equation has exactly two solutions if and only if $x^2 + a \equiv 0 \pmod 9$ has two solutions and $x^2 + a \equiv 0 \pmod{223}$ has one solution.

Let $a = 223b$. Since $223 \equiv -2 \pmod 9$ and -2 is a quadratic residue modulo 9, so $-b$ must be also a quadratic residue modulo 9. Hence $b \in \{2, 5, 8\}$, and we have

$$a = 2 \times 223 = 446, \quad a = 5 \times 223 = 1115, \quad a = 8 \times 223 = 1784.$$

Testing Questions   (A)

1. Find all the quadratic residues and quadratic non-residues modulo $7, 11, 17$ and 19 respectively.

2. Prove that there are infinitely many prime numbers of form $4k + 1$.

3. Given that p is an odd prime number. Prove that -1 is a quadratic residue modulo p if and only if $p \equiv 1 \pmod 4$.

4. Determine the number of solutions to each of the following quadratic congruence equations:

    (i) $x^2 \equiv -2 \pmod{67}$;   (ii) $x^2 \equiv 2 \pmod{67}$;   (iii) $x^2 \equiv -2 \pmod{37}$;
    (iv) $x^2 \equiv 2 \pmod{37}$; (v) $x^2 \equiv -1 \pmod{221}$; (vi) $x^2 \equiv -1 \pmod{427}$.

5. (HONG KONG/2002) Let p be a prime of form $4k + 1$. Evaluate $\displaystyle\sum_{k=1}^{p-1} \left\{ \frac{k^2}{p} \right\}$,

    where $\{x\} = x - \lfloor x \rfloor$.

6. (TURKEY/2002) Find all prime numbers p for which the number of ordered pairs of integers (x, y) with $0 \le x, y \le p$ satisfying the condition $y^2 \equiv x^3 - x \pmod p$ is exactly p.

7. Let k be a positive integer. Prove that there are infinitely many perfect squares of the form $n \cdot 2^k - 7$, where n is a positive integer.

Testing Questions   (B)

1. (BULGARIA/TST/2007) Let $p = 4k + 3$ be a prime number. Find the number of different residues mod p of $(x^2 + y^2)^2$ where $(x, p) = (y, p) = 1$.

2. (SMO/2004) Find the number of ordered pairs (a, b) of integers, where $1 \le a, b \le 2004$, such that $x^2 + ax + b = 167y$ has integer solutions in x and y. Justify your answer.

3. (TURKEY/2008) (a) Find all primes p such that $\dfrac{7^{p-1} - 1}{p}$ is a perfect square;

    (b) Find all primes p such that $\dfrac{11^{p-1} - 1}{p}$ is a perfect square.

4. (VIETNAM/TST/2008) Let m and n be positive integers. Prove that $6m \mid [(2m + 3)^n + 1]$ if and only if $4m \mid (3^n + 1)$.

5. (ROMANIA/TST/2005) Let $n \ge 0$ be an integer and let $p \equiv 7 \pmod 8$ be a prime number. Prove that $\displaystyle\sum_{k=1}^{p-1} \left\{ \frac{k^{2^n}}{p} + \frac{1}{2} \right\} = \frac{p-1}{2}$, where $\{x\} = x - \lfloor x \rfloor$.

Lecture 25

Some Important Inequalities (I)

Theorem I. *(Cauchy-Schwartz Inequality)* *For any real a_1, a_2, $\cdots$, a_n and b_1, b_2, $\cdots$, b_n,*

$$(a_1 b_1 + a_2 b_2 + \cdots + a_n b_n)^2 \leq (a_1^2 + a_2^2 + \cdots + a_n^2)(b_1^2 + b_2^2 + \cdots + b_n^2).$$

Furthermore, if a_1, a_2, $\cdots$, a_n are not all zeros, the equality holds if and only if there exists a constant k such that $b_i = k a_i$ for all $i = 1, 2, \cdots, n$.

The following consequences are actually common ways to apply the Cauchy-Schwartz inequality.

Consequence 1. $\displaystyle\sum_{i=1}^{n} \frac{a_i^2}{b_i} \geq \frac{\left(\sum_{i=1}^{n} a_i\right)^2}{\sum_{i=1}^{n} b_i}$, where $a_1, a_2, \ldots, a_n, b_1, b_2, \ldots, b_n$

≥ 0 and $\displaystyle\sum_{i=1}^{n} b_i \neq 0$.

Consequence 2. $\displaystyle\sum_{i=1}^{n} \frac{a_i}{b_i} \geq \frac{\left(\sum_{i=1}^{n} a_i\right)^2}{\sum_{i=1}^{n} a_i b_i}$, where $a_1, a_2, \ldots, a_n, b_1, b_2, \ldots, b_n$

≥ 0 and $\displaystyle\sum_{i=1}^{n} a_i b_i \neq 0$.

Consequence 3. $\displaystyle\sqrt{\sum_{i=1}^{n} a_i} \cdot \sqrt{\sum_{i=1}^{n} b_i} \geq \sum_{i=1}^{n} \sqrt{a_i b_i}$, where $a_1, a_2, \ldots, a_n, b_1,$

$b_2, \ldots, b_n \geq 0$.

Theorem II. *(Schur's Inequality)* *For all nonnegative real numbers x, y, z and a positive number $r > 0$, the following inequality always holds:*

$$x^r (x - y)(x - z) + y^r (y - z)(y - x) + z^r (z - x)(z - y) \geq 0,$$

*where the equality holds if and only if (i) $x = y = z$ or (ii) two of x, y, z are
equal and the other is zero.*

For $r = 1$, Schur's inequality can be expressed in the following forms:

(1) $x^3 + y^3 + z^3 - (x^2 y + xy^2 + x^2 z + xz^2 + y^2 z + yz^2) + 3xyz \geq 0$;

(2) $(x + y + z)^3 - 4(x + y + z)(xy + yz + zx) + 9xyz \geq 0$;

(3) $xyz \geq (x + y - z)(y + z - x)(z + x - y)$.

For $r = 2$, Schur's inequality can be expressed as the following forms:

(1) $x^2(x - y)(x - z) + y^2(y - z)(y - x) + z^2(z - x)(z - y) \geq 0$;

(2) $x^4 + y^4 + z^4 - (x^3 y + xy^3 + y^3 z + yz^3 + z^3 x + zx^3) + xyz(x + y + z) \geq 0$.

Examples

Example 1. (JAPAN/2010) Let x, y, z be positive real numbers. Prove that

$$\frac{1 + xy + xz}{(1 + y + z)^2} + \frac{1 + yz + yx}{(1 + z + x)^2} + \frac{1 + zx + zy}{(1 + x + y)^2} \geq 1.$$

Solution   The Cauchy-Schwartz inequality yields

$$\left(1 + \frac{y}{x} + \frac{z}{x}\right)(1 + xy + xz) \geq (1 + y + z)^2,$$

which implies that $\dfrac{1 + xy + xz}{(1 + y + z)^2} \geq \dfrac{x}{x + y + z}$. Similarly,

$$\frac{1 + yz + yx}{(1 + z + x)^2} \geq \frac{y}{x + y + z}, \quad \frac{1 + zx + zy}{(1 + x + y)^2} \geq \frac{z}{x + y + z}.$$

By adding up these three inequalities, the desired inequality is obtained.

Example 2. (MACEDONIA/2009) Given that the positive real numbers a, b, c
satisfy $ab + bc + ca = \dfrac{1}{3}$. Prove that

$$\frac{a}{a^2 - bc + 1} + \frac{b}{b^2 - ca + 1} + \frac{c}{c^2 - ab + 1} \geq \frac{1}{a + b + c}.$$

Solution   Since $0 \leq ab, bc, ca \leq \dfrac{1}{3}$, the denominators on the left hand side are all positive. The Cauchy-Schwartz inequality yields

$$
\frac{a}{a^2 - bc + 1} + \frac{b}{b^2 - ca + 1} + \frac{c}{c^2 - ab + 1}
$$

$$
= \frac{a^2}{a^3 - abc + a} + \frac{b^2}{b^3 - abc + b} + \frac{c^2}{c^3 - abc + c}
$$

$$
\geq \frac{(a + b + c)^2}{a^3 + b^3 + c^3 + a + b + c - 3abc}
$$

$$
= \frac{(a + b + c)^2}{(a + b + c)(a^2 + b^2 + c^2 - ab - bc - ca) + (a + b + c)}
$$

$$
= \frac{a + b + c}{a^2 + b^2 + c^2 - ab - bc - ca + 1}
$$

$$
= \frac{a + b + c}{a^2 + b^2 + c^2 + 2(ab + bc + ca)} = \frac{1}{a + b + c}.
$$

Example 3. (USAMO/2009) For $n \geq 2$ let $a_1, a_2, \ldots, a_n$ be positive real numbers such that

$$
(a_1 + a_2 + \cdots + a_n) \cdot \left(\frac{1}{a_1} + \frac{1}{a_2} + \cdots + \frac{1}{a_n} \right) \leq \left(n + \frac{1}{2} \right)^2.
$$

Prove that $\max\{a_1, a_2, \ldots, a_n\} \leq 4 \min\{a_1, a_2, \ldots, a_n\}$.

Solution   Let $m = \min\{a_1, a_2, \ldots, a_n\}$, $M = \max\{a_1, a_2, \ldots, a_n\}$. Based on symmetry we may assume that $m = a_1 \leq a_2 \leq \cdots \leq a_n = M$.

By the Cauchy-Schwartz inequality,

$$
\left(n + \frac{1}{2} \right)^2 \geq (a_1 + a_2 + \cdots + a_n) \left(\frac{1}{a_1} + \frac{1}{a_2} + \cdots + \frac{1}{a_n} \right)
$$

$$
= (m + a_2 + \cdots + a_{n-1} + M) \left(\frac{1}{M} + \frac{1}{a_2} + \cdots + \frac{1}{m} \right)
$$

$$
\geq \left(\sqrt{\frac{m}{M}} + \underbrace{1 + 1 + \cdots + 1}_{n-2 \text{ terms}} + \sqrt{\frac{M}{m}} \right)^2 \Rightarrow n + \frac{1}{2} \geq \sqrt{\frac{m}{M}} + n - 2 + \sqrt{\frac{M}{m}}
$$

$$
\Rightarrow 2(m + M) \leq 5\sqrt{mM} \Rightarrow 4m^2 + 4M^2 - 17mM \leq 0
$$

$$
\Rightarrow (4M - m)(M - 4m) \leq 0 \Rightarrow M \leq 4m.
$$

Example 4. (BOSNIA/2008) If a, b and c are positive reals prove the inequality:

$$\left(1 + \frac{4a}{b+c}\right)\left(1 + \frac{4b}{c+a}\right)\left(1 + \frac{4c}{a+b}\right) > 25.$$

Solution $\left(1 + \dfrac{4a}{b+c}\right)\left(1 + \dfrac{4b}{c+a}\right)\left(1 + \dfrac{4c}{a+b}\right) > 25$

$\Leftrightarrow (b + c + 4a)(c + a + 4b)(a + b + 4c) > 25(a + b)(b + c)(c + a)$

$\Leftrightarrow a^3 + b^3 + c^3 + 7abc > a^2b + ab^2 + b^2c + bc^2 + c^2a + ca^2.$

Schur's Inequality gives

$$a^3 + b^3 + c^3 + 3abc \geq a^2b + ab^2 + b^2c + bc^2 + c^2a + ca^2,$$

so the desired inequality holds.

Sometimes, the AM-GM inequality, Cauchy-Schwartz inequality, Schur's inequality and other inequalities are used together to solve a problem, as shown in the following examples.

Example 5. (CHNMO/TST/2008) Let x, y, z be positive real numbers. Prove that

$$\frac{xy}{z} + \frac{yz}{x} + \frac{zx}{y} > 2\sqrt[3]{x^3 + y^3 + z^3}.$$

Solution Let $\dfrac{xy}{z} = a^2, \dfrac{yz}{x} = b^2, \dfrac{zx}{y} = c^2$, where $a, b, c > 0$, then $x = ca, y = ab, z = bc$ and the original inequality becomes

$$a^2 + b^2 + c^2 > 2\sqrt[3]{a^3b^3 + b^3c^3 + c^3a^3}. \tag{25.1}$$

Since

$(25.1) \Leftrightarrow (a^2 + b^2 + c^2)^3 > 8(a^3b^3 + b^3c^3 + c^3a^3)$

$\Leftrightarrow a^6 + b^6 + c^6 + 3(a^4b^2 + a^2b^4 + b^4c^2 + b^2c^4 + c^4a^2 + c^2a^4) + 6a^2b^2c^2$

$\qquad > 8(a^3b^3 + b^3c^3 + c^3a^3), \tag{25.2}$

we only have to show (25.2). Schur's inequality gives

$$a^6 + b^6 + c^6 + 3a^2b^2c^2 > a^4b^2 + a^2b^4 + b^4cc^2 + b^2c^4 + c^4a^2 + c^2a^4. \tag{25.3}$$

The AM-GM inequality gives

$$a^4b^2 + a^2b^4 \geq 2a^3b^3, b^4c^2 + b^2c^4 \geq 2b^3c^3, c^4a^2 + c^2a^4 \geq 2c^3a^3,$$

hence,

$$a^4b^2+a^2b^4+b^4c^2+b^2c^4+c^4a^2+c^2a^4 \geq 2(a^3b^3+b^3c^3+c^3a^3), \qquad (25.4)$$

therefore $(25.3) + 3 \times (25.4)$ yields (25.2) at once.

Example 6. (TURKEY/TST/2009) In a triangle ABC, the incircle touches the sides AB, AC and BC at C_1, B_1 and A_1 respectively. Prove that $\sqrt{\dfrac{AB_1}{AB}} + \sqrt{\dfrac{BC_1}{BC}} + \sqrt{\dfrac{CA_1}{CA}} \leq \dfrac{3}{\sqrt{2}}$ is true.

Solution Let $x = AB_1, y = BC_1, z = CA_1$, then the problem is to show

$$\sqrt{\frac{x}{x+y}} + \sqrt{\frac{y}{y+z}} + \sqrt{\frac{z}{z+x}} \leq \frac{3}{\sqrt{2}},$$

namely to show $\dfrac{1}{\sqrt{1+a^2}} + \dfrac{1}{\sqrt{1+b^2}} + \dfrac{1}{\sqrt{1+c^2}} \leq \dfrac{3}{\sqrt{2}}$, where $a, b, c > 0$ and $abc = 1$.

By the Cauchy-Schwartz inequality, $\dfrac{1}{\sqrt{1+c^2}} \leq \dfrac{\sqrt{2}}{1+c}$ and

$$\frac{1}{\sqrt{1+a^2}} + \frac{1}{\sqrt{1+b^2}} \leq \sqrt{2\left(\frac{1}{1+a^2} + \frac{1}{1+b^2}\right)}.$$

Since $\dfrac{1}{1+a^2} + \dfrac{1}{1+b^2} = 1 + \dfrac{1-a^2b^2}{(1+a^2)(1+b^2)} \leq 1 + \dfrac{1-a^2b^2}{(1+ab)^2} = \dfrac{2}{1+ab}$, so by the AM-GM inequality,

$$\frac{1}{\sqrt{1+a^2}} + \frac{1}{\sqrt{1+b^2}} + \frac{1}{\sqrt{1+c^2}} \leq 2\sqrt{\frac{c}{1+c}} + \frac{\sqrt{2}}{1+c}$$

$$= \frac{\sqrt{2}}{1+c}[\sqrt{2c(1+c)} + 1] \leq \frac{\sqrt{2}}{1+c}\left(\frac{2c+c+1}{2} + 1\right) = \frac{3}{\sqrt{2}}.$$

Example 7. (IRAN/TST/2008) Let $a, b, c > 0$ and $ab + bc + ca = 1$. Prove that:

$$\sqrt{a^3 + a} + \sqrt{b^3 + b} + \sqrt{c^3 + c} \geq 2\sqrt{a + b + c}.$$

Solution Below we use $\sum$ to denote a cyclic sum. Since $\sum ab = 1$,

$$\sum \sqrt{a^3 + a} \geq 2\sqrt{\sum a} \Leftrightarrow \sum \sqrt{a(a+b)(a+c)} \geq 2\sqrt{\left(\sum a\right) \cdot \left(\sum ab\right)}$$

$$\Leftrightarrow \sum a(a+b)(a+c) + 2\sum \sqrt{a(a+b)(a+c)} \cdot \sqrt{b(b+c)(b+a)}$$

$$\geq 4\sum a \cdot \sum ab$$

$\Leftrightarrow \sum a^3 + \sum ab(a+b) + 3abc + 2 \sum \sqrt{a(a+b)(a+c)} \cdot \sqrt{b(b+c)(b+a)}$
$\geq 4 \sum ab(a+b) + 12abc$

$\Leftrightarrow \sum a^3 - 3 \sum ab(a+b) - 9abc + 2 \sum \sqrt{a(a+b)(a+c)} \cdot \sqrt{b(b+c)(b+a)}$
$\geq 0.$ (*)

The Cauchy-Schwartz inequality gives

$$\sqrt{a(a+b)(a+c)} \cdot \sqrt{b(b+c)(b+a)}$$
$$= \sqrt{(a^3 + a^2c + a^2b + abc)(ab^2 + b^2c + b^3 + abc)}$$
$$\geq a^2b + abc + ab^2 + abc = ab(a+b) + 2abc,$$

so

$$\sum \sqrt{(a^3 + a^2c + a^2b + abc)(ab^2 + b^2c + b^3 + abc)} \geq \sum ab(a+b) + 6abc,$$

therefore

$$\sum a^3 - 3 \sum ab(a+b) - 9abc + 2 \sum \sqrt{a(a+b)(a+c)} \cdot \sqrt{b(b+c)(b+a)}$$
$$\geq \sum a^3 - 3 \sum ab(a+b) - 9abc + 2 \sum ab(a+b) + 12abc$$
$$= \sum a^3 - \sum ab(a+b) + 3abc = \sum a(a-b)(a-c) \geq 0.$$

The last inequality is from Schur's inequality. Thus (*) is proven and so the original inequality is also proven.

Example 8. (GREECE/2009) Given that x, y, z are all nonnegative and $x + y + z = 2$. Prove that
$$x^2y^2 + y^2z^2 + z^2x^2 + xyz \leq 1.$$

Solution To homogenize the both sides, rewrite the given inequality in the form

$$(x + y + z)^4 \geq 16(x^2y^2 + y^2z^2 + z^2x^2) + 8xyz(x + y + z). \qquad (25.5)$$

Since $(x + y + z)^4 = x^4 + y^4 + z^4 + 4(x^3y + xy^3 + y^3z + yz^3 + z^3x + zx^3) + 6(x^2y^2 + y^2z^2 + z^2x^2) + 12xyz(x + y + z)$, so (25.5) is equivalent to

$$x^4 + y^4 + z^4 + 4(x^3y + xy^3 + y^3z + yz^3 + z^3x + zx^3) - 10(x^2y^2$$
$$+ y^2z^2 + z^2x^2) + 4xyz(x + y + z) \geq 0. \qquad (25.6)$$

Schur's inequality gives $x^2(x-y)(x-z) + y^2(y-z)(y-x) + z^2(z-x)(z-y) \geq 0$, i.e., $x^4 + y^4 + z^4 - (x^3y + xy^3 + y^3z + yz^3 + z^3x + zx^3) + xyz(x + y + z) \geq 0.$

Hence to show (25.6) it suffices to show that

$$5(x^3y + xy^3 + y^3z + yz^3 + z^3x + zx^3) - 10(x^2y^2 + y^2z^2 + z^2x^2)$$
$$+3xyz(x + y + z) \geq 0. \tag{25.7}$$

The AM-GM inequality gives

$$x^3y + xy^3 \geq 2x^2y^2, \quad y^3z + yz^3 \geq 2y^2z^2, \quad z^3x + zx^3 \geq 2z^2x^2,$$

so $5(x^3y + xy^3 + y^3z + yz^3 + z^3x + zx^3) - 10(x^2y^2 + y^2z^2 + z^2x^2) \geq 0$, and $3xyz(x + y + z) \geq 0$ is obvious, so (25.7) holds, and the equality holds if and only if two of x, y, z are equal and the third is 0.

Example 9. (TURKEY/2009) Show that for all positive real numbers a, b, c,

$$\frac{(b + c)(a^4 - b^2c^2)}{ab + 2bc + ca} + \frac{(c + a)(b^4 - c^2a^2)}{bc + 2ca + ab} + \frac{(a + b)(c^4 - a^2b^2)}{ca + 2ab + bc} \geq 0.$$

Solution If $\dfrac{(b + c)(a^4 - b^2c^2)}{ab + 2bc + ca} \geq \dfrac{a^3 + abc - b^2c - bc^2}{2}$, then by Schur's inequality,

$$\sum \frac{(b + c)(a^4 - b^2c^2)}{ab + 2bc + ca} \geq \sum \frac{a^3 + abc - b^2c - bc^2}{2}$$
$$= \frac{1}{2}\sum(a^3 + abc - a^2b - ca^2) = \frac{1}{2}\sum a(a - b)(a - c) \geq 0.$$

Since

$$\frac{(b + c)(a^4 - b^2c^2)}{ab + 2bc + ca} \geq \frac{a^3 + abc - b^2c - bc^2}{2}$$
$$\Leftrightarrow 2(b + c)(a^4 - b^2c^2) - (ab + 2bc + ca)(a^3 + abc - b^2c - bc^2) \geq 0$$
$$\Leftrightarrow (b + c)a^4 - 2bca^3 - bc(b + c)a^2 + abc(b^2 + c^2) \geq 0.$$

By the Cauchy-Schwartz inequality and the AM-GM inequality,

$$(b + c)a^3 - 2bca^2 - bc(b + c)a + bc(b^2 + c^2)$$
$$\geq \frac{4bc}{(b + c)}a^3 - 2bca^2 - bc(b + c)a + bc\frac{(b + c)^2}{2}$$
$$= \frac{bc}{2(b + c)}[8a^3 - 4(b + c)a^2 - 2(b + c)^2a + (b + c)^3]$$
$$= \frac{bc}{2(b + c)}[2a(2a - b - c)^2 + (b + c)^3] \geq 0.$$

Thus, the original inequality is proven.

Testing Questions   (A)

1. (CMC/2009) If the inequality $\sqrt{x} + \sqrt{y} \le k\sqrt{2x + y}$ holds for any positive real numbers x, y, find the range of k.

2. (SERBIA/2009) Let x, y, z be positive real numbers satisfying $xy + yz + zx = x + y + z$. Prove that

$$\frac{1}{x^2 + y + 1} + \frac{1}{y^2 + z + 1} + \frac{1}{z^2 + x + 1} \le 1.$$

3. (TURKEY/2007) a, b, c are positive numbers and $a + b + c = 3$. Prove that

$$\frac{a^2 + 3b^2}{ab^2(4 - ab)} + \frac{b^2 + 3c^2}{bc^2(4 - bc)} + \frac{c^2 + 3a^2}{ca^2(4 - ca)} \ge 4.$$

4. (CMO/2008) Let a, b, c be positive real numbers for which $a + b + c = 1$. Prove that
$$\frac{a - bc}{a + bc} + \frac{b - ca}{b + ca} + \frac{c - ab}{c + ab} \le \frac{3}{2}.$$

5   (BELARUS/2008) Prove that for nonnegative real numbers $x_1, x_2, \ldots, x_n$ the following inequality holds:

$$\frac{x_1(2x_1 - x_2 - x_3)}{x_2 + x_3} + \frac{x_2(2x_2 - x_3 - x_4)}{x_3 + x_4} + \cdots + \frac{x_{n-1}(2x_{n-1} - x_n - x_1)}{x_n + x_1}$$

$$+ \frac{x_n(2x_n - x_1 - x_2)}{x_1 + x_2} \ge 0.$$

6. (MEDITERRANEAN MO/2010) Given positive real numbers $a_1, a_2 \ldots, a_n$ such that $n > 2$ and $a_1 + a_2 + \cdots + a_n = 1$. Prove the inequality

$$\frac{a_2 a_3 \cdots a_n}{a_1 + n - 2} + \frac{a_1 a_3 \cdots a_n}{a_2 + n - 2} + \cdots + \frac{a_1 a_2 \cdots a_{n-1}}{a_n + n - 2} \le \frac{1}{(n - 1)^2}.$$

7. (KOREA/2009) Let a, b, c be the lengths of three sides of a triangle. Let

$$A = \frac{a^2 + bc}{b + c} + \frac{b^2 + ca}{c + a} + \frac{c^2 + ab}{a + b}, \quad B = \frac{1}{\sqrt{(a + b - c)(b + c - a)}}$$

$$+ \frac{1}{\sqrt{(b + c - a)(c + a - b)}} + \frac{1}{\sqrt{(c + a - b)(a + b - c)}}.$$

Prove that $AB \ge 9$.

8. (CROATIA/TST/2010) Let a, b, c be positive real numbers such that $a + b + c = 3$. Prove that

$$\frac{a^4}{b^2 + c} + \frac{b^4}{c^2 + a} + \frac{c^4}{a^2 + b} \geq \frac{3}{2}.$$

9. (VIETNAM/2008) Let x, y, z be distinct non-negative real numbers. Prove that

$$\frac{1}{(x - y)^2} + \frac{1}{(y - z)^2} + \frac{1}{(z - x)^2} \geq \frac{4}{xy + yz + zx}.$$

When does the equality hold?

10 (IRAN/2008) Let $x, y, z > 0$ and $x + y + z = 3$. Prove that:

$$\frac{x^3}{y^3 + 8} + \frac{y^3}{z^3 + 8} + \frac{z^3}{x^3 + 8} \geq \frac{1}{9} + \frac{2}{27}(xy + yz + zx).$$

Testing Questions (B)

1. (TURKEY/2008) Let a, b, c be positive reals such that their sum is 1. Prove that

$$\frac{a^2b^2}{c^3(a^2 - ab + b^2)} + \frac{b^2c^2}{a^3(b^2 - bc + c^2)} + \frac{c^2a^2}{b^3(c^2 - ca + a^2)} \geq \frac{3}{ab + bc + ca}.$$

2. (CSMO/2008) Find the largest positive number λ such that the inequality

$$|\lambda xy + yz| \leq \frac{\sqrt{5}}{2}$$

holds for all real numbers x, y, z satisfying $x^2 + y^2 + z^2 = 1$.

3. (APMO/2007) Let x, y and z be positive real numbers such that $\sqrt{x} + \sqrt{y} + \sqrt{z} = 1$. Prove that

$$\frac{x^2 + yz}{\sqrt{2x^2(y + z)}} + \frac{y^2 + zx}{\sqrt{2y^2(z + x)}} + \frac{z^2 + xy}{\sqrt{2z^2(x + y)}} \geq 1.$$

4. (UKRAINE/2008) x, y, z are nonnegative real numbers, and $x^2 + y^2 + z^2 = 3$. Prove that

$$\frac{x}{\sqrt{x^2 + y + z}} + \frac{y}{\sqrt{y^2 + z + x}} + \frac{z}{\sqrt{z^2 + x + y}} \leq \sqrt{3}.$$

5. (CGMO/2008) Let $\varphi(x) = ax^3 + bx^2 + cx + d$ be a polynomial with real coefficients. Given that $\varphi(x)$ has three positive real roots and that $\varphi(0) < 0$, prove that

$$2b^3 + 9a^2d - 7abc \leq 0.$$

Lecture 26

Some Important Inequalities (II)

Theorem I. (Rearrangement Inequality) *Let $a_1 \le a_2 \le \cdots \le a_n$ and $b_1 \le b_2 \le \cdots \le b_n$ be two groups of ordered real numbers. For a permutation $(j_1, j_2, \cdots, j_n)$ of $(1, 2, \cdots, n)$, the sums given by*

$$
\begin{aligned}
S_O &= a_1 b_1 + a_2 b_2 + \cdots + a_n b_n, \quad \textbf{(ordered sum)} \\
S_M &= a_1 b_{j_1} + a_2 b_{j_2} + \cdots + a_n b_{j_n}, \quad \textbf{(mixed sum)} \\
S_R &= a_1 b_n + a_2 b_{n-1} + \cdots + a_n b_1, \quad \textbf{(reverse sum)}
\end{aligned}
$$

must obey the inequalities : $S_R \le S_M \le S_O$.

Furthermore, $S_M = S_O$ for all S_M (or $S_M = S_R$ for all S_M) if and only if $a_1 = a_2 = \cdots = a_n$ or $b_1 = b_2 = \cdots = b_n$.

From the rearrangement theorem the following theorem can be derived easily, it has been used often in recent years in Mathematical Olympiad competitions.

Theorem II. (Chebyshev's Inequality) *Let $a_1 \le a_2 \le \cdots \le a_n$ and $b_1 \le b_2 \le \cdots \le b_n$ be two groups of ordered real numbers. Then*

$$
\frac{1}{n} \sum_{i=1}^{n} a_i b_{n+1-i} \le \left(\frac{1}{n} \sum_{i=1}^{n} a_i \right) \left(\frac{1}{n} \sum_{i=1}^{n} b_i \right) \le \frac{1}{n} \sum_{i=1}^{n} a_i b_i.
$$

The following consequences are some common ways that Chebyshev's inequality is applied.

Consequence 1. For $x_1, x_2, \ldots, x_n > 0$,

$$
\sum_{i=1}^{n} x_i^{p+q} \ge \frac{1}{n} \sum_{i=1}^{n} x_i^{p} \cdot \sum_{i=1}^{n} x_i^{q} \quad \text{when } pq > 0;
$$

$$
\sum_{i=1}^{n} x_i^{p+q} \le \frac{1}{n} \sum_{i=1}^{n} x_i^{p} \cdot \sum_{i=1}^{n} x_i^{q} \quad \text{when } pq < 0.
$$

85

Consequence 2. $x_1, x_2, \ldots, x_n > 0, r > s > 0$ and $\prod_{i=1}^{n} x_i = 1$, then

$$\sum_{i=1}^{n} x_i^r \geq \sum_{i=1}^{n} x_i^s.$$

Consequence 3. For $a_1 \leq a_2 \leq \cdots \leq a_n; b_1 \leq b_2 \leq \cdots \leq b_n$ and $m_1, m_2, \cdots m_n > 0$,

$$\sum_{i=1}^{n} m_i \cdot \sum_{i=1}^{n} a_i b_i \geq \sum_{i=1}^{n} \sqrt{m_i a_i} \cdot \sum_{i=1}^{n} \sqrt{m_i b_i}.$$

Definition 26.1. A function f defined on some interval $I = [a, b]$ is said to be **convex** (or **concave upward**) if it satisfies the following requirement :

$$f\left(\frac{x_x + x_2}{2}\right) \leq \frac{1}{2}[f(x_1) + f(x_2)], \qquad x_1, x_2 \in I. \tag{26.1}$$

Further, f is said to be **strictly convex** provided the equality holds if and only if $x_1 = x_2$.

Jensen' Inequality. For any strictly convex function $f(x)$ which is defined on $I = [a, b]$ (or $I = (a, b)$), the inequality

$$f\left(\frac{x_1 + x_2 + \cdots + x_n}{n}\right) \leq \frac{1}{n}[f(x_1) + f(x_2) + \cdots + f(x_n)] \tag{26.2}$$

holds for any $x_1, x_2, \cdots, x_n \in I$. Furthermore, the equality holds if and only if $x_1 = x_2 = \cdots = x_n$.

For continuous convex functions we have the following extended Jensen's inequality.

Theorem III. (Weighted Jensen's Inequality). *Given that f is a continuous function defined on some interval $I = [a, b]$ (or $I = (a, b)$). Then f is strictly convex on I if and only if for any positive integer $n \in \mathbb{N}$ and $\lambda_1, \lambda_2, \cdots \lambda_n > 0$ satisfying $\sum_{i=1}^{n} \lambda_i = 1$,*

$$f(\lambda_1 x_1 + \lambda_2 x_2 + \cdots + \lambda_n x_n) \leq \lambda_1 f(x_1) + \lambda_2 f(x_2) + \cdots + \lambda_n f(x_n), \tag{26.3}$$

for all $x_1, \cdots, x_n \in I$, and the equality holds if and only if $x_1 = x_2 = \cdots = x_n$.

Applying Jensen's inequality yields the following often used theorem.

Theorem IV. (Power Mean Inequality) *For any nonnegative real numbers x_1, $x_2, \cdots, x_n$ and $\alpha, \beta > 0$ with $\alpha > \beta$,*

$$\left(\frac{1}{n}\sum_{i=1}^{n} x_i^{\alpha}\right)^{\frac{1}{\alpha}} \geq \left(\frac{1}{n}\sum_{i=1}^{n} x_i^{\beta}\right)^{\frac{1}{\beta}}. \tag{26.4}$$

Examples

Example 1. (POLAND/2007) a, b, c are positive real numbers satisfying $abc = 1$. Prove that

$$\frac{c}{b} + \frac{b}{a} + \frac{a}{c} \le a^3 b + b^3 c + c^3 a.$$

Solution Since $abc = 1$, let $a = \frac{x}{y}, b = \frac{y}{z}, c = \frac{z}{x}$. Then

$$\frac{c}{b} + \frac{b}{a} + \frac{a}{c} \le a^3 b + b^3 c + c^3 a$$

$$\Leftrightarrow \frac{z^2}{xy} + \frac{y^2}{zx} + \frac{x^2}{yz} \le \frac{x^3}{y^2 z} + \frac{y^3}{z^2 x} + \frac{z^3}{x^2 y}$$

$$\Leftrightarrow x^3 + y^3 + z^3 \le \frac{x^4}{y} + \frac{y^4}{z} + \frac{z^4}{x}.$$

Since the order of x^4, y^4, z^4 in magnitude and that of $\frac{1}{x}, \frac{1}{y}, \frac{1}{z}$ are reverse, by the rearrangement inequality, any mixed sum is not less than the reverse sum, so

$$x^3 + y^3 + z^3 = \frac{x^4}{x} + \frac{y^4}{y} + \frac{z^4}{z} \le \frac{x^4}{y} + \frac{y^4}{z} + \frac{z^4}{x}.$$

Example 2. (CGMO/2006) For $x_1, x_2, \ldots, x_n > 0$ and $k \ge 1$, prove that

$$\sum_{i=1}^{n} \frac{1}{1 + x_i} \cdot \sum_{i=1}^{n} x_i \le \sum_{i=1}^{n} \frac{1}{x_i^k} \cdot \sum_{i=1}^{n} \frac{x_i^{k+1}}{1 + x_i}.$$

Solution Without loss of generality, assume that $x_1 \ge x_2 \ge \cdots \ge x_n$. Then

$$\frac{1}{x_1^k} \le \frac{1}{x_2^k} \le \cdots \le \frac{1}{x_n^k}$$

and, since $x_i^k (1 + x_j) - x_j^k (1 + x_i) = x_i x_j (x_i^{k-1} - x_j^{k-1}) + (x_i^k - x_j^k) \ge 0$ for $1 \le i < j \le n$,

$$\frac{x_1^k}{1 + x_1} \ge \frac{x_2^k}{1 + x_2} \ge \cdots \ge \frac{x_n^k}{1 + x_n}.$$

By using Chebyshev's inequality twice,

$$\sum_{i=1}^{n} \frac{1}{1 + x_i} \cdot \sum_{i=1}^{n} x_i = \sum_{i=1}^{n} x_i \cdot \sum_{i=1}^{n} \left(\frac{1}{x_i^k} \cdot \frac{x_i^k}{1 + x_i} \right)$$

$$\le \frac{1}{n} \sum_{i=1}^{n} x_i \cdot \sum_{i=1}^{n} \frac{1}{x_i^k} \cdot \sum_{i=1}^{n} \frac{x_i^k}{1 + x_i}$$

$$\le \sum_{i=1}^{n} \frac{1}{x_i^k} \cdot \sum_{i=1}^{n} \left(x_i \cdot \frac{x_i^k}{1 + x_i} \right) = \sum_{i=1}^{n} \frac{1}{x_i^k} \cdot \sum_{i=1}^{n} \frac{x_i^{k+1}}{1 + x_i}.$$

Example 3. (CSMO/2007) Let $a, b, c > 0$ and $abc = 1$. Prove that for integer $k \geq 2$,

$$\frac{a^k}{b+c} + \frac{b^k}{c+a} + \frac{c^k}{a+b} \geq \frac{3}{2}.$$

Solution   Without loss of generality we may assume that $a \geq b \geq c$. Then

$$\frac{a^k}{b+c} \geq \frac{b^k}{c+a} \geq \frac{c^k}{a+b} \quad \text{and} \quad b+c \leq c+a \leq a+b.$$

Therefore Chebyshev's inequality gives

$$\left(\frac{a^k}{b+c} + \frac{b^k}{c+a} + \frac{c^k}{a+b} \right) (b+c+c+a+a+b) \geq 3(a^k + b^k + c^k).$$

From Consequence 2 of Chebyshev's inequality, $a^k + b^k + c^k \geq a + b + c$, so

$$\frac{a^k}{b+c} + \frac{b^k}{c+a} + \frac{c^k}{a+b} \geq \frac{3(a^k + b^k + c^k)}{2(a+b+c)} \geq \frac{3}{2}.$$

Example 4. (CHNMO/TST//2006) Let $x_1, x_2, \ldots, x_n > 0$ and $x_1 + x_2 + \cdots + x_n = 1$. Prove that

$$\prod_{k=1}^{n} \frac{1 + x_k}{x_k} \geq \prod_{k=1}^{n} \frac{n - x_k}{1 - x_k}.$$

Solution   Let $f(x) = \ln \left(1 + \frac{1}{x}\right)$, $0 < x < 1$. For any $0 < x_1 < x_2 < 1$,

$$f\left(\frac{x_1 + x_2}{2}\right) \leq \frac{1}{2}[f(x_1) + f(x_2)] \Leftrightarrow 1 + \frac{2}{x_1 + x_2} \leq \sqrt{\left(1 + \frac{1}{x_1}\right)\left(1 + \frac{1}{x_2}\right)}$$

$$\Leftrightarrow \frac{4}{(x_1 + x_2)^2} + \frac{4}{x_1 + x_2} \leq \frac{1}{x_1 x_2} + \frac{1}{x_1} + \frac{1}{x_2} \Leftrightarrow (x_1 + x_2)^2 \geq 4 x_1 x_2.$$

Therefore $f(x)$ is a continuous convex function on $(0, 1)$. By the Weighted Jensen's inequality, for any fixed $k = 1, 2, \ldots, n$,

$$\frac{1}{n - 1} \sum_{\substack{i=1, \\ i \neq k}}^{n} \ln\left(1 + \frac{1}{x_i}\right) \geq \ln\left(1 + \frac{n - 1}{\sum_{\substack{i=1, \\ i \neq k}}^{n} x_i}\right) = \ln\left(\frac{n - x_k}{1 - x_k}\right),$$

i.e.,

$$\prod_{\substack{i=1, \\ i \neq k}}^{n} \left(1 + \frac{1}{x_i}\right) \geq \left[\frac{n - x_k}{1 - x_k}\right]^{n-1}.$$

Multiplying these n inequalities up, the desired inequality is obtained:

$$\prod_{k=1}^{n} \frac{1+x_k}{x_k} \geq \prod_{k=1}^{n} \frac{n-x_k}{1-x_k}.$$

Example 5. (SMO/2009) Find the largest constant C such that $\sum_{i=1}^{4}\left(x_i + \frac{1}{x_i}\right)^3 \geq$ C for all positive real numbers x_1, x_2, x_3, x_4 such that

$$x_1^3 + x_3^3 + 3x_1x_3 = x_2 + x_4 = 1.$$

Solution Note that $0 = x_1^3 + x_3^3 + 3x_1x_3 - 1 = x_1^3 + x_3^3 + (-1)^3 - 3x_1x_3(-1)$ implies that

$$\frac{1}{2}(x_1 + x_3 - 1)[(x_1 - x_3)^2 + (x_1 + 1)^2 + (x_3 + 1)^2] = 0 \Leftrightarrow x_1 + x_3 = 1$$

as $x_1, x_3 > 0$. Thus, it suffices to find the maximum C such that $\sum_{i=1}^{2}\left(y_i + \frac{1}{y_i}\right)^3$ $\geq \frac{C}{2}$ under the restrictions $y_1, y_2 > 0$ and $y_1 + y_2 = 1$.

Consider the function $F(x) = \left(x + \frac{1}{x}\right)^3, 0 < x < 1$. It is easy by induction on n to show that $f(x) = x^n, x > 0$ is an increasing convex function for any $n \in \mathbb{N}$. Since $g_1(x) = x$ and $g_2(x) = \frac{1}{x}$ are both convex on $x > 0$, so their sum $g(x) = g_1(x) + g_2(x)$ is also convex on $x > 0$. Then $F(x) = f(g(x))$ is also convex on $x > 0$ since for $0 < x_1 < x_2$,

$$\frac{1}{2}[F(x_1) + F(x_2)] = \frac{1}{2}[f(g(x_1)) + f(g(x_2))] \geq f\left(\frac{g(x_1) + g(x_2)}{2}\right)$$

$$\geq f\left(g\left(\frac{x_1 + x_2}{2}\right)\right) = F\left(\frac{x_1 + x_2}{2}\right).$$

Then

$$\frac{1}{2}[F(x) + F(1-x)] \geq F\left(\frac{x}{2} + \frac{1-x}{2}\right) = F\left(\frac{1}{2}\right)$$

implies that $\frac{C_{\max}}{2} = 2F\left(\frac{1}{2}\right) = 2 \cdot \left(\frac{5}{2}\right)^3 = \frac{125}{4}$, hence $C_{\max} = \frac{125}{2}$.

Example 6. (CROATIA/2008) a, b, c are positive numbers such that $a + b + c = 1$. Prove that

$$\frac{1}{bc + a + \dfrac{1}{a}} + \frac{1}{ca + b + \dfrac{1}{b}} + \frac{1}{ab + c + \dfrac{1}{c}} \leq \frac{27}{31}.$$

Solution $\dfrac{1}{bc + a + \frac{1}{a}} + \dfrac{1}{ca + b + \frac{1}{b}} + \dfrac{1}{ab + c + \frac{1}{c}} \le \dfrac{27}{31}$ is equivalent to

$$\frac{9a^2 + 9abc + 9 - 31a}{a^2 + abc + 1} + \frac{9b^2 + 9abc + 9 - 31b}{b^2 + abc + 1} + \frac{9c^2 + 9abc + 9 - 31c}{c^2 + abc + 1} \ge 0.$$

We may assume $a \ge b \ge c$. It is clear that $9(a + b) < 31$, so

$$9a^2 + 9abc + 9 - 31a \le 9b^2 + 9abc + 9 - 31b \le 9c^2 + 9abc + 9 - 31c.$$

Since $a^2 \ge b^2 \ge c^2 \Rightarrow \dfrac{1}{a^2 + abc + 1} \le \dfrac{1}{b^2 + abc + 1} \le \dfrac{1}{c^2 + abc + 1}$, by Chebyshev's inequality,

$$3 \sum_{a,b,c} \frac{9a^2 + 9abc + 9 - 31a}{a^2 + abc + 1}$$

$$\ge \left[\sum_{a,b,c} (9a^2 + 9abc + 9 - 31a) \right] \cdot \left[\sum_{a,b,c} \frac{1}{a^2 + abc + 1} \right].$$

Thus, it suffices to show that

$$\sum_{a,b,c} (9a^2 + 9abc + 9 - 31a) \ge 0$$
$$\Leftrightarrow 9(a^2 + b^2 + c^2) + 27abc + 27 - 31(a + b + c) \ge 0$$
$$\Leftrightarrow 9(a^2 + b^2 + c^2) + 27abc - 4 \ge 0$$
$$\Leftrightarrow 9(a^2 + b^2 + c^2)(a + b + c) + 27abc - 4(a + b + c)^3 \ge 0$$

which is equivalent to

$$5(a^3 + b^3 + c^3) - 3(a^2b + ab^2 + b^2c + bc^2 + c^2a + ca^2) + 3abc \ge 0. \qquad (*)$$

Schur's inequality gives

$$a^3 + b^3 + c^3 + 3abc \ge a^2b + ab^2 + b^2c + bc^2 + c^2a + ca^2 \qquad (**)$$

and the AM-GM inequality gives

$$a^3 + b^3 + c^3 \ge 3abc. \qquad (***)$$

By $3 \times (**) + 2 \times (***)$, we obtain the inequality $(*)$.

Testing Questions   (A)

1. Let $a, b > 0$ and $p, q > 0$. Prove that $a^{p+q} b^{p+q} \geq a^p b^q + a^q b^p$, and the equality holds if and only if $a = b$.

2. Given that a, b, c are the lengths of three sides of a triangle ABC. Prove that
$$a^2(b + c - a) + b^2(c + a - b) + c^2(a + b - c) \leq 3abc.$$

3. (BALKAN/2007) When $a, b, c > 0$ and $\dfrac{1}{a+b+1} + \dfrac{1}{b+c+1} + \dfrac{1}{c+a+1} \geq 1$, prove that
$$a + b + c \geq ab + bc + ca.$$

4. (SERBIA/2007) Let k be a given natural number. Prove that for any positive numbers x, y, z with sum 1 the following inequality holds:
$$\frac{x^{k+2}}{x^{k+1} + y^k + z^k} + \frac{y^{k+2}}{y^{k+1} + z^k + x^k} + \frac{z^{k+2}}{z^{k+1} + x^k + y^k} \geq \frac{1}{7}.$$

   When does equality occur?

5. (KOREA/2010) Given an arbitrary triangle ABC, denote by P, Q, R the intersections of the incircle with sides BC, CA, AB respectively. Let the area of triangle ABC be T, and its perimeter be L. Prove that the inequality
$$\left(\frac{AB}{PQ}\right)^3 + \left(\frac{BC}{QR}\right)^3 + \left(\frac{CA}{RP}\right)^3 \geq \frac{2}{\sqrt{3}} \cdot \frac{L^2}{T}$$
   holds.

6. (**Höder's Inequality**) Prove that for any $x_i, y_i > 0$, $i = 1, 2, \cdots, n$ and $p, q > 1$ with $\dfrac{1}{p} + \dfrac{1}{q} = 1$,
$$\sum_{i=1}^{n} x_i y_i \leq \left(\sum_{i=1}^{n} x_i^p\right)^{\frac{1}{p}} \cdot \left(\sum_{i=1}^{n} y_i^q\right)^{\frac{1}{q}}.$$

7. (IRELAND/2008) Let x, y, z be positive real numbers such that $xyz = 1$. Prove that

   (i) $27 \leq (1+x+y)^2 + (1+y+z)^2 + (1+x+z)^2$, where the equality holds if and only if $x = y = z = 1$.

   (ii) $(1+x+y)^2 + (1+y+z)^2 + (1+x+z)^2 \leq 3(x+y+z)^2$, where the equality holds if and only if $x = y = z = 1$.

Testing Questions   (B)

1. (CHNMO/TST/2007) $a_1, a_2, \ldots, a_n$ are positive real numbers satisfying $a_1 + a_2 + \cdots + a_n = 1$. Prove that

$$\sum_{i=1}^{n} a_i a_{i+1} \cdot \sum_{i=1}^{n} \frac{a_i}{a_{i+1}^2 + a_{i+1}} \ge \frac{n}{n+1}.$$

2. (THAILAND/2010) a, b, c are real numbers and $a + b + c = 3$. Prove that

$$\frac{1}{\sqrt{2(a^2 + bc)}} + \frac{1}{\sqrt{2(b^2 + ca)}} + \frac{1}{\sqrt{2(c^2 + ab)}}$$

$$\ge \frac{1}{a + bc} + \frac{1}{b + ca} + \frac{1}{c + ab}.$$

3. (TURKEY/2010) For positive numbers a, b, c prove that

$$\sum_{cyc} \sqrt[4]{\frac{(a^2 + b^2)(a^2 - ab + b^2)}{2}} \le \frac{2}{3} \sum_{cyc} a^2 \cdot \sum_{cyc} \frac{1}{a + b}.$$

4. (IRAN/TST/2008) Let $a, b, c > 0$ and $ab + bc + ca = 1$. Prove that:

$$\sqrt{a^3 + a} + \sqrt{b^3 + b} + \sqrt{c^3 + c} \ge 2\sqrt{a + b + c}.$$

5. (LITHUANIA/TST/2009-2010) Let a, b, c be the lengths of three sides of a triangle, and let

$$A = \sum_{cyc} \frac{a^2 + bc}{b + c}, \qquad B = \sum_{cyc} \frac{1}{\sqrt{(a + b - c)(b + c - a)}}.$$

Prove that $AB \ge 9$.

Lecture 27

Some Methods For Solving Inequalities

In this lecture, the following common methods for solving inequalities are introduced along with some examples:

Algebraic Manipulation

In many problems of involving inequalities, the first step is "to use algebraic manipulation to simplify both side of the given inequality such that solving it becomes much easier, or such that we have a new foundation for further progress.

Example 1. (SAUDI ARABIA/2010) Let a, b, c be positive real numbers and $abc = 8$. Prove that

$$\frac{a-2}{a+1} + \frac{b-2}{b+1} + \frac{c-2}{c+1} \leq 0.$$

Solution $\frac{a-2}{a+1} + \frac{b-2}{b+1} + \frac{c-2}{c+1} = 3 - 3\left(\frac{1}{a+1} + \frac{1}{b+1} + \frac{1}{c+1}\right)$, so

$$\frac{a-2}{a+1} + \frac{b-2}{b+1} + \frac{c-2}{c+1} \leq 0 \Leftrightarrow \frac{1}{a+1} + \frac{1}{b+1} + \frac{1}{c+1} \geq 1$$

$$\Leftrightarrow \frac{ab+bc+ca+2a+2b+2c+3}{9+ab+bc+ca+a+b+c} \geq 1 \Leftrightarrow a+b+c \geq 6.$$

By the AM-GM inequality, $a+b+c \geq 3\sqrt[3]{abc} = 6$, so the conclusion is proven.

Example 2. (CROATIA/2010) Let a, b, c be positive real numbers satisfying $a^2 + b^2 + c^2 = \frac{1}{2}$. Prove that

$$\frac{1-a^2+c^2}{c(a+2b)} + \frac{1-b^2+a^2}{a(b+2c)} + \frac{1-c^2+b^2}{b(c+2a)} \geq 6.$$

Solution   By substituting the given equality into the inequality, the original inequality is equivalent to

$$\frac{a^2 + 2b^2 + 3c^2}{ac + 2bc} + \frac{b^2 + 2c^2 + 3a^2}{ba + 2ca} + \frac{c^2 + 2a^2 + 3b^2}{cb + 2ab} \geq 6.$$

$a^2 + 2b^2 + 3c^2 = (a^2 + c^2) + 2(b^2 + c^2) \geq 2ac + 4bc = 2(ac + 2bc)$ implies that

$$\frac{a^2 + 2b^2 + 3c^2}{ac + 2bc} \geq 2,$$

and similarly, $\dfrac{b^2 + 2c^2 + 3a^2}{ba + 2ca} \geq 2, \dfrac{c^2 + 2a^2 + 3b^2}{cb + 2ab} \geq 2.$ By adding up them, we have

$$\frac{a^2 + 2b^2 + 3c^2}{ac + 2bc} + \frac{b^2 + 2c^2 + 3a^2}{ba + 2ca} + \frac{c^2 + 2a^2 + 3b^2}{cb + 2ab} \geq 6.$$

Substitution of Variables or Expressions

Substitution of variables or expressions is a natural and useful method for simplifying inequalities, as shown in the following examples.

Example 3. (HUNGARY-ISRAEL/2009) Let x, y and z be non negative numbers. Prove that

$$\frac{x^2 + y^2 + z^2 + xy + yz + zx}{6} \leq \frac{x + y + z}{3} \cdot \sqrt{\frac{x^2 + y^2 + z^2}{3}}.$$

Solution   When $x = y = z = 0$, the inequality is obviously true. Below we assume that (x, y, z) are not all zeros. Then

$$\frac{x^2 + y^2 + z^2 + xy + yz + zx}{6} \leq \frac{x + y + z}{3} \cdot \sqrt{\frac{x^2 + y^2 + z^2}{3}}$$

$$\Leftrightarrow \frac{(x^2 + y^2 + z^2 + xy + yz + zx)^2}{4} \leq \frac{(x + y + z)^2 (x^2 + y^2 + z^2)}{3}$$

$$\Leftrightarrow \frac{4(x + y + z)^2}{x^2 + y^2 + z^2 + xy + yz + zx} \geq \frac{3(x^2 + y^2 + z^2 + xy + yz + zx)}{x^2 + y^2 + z^2}$$

$$\Leftrightarrow 4\left(1 + \frac{xy + yz + zx}{x^2 + y^2 + z^2 + xy + yz + zx}\right) \geq 3\left(1 + \frac{xy + yz + zx}{x^2 + y^2 + z^2}\right).$$

If $xy+yz+zx = 0$, then the last inequality is clear. When $xy+yz+zx \neq 0$, let $\dfrac{x^2 + y^2 + z^2}{xy + yz + zx} = t$. Then the original inequality is equivalent to

$$4\left(1 + \frac{1}{t+1}\right) \geq 3\left(1 + \frac{1}{t}\right) \Leftrightarrow 1 + \frac{4}{t+1} \geq \frac{3}{t}$$
$$\Leftrightarrow t^2 + 2t - 3 \geq 0 \Leftrightarrow (t+3)(t-1) \geq 0.$$

The last inequality is clear since $t \geq 1$, so the original inequality is proven.

Example 4. (IMO/Shortlist/2007) Let n be a positive integer, and let x and y be positive real numbers such that $x^n + y^n = 1$. Prove that

$$\left(\sum_{k=1}^{n} \frac{1+x^{2k}}{1+x^{4k}}\right)\left(\sum_{k=1}^{n} \frac{1+y^{2k}}{1+y^{4k}}\right) < \frac{1}{(1-x)(1-y)}.$$

Solution For each real $t \in (0,1)$,

$$\frac{1+t^2}{1+t^4} = \frac{1}{t} - \frac{(1-t)(1-t^3)}{t(1+t^4)} < \frac{1}{t}.$$

Substituting $t = x^k$ and $t = y^k$, the inequality above gives

$$0 < \sum_{k=1}^{n} \frac{1+x^{2k}}{1+x^{4k}} < \sum_{k=1}^{n} \frac{1}{x^k} = \frac{1-x^n}{x^n(1-x)}$$

and

$$0 < \sum_{k=1}^{n} \frac{1+y^{2k}}{1+y^{4k}} < \sum_{k=1}^{n} \frac{1}{y^k} = \frac{1-y^n}{y^n(1-y)}.$$

Since $1 - y^n = x^n$ and $1 - x^n = y^n$,

$$\frac{1-x^n}{x^n(1-x)} = \frac{y^n}{x^n(1-x)}, \qquad \frac{1-y^n}{y^n(1-y)} = \frac{x^n}{y^n(1-y)}$$

and therefore

$$\left(\sum_{k=1}^{n} \frac{1+x^{2k}}{1+x^{4k}}\right)\left(\sum_{k=1}^{n} \frac{1+y^{2k}}{1+y^{4k}}\right) < \frac{y^n}{x^n(1-x)} \cdot \frac{x^n}{y^n(1-y)} = \frac{1}{(1-x)(1-y)}.$$

Enlargement and Compression of One side of an Inequality

When proving an inequality, if it's impossible or very difficult to do a direct comparison of two sides at the beginning, enlarging the smaller side or compressing the bigger side one or more times is a common technique. Applying this technique, the direct comparison of two sides becomes much easier or obvious, as shown in the following examples.

Example 5. (CROATIA/TST/2010) Let a, b, c be positive real numbers, and $a + b + c = 3$. Prove that

$$\frac{a^4}{b^2 + c} + \frac{b^4}{c^2 + a} + \frac{c^4}{a^2 + b} \geq \frac{3}{2}.$$

Solution   By the Cauchy-Schwartz inequality,

$$(b^2 + c + c^2 + a + a^2 + b) \cdot \left(\frac{a^4}{b^2 + c} + \frac{b^4}{c^2 + a} + \frac{c^4}{a^2 + b} \right) \geq (a^2 + b^2 + c^2)^2,$$

So

$$\frac{a^4}{b^2 + c} + \frac{b^4}{c^2 + a} + \frac{c^4}{a^2 + b} \geq \frac{(a^2 + b^2 + c^2)^2}{a^2 + b^2 + c^2 + 3}.$$

Let $a^2 + b^2 + c^2 = x$. Then $3(a^2 + b^2 + c^2) \geq (a + b + c)^2 \Rightarrow x \geq 3$. Therefore

$$\frac{x^2}{3 + x} \geq \frac{3}{2} \Leftrightarrow 2x^2 \geq 9 + 3x \Leftrightarrow 2x^2 - 3x - 9 \geq 0 \Leftrightarrow (2x + 3)(x - 3) \geq 0.$$

The last inequality is obvious since $x \geq 3$. Thus, the original inequality is proven.

Example 6. (RUSMO/2008) Let $x_1 \geq x_2 \geq \cdots \geq x_n \geq 0$ and $\displaystyle\sum_{i=1}^{n} \frac{x_i}{\sqrt{i}} = 1$.
Prove that

$$\sum_{i=1}^{n} x_i^2 \leq 1.$$

Solution   Since $0 \leq x_n \leq x_{n-1} \leq \cdots \leq x_1$, so for any $k = 1, 2, \ldots, n$,

$$1 \geq \sum_{i=1}^{k} \frac{x_i}{\sqrt{i}} \geq \left(\sum_{i=1}^{k} \frac{1}{\sqrt{i}} \right) x_k \geq \frac{k}{\sqrt{k}} x_k = \sqrt{k} x_k,$$

therefore

$$x_k^2 \leq \frac{x_k}{\sqrt{k}} \Rightarrow \sum_{k=1}^{n} x_k^2 \leq \sum_{k=1}^{n} \frac{x_k}{\sqrt{k}} = 1.$$

Example 7. (GREECE/2008) Let x, y, z be positive real numbers satisfying $x, y, z <$
2 and $x^2 + y^2 + z^2 = 3$. Prove that

$$\frac{3}{2} < \frac{1+y^2}{2+x} + \frac{1+z^2}{2+y} + \frac{1+x^2}{2+z} < 3.$$

Solution For the left inequality, since $x, y, z < 2$, so

$$\frac{1+y^2}{2+x} + \frac{1+z^2}{2+y} + \frac{1+x^2}{2+z} > \frac{1+y^2}{4} + \frac{1+z^2}{4} + \frac{1+x^2}{4} = \frac{3}{2}.$$

For the right inequality, since $x, y, z > 0$,

$$\frac{1+y^2}{2+x} + \frac{1+z^2}{2+y} + \frac{1+x^2}{2+z} < \frac{1+y^2}{2} + \frac{1+z^2}{2} + \frac{1+x^2}{2} = 3.$$

Localization of An Inequality

Sometimes, to prove an inequality, it is possible to partition the inequality into a few or several parts. Each part can be dealt with individually, and the entire inequality can be proven by simply combining the results of the parts.

Example 8. (IMO/2005) Let x, y and z be positive real numbers such that $xyz \geq$ 1. Prove the inequality

$$\frac{x^5 - x^2}{x^5 + y^2 + z^2} + \frac{y^5 - y^2}{y^5 + z^2 + x^2} + \frac{z^5 - z^2}{z^5 + x^2 + y^2} \geq 0.$$

Solution To compress each term, we notice that in the first term on the left hand side, the numerator contains the factor $(x^3 - 1)$, and $x^3(x^2 + y^2 + z^2) - (x^5 + y^2 + z^2) = (x^3 - 1)(y^2 + z^2)$ also contains the factor $(x^3 - 1)$. Therefore the denominator of the first term can be changed to $x^3(x^2 + y^2 + z^2)$ as follows:

$$\frac{x^5 - x^2}{x^5 + y^2 + z^2} - \frac{x^5 - x^2}{x^3(x^2 + y^2 + z^2)} = \frac{x^2(x^3 - 1)^2(y^2 + z^2)}{x^3(x^5 + y^2 + z^2)(x^2 + y^2 + z^2)} \geq 0$$

$$\Rightarrow \frac{x^5 - x^2}{x^5 + y^2 + z^2} \geq \frac{x^5 - x^2}{x^3(x^2 + y^2 + z^2)}.$$

By considering the other two terms similarly, we obtain

$$\sum \frac{x^5 - x^2}{x^5 + y^2 + z^2} \geq \sum \frac{x^5 - x^2}{x^3(x^2 + y^2 + z^2)} = \frac{1}{x^2 + y^2 + z^2} \sum \left(x^2 - \frac{1}{x} \right)$$

$$\geq \frac{1}{x^2 + y^2 + z^2} \sum (x^2 - yz) = \frac{1}{2(x^2 + y^2 + z^2)} \sum (x - y)^2 \geq 0.$$

The equality holds if and only if $x = y = z = 1$.

Example 9. (IMO/Shortlist/2006) Let a, b, c be the sides of a triangle. Prove that

$$\frac{\sqrt{b+c-a}}{\sqrt{b}+\sqrt{c}-\sqrt{a}} + \frac{\sqrt{c+a-b}}{\sqrt{c}+\sqrt{a}-\sqrt{b}} + \frac{\sqrt{a+b-c}}{\sqrt{a}+\sqrt{b}-\sqrt{c}} \le 3.$$

Solution   Note first that the denominators are all positive, e.g. $\sqrt{a} + \sqrt{b} > \sqrt{a+b} > \sqrt{c}$.

Let $x = \sqrt{b} + \sqrt{c} - \sqrt{a}$, $y = \sqrt{c} + \sqrt{a} - \sqrt{b}$, and $z = \sqrt{a} + \sqrt{b} - \sqrt{c}$. Then

$$b + c - a = \left(\frac{x+z}{2}\right)^2 + \left(\frac{x+y}{2}\right)^2 - \left(\frac{y+z}{2}\right)^2 = \frac{x^2 + xy + xz - yz}{2}$$

$$= x^2 - \frac{1}{2}(x-y)(x-z)$$

and

$$\frac{\sqrt{b+c-a}}{\sqrt{b}+\sqrt{c}-\sqrt{a}} = \sqrt{1 - \frac{(x-y)(x-z)}{2x^2}} \le 1 - \frac{(x-y)(x-z)}{4x^2},$$

where we applied the inequality $\sqrt{1+2u} \le 1 + u$ to obtain the last inequality. Similarly, we obtain

$$\frac{\sqrt{c+a-b}}{\sqrt{c}+\sqrt{a}-\sqrt{b}} \le 1 - \frac{(y-x)(y-z)}{4y^2}, \quad \frac{\sqrt{a+b-c}}{\sqrt{a}+\sqrt{b}-\sqrt{c}} \le 1 - \frac{(z-y)(z-x)}{4z^2}.$$

Substituting these quantities into the original inequality, it is enough to prove that

$$x^{-2}(x-y)(x-z) + y^{-2}(y-z)(y-x) + z^{-2}(z-x)(z-y) \ge 0. \quad (27.1)$$

By symmetry we can assume $x \le y \le z$. Then, by adding up,

$$\frac{(x-y)(x-z)}{x^2} + \frac{(y-z)(y-x)}{y^2} = \frac{(y-x)(z-x)}{x^2} + \frac{(y-z)(y-x)}{y^2}$$

$$\ge \frac{(y-x)(z-y)}{y^2} + \frac{(y-z)(y-x)}{y^2} = 0$$

and

$$\frac{(z-x)(z-y)}{z^2} \ge 0,$$

(27.1) is obtained.

Mathematical Induction

Readers can see some examples of proving inequalities by mathematical induction in Lecture 16. Here, we only give two more problems which use mathematical induction (Testing Questions A Q9 and Testing Questions B Q5) for the reader's practice.

Testing Questions   (A)

1. (TAIWAN/TST/2007)   Given $0 < a, b \le 1$, prove that
$$\frac{1}{\sqrt{a^2 + 1}} + \frac{1}{\sqrt{b^2 + 1}} \le \frac{2}{\sqrt{1 + ab}}.$$

2. (TURKEY/TST/2007) Let a, b, c be positive reals such that their sum is 1. Prove that
$$\frac{1}{ab + 2c^2 + 2c} + \frac{1}{bc + 2a^2 + 2a} + \frac{1}{ca + 2b^2 + 2b} \ge \frac{1}{ab + bc + ca}.$$

3. (IRE/2007) Let a, b, c be positive real numbers. Prove that
$$\frac{a + b + c}{3} \le \sqrt{\frac{a^2 + b^2 + c^2}{3}} \le \frac{\frac{ab}{c} + \frac{bc}{a} + \frac{ca}{b}}{3}.$$
When do the equalities hold?

4. (ESTONIA//TST/2009) For any group of three distinct positive real numbers a, b, c, prove that
$$\frac{(a^2 - b^2)^3 + (b^2 - c^2)^3 + (c^2 - a^2)^3}{(a - b)^3 + (b - c)^3 + (c - a)^3} > 8abc.$$

5. (BULGARIA/2007) If $x, y, z > 0$, prove that
$$\frac{(x + 1)(y + 1)^2}{3\sqrt[3]{z^2 x^2} + 1} + \frac{(y + 1)(z + 1)^2}{3\sqrt[3]{x^2 y^2} + 1} + \frac{(z + 1)(x + 1)^2}{3\sqrt[3]{y^2 z^2} + 1} \ge x + y + z + 3.$$

6. (GREECE/2007) Let a, b, c be sides of a triangle. Show that
$$\frac{(a + c - b)^4}{a(a + b - c)} + \frac{(a + b - c)^4}{b(b + c - a)} + \frac{(b + c - a)^4}{c(c + a - b)} \ge ab + bc + ca.$$

7. (SSSMO/2008) Let $a, b, c > 0$. Prove that

$$\frac{(1 + a^2)(1 + b^2)(1 + c^2)}{(1 + a)(1 + b)(1 + c)} \geq \frac{1}{2}(1 + abc).$$

8. (CROATIA/2009) Let positive real numbers x, y, z satisfy $xyz = 1$. Prove that

$$\frac{x^3 + y^3}{x^2 + xy + y^2} + \frac{y^3 + z^3}{y^2 + yz + z^2} + \frac{z^3 + x^3}{z^2 + zx + x^2} \geq 2.$$

9. (CHINA/2003) Prove the inequality

$$\left(\frac{n}{2}\right)^n > n! > \left(\frac{n}{3}\right)^n, \qquad \text{for } n = 6, 7, 8, \ldots .$$

Testing Questions   (B)

1. (BALKAN/2010) Let a, b, c be positive real numbers. Prove that

$$\frac{a^2 b(b - c)}{a + b} + \frac{b^2 c(c - a)}{b + c} + \frac{c^2 a(a - b)}{c + a} \geq 0.$$

2. (GERMANY/2010) Let a, b, c be three distinct positive real numbers. Prove that

$$\left(\frac{2a - b}{a - b}\right)^2 + \left(\frac{2b - c}{b - c}\right)^2 + \left(\frac{2c - a}{c - a}\right)^2 \geq 5.$$

3. (CROATIA/TST/2010) Find the minimum value of D such that

$$\frac{a + b}{a + 2b} + \frac{b + c}{b + 2c} + \frac{c + a}{c + 2a} < D$$

   always holds for any positive real numbers a, b, c.

4. (APMO/2005) Let a, b and c be positive real numbers such that $abc = 8$. Prove that

$$\frac{a^2}{\sqrt{(1 + a^3)(1 + b^3)}} + \frac{b^2}{\sqrt{(1 + b^3)(1 + c^3)}} + \frac{c^2}{\sqrt{(1 + c^3)(1 + a^3)}} \geq \frac{4}{3}.$$

5. (High-School Mathematics/2011) Given positive integers $a_1 < a_2 < \cdots < a_n$ and real number $k \geq 1$. Prove that

$$\sum_{i=1}^{n} a_i^{2k+1} \geq \left(\sum_{i=1}^{n} a_i^k\right)^2.$$

Lecture 28

Some Basic Methods in Counting (I)

In this lecture two kinds of methods of counting are introduced:

(1) The applications of the three fundamental principles (Addition Principle, Multiplication Principle and Inclusion-Exclusion Principle) and Permutations and Combinations;

(2) The applications of the Pairing Method and Correspondence Principle.

Applications of Three Principles, Permutations and Combinations

Example 1. (SMO/2006) How many integers are there between 0 and 10^5 having the digit sum equal to 8?

Solution The question is equivalent to the following question:

How many permutations $(x_1, x_2, x_3, x_4, x_5)$ of five non-negative integers are there such that

$$x_1 + x_2 + x_3 + x_4 + x_5 = 8?$$

Therefore the answer is $\binom{12}{4} = 495$.

Example 2. (CMC/2009) Let $A_1 A_2 \cdots A_{200}$ be a regular 200-sided convex polygon. Make the diagonals $A_i A_{i+9}, i = 1, 2, \ldots, 200$, where $A_{i+200} = A_i$ for $1 \le i \le 9$. How many distinct points of intersection are formed inside the polygon by these 200 diagonals?

Solution For each diagonal $A_i A_{i+9}$, the two diagonals $A_{i+j} A_{i+j+9}$ and $A_{i+j-9} A_{i+j}$ for $j = 1, 2, \cdots, 8$, intersect $A_i A_{i+9}$, where we have as a convention $A_m = A_{m+200}$ for $-9 \le m \le -1$. Thus, the number of points of intersection on the diagonal $A_i A_{i+9}$ is $2 \times 8 = 16$.

Since the lengths of these 200 diagonals are equal, they all are tangent to a same circle. Since from any point outside the circle exactly two tangent lines to the circle can be introduced, so any three of the 200 diagonals cannot intersect at one common point. Thus, the total number of interior point of intersection of the 200 diagonals is $\dfrac{200 \cdot 16}{2} = 1600$.

Example 3. (CMC/2009) Find the number of ordered 6-tuples $(x_1, x_2, x_3, x_4, x_5, x_6)$ of positive integers satisfying

$$x_1 + x_2 + x_3 + 3x_4 + 3x_5 + 5x_6 = 21.$$

Solution   Let $x_1 + x_2 + x_3 = x, x_4 + x_5 = y, x_6 = z$, then $x \geq 3, y \geq 2, z \geq 1$.

Consider the Diophantine equation $x + 3y + 5z = 21$ with $x \geq 3, y \geq 2, z \geq 1$. Then

$$5z = 21 - x - 3y \leq 12 \Rightarrow z \leq 2.$$

When $z = 1$, then $x + 3y = 16 \Rightarrow x = 1 + 3t, y = 5 - t$, so $(x, y) = (10, 2), (7, 3), (4, 4)$.

When $z = 2$, then $x + 3y = 11 \Rightarrow x = 2 + 3t, y = 3 - t$, so $(x, y) = (5, 2)$ only.

Thus, $(x, y, z) = (10, 2, 1), (7, 3, 1), (4, 4, 1), (5, 2, 2)$.

For the equation $x_1 + x_2 + x_3 = x$, the number of solutions is $\begin{pmatrix} x-1 \\ 2 \end{pmatrix}$; the equation $x_4 + x_5 = y$ has $y - 1$ distinct solutions, so the number of solutions for $(x_1, x_2, x_3, x_4, x_5, x_6)$ is given by

$$\begin{pmatrix} 9 \\ 2 \end{pmatrix} \cdot (2-1) + \begin{pmatrix} 6 \\ 2 \end{pmatrix}(3-1) + \begin{pmatrix} 3 \\ 2 \end{pmatrix}(4-1) + \begin{pmatrix} 4 \\ 2 \end{pmatrix}(2-1) = 36 + 30 + 9 + 6 = 81.$$

Example 4. (SMO/2008) Determine the number of 4-element subsets $\{a, b, c, d\}$ of $\{1, 2, 3, 4, \ldots, 20\}$ such that $a + b + c + d$ is divisible by 3.

Solution   Partition the set $S = \{1, 2, \ldots, 20\}$ into three "subsets according to their remainder mod 3: C_0, C_1, C_2, and $|C_0| = 6, |C_1| = |C_2| = 7$.

(i) When a, b, c, d all come from C_0, the number of the 4-element subsets is $\binom{6}{4} = 15$;

(ii) When 2 of $\{a, b, c, d\}$ are from C_0, the third from C_1 and the fourth from C_2, then the number of the 4-element subsets is $\begin{pmatrix} 6 \\ 2 \end{pmatrix} \cdot \begin{pmatrix} 7 \\ 1 \end{pmatrix}^2 = 735$.

(iii) When three of a, b, c, d are from C_1 or are from C_2, and the fourth element is from C_0, then the number of these 4-element subsets is

$$\binom{7}{3} \cdot \binom{6}{1} + \binom{7}{3} \cdot \binom{6}{1} = 420;$$

(iv) When 2 of $\{a, b, c, d\}$ are from C_1 and other 2 are from C_2, then the number of the 4-element subsets is $\binom{7}{2}^2 = 21^2 = 441$;

Thus, the total number of the 4-element subsets is $15 + 735 + 420 + 441 = 1611$.

Applications of Correspondence And Bijection

Theorem I. *Let M and N be two finite sets. If there exists an injection from M into N, then $|M| \leq |N|$; if there exists a surjection from M to N, then $|M| \geq |N|$; if there exists a bijection from M to N, then $|M| = |N|$.*

Example 5. (SMO/2006) For any non-empty finite set A of real numbers, let $s(A)$ be the sum of the elements in A. There are exactly 61 of 3-element subsets A of $\{1, \ldots, 23\}$ with $s(A) = 36$. Find the number of 3-element subsets of $\{1, \ldots, 23\}$ with $s(A) < 36$.

Solution For each 3-element subset $x = \{a, b, c\}$ of the set $M = \{1, 2, 3, \cdots, 23\}$ with $a + b + c < 36$, define the map $f(x) = \{24 - a, 24 - b, 24 - c\}$. Then f is a bijection between the set C_1 of all three element subsets of M with $a + b + c < 36$ and the set C_2 of all three element subsets of M with $a + b + c > 36$. Hence $|C_1| = |C_2|$.

Since there are a total of $\binom{23}{3} = 1771$ different 3-element subsets of M, so

$$|C_1| = \frac{1}{2}(1771 - 61) = 855.$$

Example 6. Partition each side of an equilateral triangle XYZ of side n into n equal parts. By introducing lines parallel to the sides of XYZ which pass through these partition points we obtain a grid formed by n^2 equal equilateral triangle of side 1, as shown in the following diagram. Find the number of parallelograms contained in the grid.

Solution Use S_{yz} to denote the set of all parallelograms in the grid that have sides which are parallel to the lines XY or XZ. Similarly, we define the sets S_{zx} and S_{xy}. By symmetry we have $|S_{yz}| = |S_{xy}| = |S_{zx}|$, so the answer is $3|S_{yz}|$.

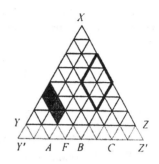

To get $|S_{yz}|$, we extend the ray XY one unit to Y' and the ray XZ one unit to Z' respectively. Let p be a parallelogram in S_{yz}. Then by extending the sides of p so that we have four points of intersection on $Y'Z'$. Conversely, every four points on $Y'Z'$ determine unique parallelogram p in S_{yz}.

For example, in the diagram, the shaded parallelogram corresponds to $\{Y', A, F, B\}$, and the parallelogram with bold sides corresponding to $\{A, B, C, Z'\}$. Thus, we get a bijection between S_{yz} and the set of groups of four points on $Y'Z'$. Therefore the answer is $3\binom{n+2}{4}$.

Example 7. There are n ($n \geq 6$) given points on a circle, and each two points are connected by a segment. Suppose that any three segments are not concurrent, so any three intersecting segments form a triangle inside the circle. Find the number of triangles formed.

Solution   We call a point on the circle an exterior point and a point inside the circle an internal point. Then the triangles can be classified into four classes:

C_1:   all the three vertices are exterior points. Let $|C_1|$ be I_1;
C_2:   two vertices are exterior points, and one vertex is internal point. Let $|C_2|$ be I_2;
C_3:   one vertex is exterior and other two are internal. Let $|C_3|$ be I_3;
C_4:   all the three vertices are internal points. Let $|C_4|$ be I_4;

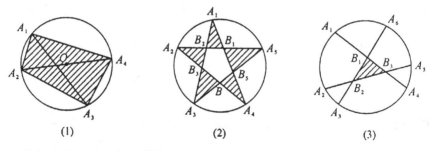

It is obvious that $I_1 = {}^nC_3$.

Next, as shown in the diagram (1) above, every four points on the circle determine four triangles in C_2, and *vice versa*. Thus, $I_2 = 4 \cdot {}^nC_4$.

Similarly, as shown in the diagram (2) above, every five points on the circle determine five triangles in C_3, and vice versa. So $I_3 = 5 \cdot {}^nC_5$.

Finally, as shown in the diagram (3) above, $I_4 = {}^nC_6$.
Thus the total number of triangles is ${}^nC_3 + 4 \cdot {}^nC_4 + 5 \cdot {}^nC_5 + {}^nC_6$.

Example 8. (CMC/2009) Let M be a set with $|M| = n$. If M has k distinct subsets $A_1, A_2, \ldots, A_k$ such that $A_i \cap A_j \neq \emptyset$ for any $i, j \in \{1, 2, \cdots, k\}$. Find the maximum value of k.

Solution   (i)   For $M = \{a_1, a_2, \cdots, a_n\}$ we first show that there are 2^{n-1} subsets of M such that any two of them have a non-empty intersection. This is because M has a total of 2^{n-1} distinct subsets containing a_i for each fixed i in $\{1, 2, \ldots, n\}$.

(ii)   Next we show that for any $2^{n-1} + 1$ distinct subsets of M, there must be two which are disjoint.

Let $B_1, B_2, \ldots, B_{2^{n-1}}$ be 2^{n-1} distinct subsets of M such that each contains the element a_n, and let B_i' be its complementary set, i.e., $B_i' = M \setminus B_i, i = 1, 2, \ldots, 2^{n-1}$. Since different B_i has different B_i', the subsets $B_1, B_2, \ldots, B_{2^{n-1}}$; $B_1', B_2', \ldots, B_{2^{n-1}}'$ consist of all 2^n subsets of M.

We can form the 2^{n-1} pairs $(B_i, B_i'), i = 1, 2, \ldots, 2^{n-1}$ by using these 2^n subsets. For any given distinct $2^{n-1} + 1$ subsets, by the pigeonhole principle, there must exist two subsets coming from the same pair, so these two must be disjoint.

Testing Questions   (A)

1.   (SMO/2006) There are 10! permutations $s_0 s_1 \ldots s_9$ of $0, 1, \ldots, 9$. How many of them satisfy $s_k \geq k - 2$ for $k = 0, 1, \ldots, 9$?

2.   (CMC/2009) "Consider all positive integers consisting of only nonzero digits which have digit sum 7. Find the times of the digit 3 appeared in all such integers.

3.   (SMO/2007) Let $S = \{1, 2, 3, 4, \ldots, 50\}$. A 3-element subset $\{a, b, c\}$ of S is said to be *good* if $a + b + c$ is divisible by 3. Determine the number of 3-elements of S which are good.

4.   (CMC/2009) x, y, z be natural numbers not greater than 6. How many ordered triples (x, y, z) are there such that the product xyz is divisible by 10?

5.   (SMO/2006) Let P be a 30-sided polygon inscribed in a circle. Find the number of triangles whose vertices are the vertices of P such that any two vertices of each triangle are separated by at least three other vertices of P.

6. (CMC/2010) Colour the sides and diagonals of an n-sided convex polygon $A_1 A_2 \cdots A_n$ with either red or blue, such that there is no triangle with three sides of same colour. If by $b_k, k = 1, 2, \ldots, n$ we denote the total number of blue sides and diagonals emitting from A_k, prove that

$$b_1 + b_2 + \cdots + b_n \leq \frac{n^2}{2}.$$

7. (CMC/2009) There are six distinct books: one is a mathematical book, two are English books and three are music books. How many ways are there to arrange these books in a row such that any two books of the same kind are separated??

8. (BMO/2009-2010) Isaac attempts all six questions on an Olympiad paper in order. Each question is marked on a scale from 0 to 10. He never scores more in a later question than in any earlier question. How many different possible sequences of six marks can he achieve?

9. When n objects are arranged in a row, a subset of these objects is said to be distant if any two objects in the subset are not adjacent. Prove that the number of distant subsets containing k objects is is $\binom{n-k+1}{k}$.

Testing Questions   (B)

1. (SMO/2008) Find the number of 10-letter permutations comprising 4 a's, 3 b's, 3 c's such that no two adjacent letters are identical.

2. (RUSMO/2005) 100 people from 50 countries, two from each country, sit in a circle. Prove that one may partition them into 2 groups in such a way that neither two countrymen, nor three consecutive people on a circle, are in the same group.

3. (USAMO/2004) Alice and Bob play a game on a 6 by 6 grid. On his or her turn, a player chooses a rational number not yet appearing in the grid and writes it in an empty square of the grid. Alice goes first and then the players alternate. When all squares have numbers written in them, in each row, the square with the greatest number in that row is colored black. Alice wins if she can then draw a line from the top of the grid to the bottom of the grid that stays in black squares, and Bob wins if she can't. (If two squares share a vertex, Alice can draw a line from one to the other that stays in those two squares.) Find, with proof, a winning strategy for one of the players.

4. (CNMO/2009) The regular triangular array consists of $\dfrac{n(n+1)}{2}$ points. If $f(n)$ denotes the number of equilateral triangles formed by taking three of their points as vertices, find an expression of $f(n)$ in terms of n.

5. (USAMO/2008) At a certain mathematical conference, every pair of mathematicians are either friends or strangers. At mealtime, every participant eats in one of two large dining rooms. Each mathematician insists upon eating in a room which contains an even number of his or her friends. Prove that the number of ways that the mathematicians may be split between the two rooms is a power of two (i.e., is of the form 2^k for some positive integer k).

Lecture 29

Some Basic Methods in Counting (II)

In this lecture two kinds of methods of counting are introduced:
  (1) Applications of the *Counting in Two Ways* method;
  (2) Applications of the *Recurrence method*.

Applications of Counting by Two Ways

Examples

Example 1. (Fubini's Principle) For finite sets $A = \{a_1, a_2, \ldots, a_m\}$ and $B = \{b_1, b_2, \ldots, b_n\}$ define their *Cartesian product* $A \times B$ as the set of ordered pairs given by

$$A \times B = \{(a_i, b_j) \mid 1 \leq i \leq m, 1 \leq j \leq n\}.$$

Prove that $|A \times B| = \sum_{i=1}^{m} |C_i| = \sum_{j=1}^{n} |D_j|$, where

$$C_i = \{(a_i, b) \mid b \in B\}, \quad D_j = \{(a, b_j) \mid a \in A\}, \quad 1 \leq i \leq m, 1 \leq j \leq n.$$

Solution We count $|A \times B|$ by partitioning $A \times B$ in two ways: (i) according to the first component of the pairs (a_i, b_j) to classify $A \times B$ and (ii) according to the second component of the pairs (a_i, b_j) to classify $A \times B$.

Therefore this example is a typical example of counting in two ways.

Example 2. (IMO/2001/Q4) Let n be an odd integer greater than 1, and let $k_1, k_2, \cdots, k_n$ be given integers. For each of the $n!$ permutations $a = (a_1, a_2, \cdots, a_n)$ of $1, 2, \cdots, n$, let

$$S(a) = \sum_{i=1}^{n} k_i a_i.$$

Prove that there are two permutations b and c, $b \neq c$, such that $n!$ is a divisor of $S(b) - S(c)$.

Solution   Let $\sum S(a)$ be the sum of $S(a)$ over all $n!$ permutations $a = (a_1, a_2, \cdots, a_n)$. We compute $\sum S(a) \bmod n!$ two ways, one of which depends on the desired conclusion being false, and reach a contradiction when n is odd.

First way. In $\sum S(a)$, k_1 is multiplied by each of $i \in \{1, 2, \cdots, n\}$ a total of $(n-1)!$ times, once for each permutation of $\{1, 2, \cdots, n\}$ in which $a_1 = i$. Thus the coefficient of k_1 in $\sum S(a)$ is

$$(n-1)!(1 + 2 + \cdots + n) = \frac{(n+1)!}{2}.$$

The same is true for all k_i, so

$$\sum S(a) = \frac{(n+1)!}{2} \sum_{i=1}^{n} k_i. \tag{29.1}$$

Second way. If $n!$ is not a divisor of $S(b) - S(c)$ for any $b \neq c$, then each $S(a)$ must have a different remainder mod $n!$. Since there are $n!$ permutations, these remainders must be precisely the numbers $0, 1, 2, \cdots, n! - 1$. Thus

$$\sum S(a) \equiv \frac{(n! - 1)(n!)}{2} \pmod{n!} \tag{29.2}$$

Combining (29.1) and (29.2), we get

$$\frac{(n+1)!}{2} \sum_{i=1}^{n} k_i \equiv \frac{(n! - 1)(n!)}{2} \pmod{n!} \tag{29.3}$$

Now for odd n, the left side of (29.3) is congruent to $0 \bmod n!$, while for odd $n > 1$, the right side is not congruent to $0 \bmod n!$ ($n! - 1$ is odd). We have a contradiction.

Example 3. (Erdös-Ko-Rado Theorem) Let $2 \leq r \leq \dfrac{n}{2}$, $Z_n = \{1, 2, \cdots, n\}$, $\mathcal{A}$ be a family of sets consisting of r-element subsets of Z_n. If any two elements in $\mathcal{A}$ have non-empty intersection, prove that $|\mathcal{A}| \leq \dbinom{n-1}{r-1}$, where the equality holds if

$$\mathcal{A} = \{A \mid A \text{ is an } r\text{-element set of } Z_n \text{ and contains a fixed element } x \text{ of } Z_n\}.$$

Solution There are $(n-1)!$ ways to arrange the elements in Z_n at a circle. For each permutation π_j $(j = 1, 2, \cdots, (n-1)!)$, if all the elements of $A_i \in \mathcal{A}$ appear consecutively in π_j, then make up a pair (A_i, π_j). By S we denote the set of all the pairs.

Since any two elements in $\mathcal{A}$ have non-empty intersection, and $2 \le r \le \frac{n}{2}$, π_j appears at most r times in S, therefore

$$|S| \le r \cdot [(n-1)!]. \tag{29.4}$$

On the other hand, for each $A_i \in \mathcal{A}$, there are $(r!)[(n-r)!]$ permutations π_j, such that all the elements of $A_i \in \mathcal{A}$ appear consecutively in π_j, therefore

$$|S| = |\mathcal{A}| \cdot (r!) \cdot [(n-r)!]. \tag{29.5}$$

Combining (29.4) and (29.5), we obtain

$$|\mathcal{A}| \cdot (r!) \cdot [(n-r)!] \le r \cdot [(n-1)!],$$

$$\therefore |\mathcal{A}| \le \frac{(n-1)!}{[(r-1)!][(n-r)!]} = \binom{n-1}{r-1}.$$

Next, we have $|\mathcal{A}| = \binom{n-1}{r-1}$ if

$$\mathcal{A} = \{A \,|\, A \text{ is an } r\text{-element subset of } Z \text{ and contains a fixed element of } Z\},$$

so the equality holds.

Example 4. (CWMO/2008) Let P be an arbitrary point inside the regular n-sided polygon $A_1 A_2 \cdots A_n$, and let the lines $A_i P$ intersect the boundary of $A_1 A_2 \cdots A_n$ at B_i, $i = 1, 2, \ldots, n$, respectively. Prove that $\displaystyle\sum_{i=1}^{n} PA_i \ge \sum_{i=1}^{n} PB_i$.

Solution Let $t = \left\lfloor \dfrac{n}{2} \right\rfloor$ and $A_{n+j} = A_j$, $j = 1, 2, \ldots, n$. Let d be the length of the longest diagonal of $A_1 A_2 \cdots A_n$, then

$$A_i P + PB_i = A_i B_i \le d, \qquad i = 1, 2, \ldots, n. \tag{29.6}$$

On the other hand, from the triangle inequality,

$$A_i P + PA_{i+t} \ge A_i A_{i+t} = d, \qquad i = 1, 2, \ldots, n. \tag{29.7}$$

For each of (29.6) and (29.7), by taking summation on i from 1 to n, it follows that

$$\sum_{i=1}^{n}(A_i P + PA_{i+t}) \geq nd \geq \sum_{i=1}^{n}(A_i P + PB_i)$$

$$\Rightarrow 2\sum_{i=1}^{n} PA_i \geq \sum_{i=1}^{n} PA_i + \sum_{i=1}^{n} PB_i \Rightarrow \sum_{i=1}^{n} PA_i \geq \sum_{i=1}^{n} PB_i.$$

Applications of the Recurrence Method in Counting

Example 5. (SMO/2006) Three people A, B and C play a game of passing a basketball from one to another. Find the number of ways of passing the ball starting with A and reaching A again on the 11th pass. For example, one possible sequence of passing is

$$A \to B \to A \to B \to C \to A \to B \to C \to B \to C \to B \to A.$$

Solution   For any positive integer $n \geq 1$ let s_n be the number of ways that the basketball reaches A at nth pass. Then $s_1 = 0$. Since there are a total of 2^n different ways to complete n passes without restriction, so

$$s_{n+1} = 2^n - s_n.$$

Since $s_1 = 0, s_2 = 2$, we have

n	1	2	3	4	5	6	7	8	9	10	11
s_n	0	2	2	6	10	22	42	86	170	342	682

Thus the answer is 682.

Example 6. (CMC/2010) Let $a_1, a_2, \cdots, a_n$ be a permutation of $1, 2, \cdots, n$ that satisfies

(1) $a_1 = 1$,   (2) $|a_i - a_{i-1}| \leq 2, i = 2, 3, \ldots, n$.

If we let $f(n)$ denote the number of such permutations, find the remainder of $f(2010)$ when it is divided by 3.

Solution   It can be verified that $f(1) = f(2) = 1, f(3) = 2$. For $n \geq 4$, $a_1 = 1 \Rightarrow a_2 = 2$ or 3.

When $a_2 = 2$, the number of permissible permutations is $f(n-1)$, since we can obtain a permissible permutation by removing a_1 and using $a_i - 1$ to replace a_i for $2 \leq i \leq n$, and this correspondence is one-to-one.

When $a_2 = 3$, then $a_3 = 2$ or $s_3 \neq 2$. If $a_3 = 2$, then that the number of permissible permutations is $f(n-3)$. If $a_3 \neq 2$, then 2 must be arranged behind 4, and after the increasing arrangement of odd numbers the decreasing arrangement of even numbers is followed, i.e, there is exact one such permutation. Thus,

$$f(n) = f(n-1) + f(n-3) + 1.$$

Let $r(n)$ be the remainder of $f(n)$ modulo 3 for $n = 1, 2, \cdots$. Then $r(1) = r(2) = 1, r(3) = 2$, and for $n \geq 4$,

$$r(n) \equiv r(n-1) + r(n-3) + 1 \pmod 3.$$

For $n = 1, 2, 3, \ldots, 11$, they are $1, 1, 2, 1, 0, 0, 2, 0, 1, 1, 2$, therefore the sequence $\{r(n)\}$ is periodic and 8 is its minimum period. Thus, since $2010 \equiv 2 \pmod 8$,

$$r(2010) = r(2) = 1, \quad \text{namely the remainder of } f(2010) \bmod 3 \text{ is } 1.$$

Example 7. If we put n balls labeled 1 to n into n boxes labeled 1 to n such that each box has one ball, how many ways are there such that in each box the label of the ball is not equal to the label of the box?

Solution Suppose that the boxes are arranged in ascending order, only consider the arrangements of the n balls below. When the number of balls and the boxes are both n, let A_n be the set of all arrangements of the n balls satisfying the requirement, and $|A_n| = a_n$.

Now consider the recursive relation among a_{n-1}, a_n and a_{n+1} as follows. For each arrangement in A_n, we can put the ball with label $n+1$ into box k and the ball in box k into box $n+1$ for $1 \leq k \leq n$, then these na_n arrangements are A_{n+1}.

Among all the arrangements for n balls, if only ball k is in box k and the other $n-1$ balls not in the boxes with same label as ball, the number of such arrangements is a_{n-1}. By putting ball $n+1$ into box k and putting ball k into box $n+1$, a total of na_{n-1} another arrangements in A_{n+1} is obtained. Thus,

$$a_{n+1} = na_n + na_{n-1}, n = 2, 3, \ldots.$$

It is clear that $a_1 = 0, a_2 = 1$. Let $a_n = (n!)b_n, n \geq 1$, then $b_1 = 0, b_1 = \frac{1}{2}$ and

$$(n+1)b_{n+1} = nb_n + b_{n-1} \Rightarrow b_{n+1} - b_n = -\frac{1}{n+1}(b_n - b_{n-1}),$$

hence

$$\prod_{k=2}^{n-1}(b_{k+1} - b_k) = \prod_{k=2}^{n-1}\left[-\frac{1}{k+1}(b_k - b_{k-1})\right]$$

$$\Rightarrow b_n - b_{n-1} = (-1)^{n-2} \cdot \frac{2}{n!}(b_2 - b_1) = \frac{(-1)^n}{n!}.$$

$$\Rightarrow b_n = \sum_{i=2}^{n}(b_i - b_{i-1}) = \sum_{i=2}^{n}(-1)^i \cdot \frac{1}{i!} = \sum_{i=0}^{n}(-1)^i \frac{1}{i!}.$$

Thus, $a_n = n! \sum_{i=0}^{n} (-1)^i \frac{1}{i!}$.

Example 8. (SMO/2011) A collection of 2011 of circles divide the plane into N regions in such a way that any pair of circles intersects at two points and no point lies on three circles. Find the last four digits of N.

Solution   Let a_n be the number of regions obtained by n circles divided in such a way that any pair of circles intersects at two points and no point lies on three circles. When the number of circles from n increases to $n + 1$, since there are $2n$ points of intersection on the $n + 1$th circle, the circle is partitioned in $2n$ arcs, and each arc partitions one region into two, so $a_1 = 2$ and

$$a_{n+1} = a_n + 2n \Rightarrow a_n = \sum_{k=1}^{n-1}(a_{k+1} - a_k) = 2\sum_{k=1}^{n-1} k \Rightarrow a_n = 2 + (n-1)n.$$

Thus,

$$N = a_{2011} = 2 + 2010 \cdot 2011 = 4042112 \Rightarrow \text{its last four digits are 2112.}$$

Testing Questions   (A)

1. (CMC/2000) There are totally n persons. Given that there is at most one telephone call taken between any two of them, and the total number of calls taken by any $n - 2$ of them is equal 3^k (where k is a positive integer). Find all the possible value of n.

2. Given that a natural number n is formed by α_1 digits of 1, α_2 digits of 2, $\cdots$, α_9 digits of 9. Prove that

    $$2^{\alpha_1} 3^{\alpha_2} \cdots 9^{\alpha_8} 10^{\alpha_9} \leq n + 1.$$

3. (IMO/1998/Q2) In a competition, there are a contestants and b judges, where $b \geq 3$ is an odd integer. Each judge rates each contestant as either "pass" or "fail". Suppose k is a number such that, for any two judges, their ratings coincide for at most k contestants. Prove that

    $$\frac{k}{a} \geq \frac{b-1}{2b}.$$

4. A circular disk is divided into n (≥ 2) sectors $S_1, S_2, \cdots, S_n$ and each sector is coloured with one of m (≥ 2) colours such that the colours of any two adjacent sectors are different. How many distinct colourings are there?

5. (CHNMO/2001) A circle with perimeter 24 is divided into 24 equal arcs. How many ways are there to choose 8 points from the 24 points of division such that the length of the arc between any two of them is neither 3 nor 8?

6. (CSMO/2008) Let n be a positive integer, and let $f(n)$ denote the number of n digit number $\overline{a_1 a_2 \cdots a_n}$, called a *wave number*, satisfying the following conditions:

    (i) $a_i \in \{1, 2, 3, 4\}$ and $a_i \neq a_{i+1}, i = 1, 2, \ldots$;

    (ii) When $n \geq 3$, the numbers $a_i - a_{i+1}$ and $a_{i+1} - a_{i+2}$ have opposite signs for $i = 1, 2, \ldots$.

    Find (1) the value of $f(10)$;   (2) the remainder of $f(2008)$ modulo 13.

7. (MOSP/2003) Given n collinear points, consider the distances between any two of them. If each distance appears at most twice, prove that at least $\left\lfloor \dfrac{n}{2} \right\rfloor$ distances appeared only once.

8. (SMO/2007) For each positive integer n, let a_n denote the number of n-digit integers formed by some or all of the digits $0, 1, 2,$ and 3 which contain neither a block of 12 nor a block of 21. Evaluate a_9.

9. (IMO/Shortlist/2006) Let S be a finite set of points in the plane such that no three of them are on a line. For each convex polygon P whose vertices are in S, let $a(P)$ be the number of vertices of P, and let $b(P)$ be the number of points of S which are outside P. Prove that for every real number x:

$$\sum_P x^{a(P)} (1 - x)^{b(P)} = 1,$$

    where the sum is taken over all convex polygons with vertices in S.

    Remark. A line segment, a point, and the empty set are considered as convex polygons of 2, 1, and 0 vertices respectively.

Testing Questions   (B)

1. (IMO/Shortlist/2004) There are 10001 students at a university. Some students join together to form several clubs (a student may belong to different clubs). Some clubs join together to form several societies (a club may belong to different societies). There are a total of k societies. Suppose that the following conditions hold:

(i)   Each pair of students are in exactly one club.

(ii)   For each student and each society, the student is in exactly one club of the society.

(iii) Each club has an odd number of students In addition, a club with $2m + 1$ students (m is a positive integer) is in exactly m societies.

Find all possible values of k.

2.   (BULGARIA/Spring MC/2007) Given that $m, n > 1$ are odd numbers. In each square of a rectangular table of dimension $m \times n$ fill in a distinct real number. A number is called "good number" if

(i)   if it is the maximum number in its row (or in its column); and

(ii)   it is the median in its column (or in its row). Then median means half of the rest numbers in its row (or column)is greater than it and other half is less than it.

Find the maximum value of the good numbers.

3.   (SMO/2006) Find the number of 7-digit integers formed by some or all of the five digits, namely, $0, 1, 2, 3$, and 4, such that these integers contain none of the three blocks $22, 33$ and 44.

4.   (CMC/2010) On the encrypted lock, the code is made by giving one of digit 0 or 1 at each vertex of an n-sided regular polygon, and color each vertex red or blue, such that any two adjacent vertices have same color or same digit. Find the number of different codes.

5.   (CANADA/2009) Given an $m \times n$ grid with squares coloured either black or white, we say that a black square in the grid is *stranded* if there is some square to its left in the same row that is white and there is some square above it in the same column that is white (see Figure below).

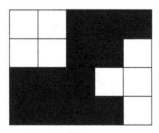

Figure.   A 4×5 grid with no stranded black squares.

Find a closed formula for the number of $2 \times n$ grids with no stranded black squares.

Lecture 30

Introduction to Functional Equations

Definition 30.1. An equation containing one or more unknown functions is called a **functional equation**, if the operations in the equation are addition, subtraction, multiplication, division and composition operations of the functions only.

Any function which causes the equation to become an identity on the domain of the function is called a **solution** of the equation.

The process for finding some or all solutions is called **solving the equation**.

Although Cauchy investigated the famous "Cauchy's Problem" in about 150 years ago, today we still do nott have a clear and complete description of the ideal theory and methods for functional equation. New problems and new methods are appearing continuously, like in Mathematical Olympiad competitions. In spite of this, we are sure that the following listed methods and tools are essential for finding the unknown function based on the given conditions in a problem.

(1) *Cauchy's Method.* Determine a function step by step: on $\mathbb{N}$ first, then on $\mathbb{Z}$, then on $\mathbb{Q}$, then finally on $\mathbb{R}$.

(2) *Evaluation of Variables.* When the given equation contains a single variable x, by letting x be certain special value, like 0, ± 1 etc., one can determine the values of the unknown function on points $0, 1$ etc., which is often important to simplify the given equation. In the multivariate case, evaluating the values of some variables is sometimes necessary to reduce the number of variables or to simplify the structure of the equation.

(3) *Substitution of Variables or Function.* It is a useful tool to simplify the structure of given equation, but the use of this tool is very flexible.

(4) *Applications of properties of unknown function.* It may be easier to determine the unknown function for determining a function if there is information on the boundedness, symmetry, continuity, momotonicity, periodicity, injectivity, surjectivity and bijectivity.

(5) *Recurrence Method.* The recursive model is a useful method for determining a function.

(6)   *Mathematical Indiction.* In recent years, many functional equation problems indicated that they could be solved by using mathematical induction.

Examples

Example 1. (Cauchy's Problem) Find all monotone functions $f(x)$, $x \in \mathbb{R}$, satisfying

$$f(x + y) = f(x) + f(y). \quad \text{(which is known as the \textbf{Cauchy's equation}.)}$$
$$(30.1)$$

Solution By using (30.1) repeatedly, it is easy to prove that for any $n \in \mathbb{N}$ and real values $x_1, x_2, \cdots, x_n$,

$$f(x_1 + x_2 + \cdots + x_n) = f(x_1) + f(x_2) + \cdots + f(x_n).$$

Then letting $x_1 = x_2 = \cdots = x_n = x$ yields

$$f(nx) = nf(x) \qquad \forall\, n \in \mathbb{N}, \ x \in \mathbb{R}. \qquad (30.2)$$

Using $\dfrac{x}{n}$ as x in (30.2) then leads to

$$f\left(\frac{1}{n}x\right) = \frac{1}{n}f(x),$$

and substituting mx into it as x leads to

$$f\left(\frac{m}{n}x\right) = \frac{m}{n}f(x) \qquad \forall\, n, m \in \mathbb{N}, \ x \in \mathbb{R}. \qquad (30.3)$$

Substituting $x = y = 0$ into (30.1) gives $f(0) = 2f(0)$, i.e., $f(0) = 0$. Then substituting $y = -x$ into (30.1) yields $f(-x) = -f(x)$, therefore

$$f\left(-\frac{m}{n}x\right) = -\frac{m}{n}f(x) \qquad \forall\, n, m \in \mathbb{N}, \ x \in \mathbb{R}. \qquad (30.4)$$

Combining (30.3) and (30.4), we obtain

$$f(rx) = rf(x) \qquad \forall\, r \in \mathbb{Q}, \ x \in \mathbb{R}. \qquad (30.5)$$

From (30.5), by letting $x = 1$, we have $f(r) = Cr$, where $C = f(1)$.

Now for any irrational number λ, there are rational numbers $r_1 < r_2$ such that $r_1 < \lambda < r_2$. If $f(x)$ is an non-decreasing function, then $f(r_1) \leq f(\lambda) \leq f(r_2)$, i.e.,

$$Cr_1 \leq f(\lambda) \leq Cr_2. \qquad (30.6)$$

Since we can take an increasing sequence of rational numbers, $\{r_1^n, \ n = 1, 2, \cdots\}$, and an decreasing sequence of rational numbers, $\{r_1^n, \ n = 1, 2, \cdots\}$, such that

$$\lim_{n \to +\infty} r_1^n = \lim_{n \to +\infty} r_2^n = \lambda,$$

we obtain

$$f(\lambda) = C\lambda, \qquad \forall \lambda \in \mathbb{R}. \tag{30.7}$$

If f is non-increasing, the discussion is similar, and we can obtain the same result (30.7). It is clear that $f(x) = Cx \ (x \in \mathbb{R})$ satisfies (1), therefore, the function given by

$$f(x) = Cx \qquad \forall x \in \mathbb{R}$$

is the unique monotone solution of (30.1). The discussion is similar when f is decreasing.

Note: In fact, if a function $f(x)$ satisfies (30.1) and any one of the following conditions:

(A) $f(x)$ is bounded at some interval (a, b), regardless how small the interval is;
(B)   There exists a interval $[0, \epsilon]$, where $\epsilon > 0$, such that $f(x) \geq 0$ on the interval, or $f(x) \leq 0$ on the interval;
(C) $f(x)$ is continuous at some point x_0,
then it must be the function $f(x) = Cx$. (see Appendix E for the proof.)

Example 2. (SMO/2011) A real-valued function f satisfies the relation

$$f(x^2 + x) + 2f(x^2 - 3x + 2) = 9x^2 - 15x \tag{30.8}$$

for all real values of x. Find $f(2011)$.

Solution   By replacing x with $1 - x$ in (30.8), it follows that

$$f(x^2 - 3x + 2) + 2f(x^2 + x) = 9x^2 - 3x - 6. \tag{30.9}$$

By eliminating $f(x^2 - 3x + 2)$ from the two simultaneous equations (30.8) and (30.9), we obtain

$$3f(x^2 + x) = 9x^2 + 9x - 12 = 9(x^2 + x) - 12,$$

i.e., $f(x^2 + x) = 3(x^2 + x) - 4, x \in \mathbb{R}$. Since the equation $x^2 + x = 2011$ has real roots, so there is x_0 such that $x_0^2 + x_0 = 2011$. Hence,

$$f(2011) = f(x_0^2 + x_0) = 3(2011) - 4 = 6029.$$

Example 3. (COLOMBIA/2009) Find all functions $f : \mathbb{Z} \to \mathbb{Z}$ satisfying
(i) $f(n)f(-n) = f(n^2)$;
(ii) $f(m + n) = f(m) + f(n) + 2mn$.

Solution   Let $g(n) = f(n) - n^2, n = 1, 2, \ldots$. Then condition (ii) gives

$$
\begin{aligned}
g(m+n) &= f(m+n) - (m+n)^2 = f(m) + f(n) + 2mn - (m+n)^2 \\
&= (f(m) - m^2) + (f(n) - n^2) = g(m) + g(n).
\end{aligned}
$$

Since g satisfies the Cauchy's equation, so $g(n) = cn, n \in \mathbb{Z}$, where c is a constant.

Condition (i) now yields

$$
\begin{aligned}
n^4 - c^2 n^2 &= (n^2 + cn)(n^2 - cn) = (n^2 + g(n))[(-n)^2 + g(-n)] \\
&= f(n)f(-n) = f(n^2) = n^4 + g(n^2) = n^4 + cn^2 \\
\Rightarrow\quad & c(c+1) = 0 \Rightarrow c = 0 \text{ or } -1.
\end{aligned}
$$

Returning to f, we have $f(n) = n^2$ or $f(n) = n^2 - n$. By checking, both satisfy the conditions in the question.

Example 4. (SMO/2009) A function $f : \mathbb{R} \to \mathbb{R}$ satisfies the relation

$$ f(x)f(y) = f(2xy + 3) + 3f(x+y) - 3f(x) + 6x, \qquad x, y \in \mathbb{R}. \quad (30.10) $$

Find the value of $f(2009)$.

Solution   Considering that x and y are symmetric in almost all terms of (30.10), if we interchange x and y, it follows that

$$ f(x)f(y) = f(2xy + 3) + 3f(x+y) - 3f(y) + 6y, \qquad x, y \in \mathbb{R}. \quad (30.11) $$

A comparison of (30.10) and (30.11) yields $-3f(x) + 6x = -3f(y) + 6y$ for all $x, y \in \mathbb{R}$, so $-3f(x) + 6x = -3C$ for some constant C, i.e.,

$$ f(x) = 2x + C, $$

where $C \in \mathbb{Z}$. To determine C, substituting the expression for f into (30.10), it follows that

$$
\begin{aligned}
&(2x + C)(2y + C) = 2(2xy + 3) + C + 3[2(x+y) + C] - 3(2x + c) + 6x \\
&\Rightarrow C^2 - C - 6 = 6(x+y) - 2C(x+y) \\
&\Rightarrow (C - 3)(C + 2) = 2(3 - C)(x+y) \\
&\Rightarrow (C - 3)(C + 2 + 2x + 2y) = 0 \text{ for all } x, y \in \mathbb{R} \Rightarrow C = 3.
\end{aligned}
$$

Thus, $f(2009) = 2 \times 2009 + 3 = 4021$.

Example 5. (ROMANIA/2008) Find functions $f : \mathbb{N}_0 \to \mathbb{N}_0$, such that

$$ f(x^2 + f(y)) = xf(x) + y \qquad \text{for all } x, y \in \mathbb{N}_0. \quad (30.12) $$

Solution   In (30.12) let $x = 0$, then

$$f(f(y)) = y, \qquad \forall y \in \mathbb{N}_0. \tag{30.13}$$

Let $x = 1$, $y = 0$ and use (30.13) and (30.12), then

$$1 + f(0) = f(f(1 + f(0))) = f(f(1) + 0) = 1 \Rightarrow f(0) = 0.$$

Denote $f(1)$ by a and let $x = 1$, $y = f(z)$ for $z \in \mathbb{N}_0$, then

$$f(1 + z) = f(1 + f(f(z))) = f(1) + f(z) = f(z) + a, \quad z \in \mathbb{N}_0.$$

By induction on z, we obtain

$$f(n) = an, \qquad n \in \mathbb{N}_0. \tag{30.14}$$

Substituting (30.14) into (30.12) yields $a(x^2 + ay) = ax^2 + y$, so $a^2 y = y$, $y \in \mathbb{N}_0$. Therefore $a = \pm 1$. Since $a \geq 0$, so $a = 1$. Thus, $f(n) = n, n \in \mathbb{N}_0$. f obviously satisfies (30.12).

Example 6. (THAILAND/2010) Find all functions $f : \mathbb{R}^+ \to \mathbb{R}^+$ satisfying

$$f(x)f(yf(x)) = f(x + y) \qquad \text{for all } x, y > 0. \tag{30.15}$$

Solution   If $f(x_0) > 1$ at some point $x_0 > 0$, let $y_0 = \dfrac{x_0}{f(x_0) - 1}$, then $y_0 > 0$ and

$$f(x_0 + y_0) = f(x_0)f(y_0 f(x_0)) = f(x_0)f(x_0 + y_0) \Rightarrow f(x_0) = 1,$$

which is a contradiction, so $f(x) \leq 1$, $x \in \mathbb{R}^+$ and $f(x) \geq f(x+y)$ for $x, y > 0$. Therefore f is a non-increasing function on $\mathbb{R}^+$.

If there exists $x_1 > 0$ such that $f(x_1) = 1$, then $f(x) = 1$ on $(0, x_1]$. Let $x = x_1$, $y = kx_1$, then (30.15) yields $f((k + 1)x_1) = f(kx_1)$. Thus, $f(kx_1) = f(x_1) = 1$ for $k = 1, 2, 3, \cdots$. By the monotonicity of f, we have $f(x) = 1$ identically on $\mathbb{R}^+$.

When $f(x) < 1$ on $\mathbb{R}^+$, then f is strictly decreasing, and f is injection. Let $x = 1$ and use $x + y - 1$ to replace y, then (30.15) yields

$$f(1)f((x + y - 1)f(1)) = f(1 + (x + y - 1)) = f(x + y) = f(x)f(yf(x)).$$

Letting $y = \frac{1}{f(x)}$, then above equality yields

$$f(1)f\left(\left(x + \frac{1}{f(x)} - 1\right)f(1)\right) = f(x)f(1).$$

Since f is an injection,

$$\left(x + \frac{1}{f(x)} - 1\right) f(1) = x \Rightarrow f(x) = \frac{1}{1 + ax},$$

where $a = \dfrac{1 - f(1)}{f(1)}$. Thus, $f(x) = \dfrac{1}{1 + ax}, x > 0$, where $a \geq 0$ is a constant.
By substituting this function into (30.15), it is easy to check that this is indeed
asolution.

Example 7. (BULGARIA/2010) Let $f : \mathbb{N} \to \mathbb{N}$ be a function such that $f(1) = 1$ and

$$f(n) = n - f(f(n-1)), \qquad n \geq 2.$$

Prove that $f(n + f(n)) = n$ for each positive integer $n \in \mathbb{N}$.

Solution   First we prove by induction on n that for each $n \in \mathbb{N}$

$$f(n) \leq f(n+1) \leq f(n) + 1. \tag{30.16}$$

When $n = 1$, $f(2) = 2 - f(f(1)) = 2 - 1 = 1 = f(1)$, so (30.16) is true.
When $n = 2$, $f(3) = 3 - f(f(2)) = 3 - 1 = 2$, so (30.16) is true.
Assume that (30.16) is true for all $n \leq k$ $(k \geq 2)$. Then for $n = k + 1$, from
(30.16), $f(n) < n, n \in \mathbb{N}$. Since $f(k) \leq f(k+1) \leq f(k) + 1 \leq k$, therefore
$f(k+1) = f(k)$ or $f(k+1) = f(k) + 1$. By the induction assumption,

$$f(f(k)) \leq f(f(k+1)) \Leftrightarrow k + 1 - f(k+1) \leq k + 2 - f(k+2)$$
$$\Leftrightarrow f(k+2) \leq f(k+1) + 1,$$
$$f(f(k+1)) \leq f(f(n)+1) \leq f(f(n)) + 1$$
$$\Rightarrow k + 2 - f(k+2) \leq k + 1 - f(k+1) + 1 \Rightarrow f(k+1) \leq f(k+2).$$

so (30.16) is true also for $n = k + 1$. By induction, (30.16) is true for all $n \in \mathbb{N}$.
Next, we prove by induction on n that for each $n \in \mathbb{N}$

$$f(n + f(n)) = n. \tag{30.17}$$

For $n = 1$, $f(1 + f(1)) = f(2) = 1$, (30.17) is proven.
Assume that (30.17) is true for $n = k$ $(k \geq 1)$, i.e., $f(k + f(k)) = k$, then

$$f(f(k + f(k))) = f(k) \quad \Rightarrow \quad k + f(k) + 1 - f(k + f(k) + 1) = f(k)$$
$$\Rightarrow \quad f(k + 1 + f(k)) = k + 1.$$

If $f(k+1) = f(k)$, then (30.17) holds for $n = k + 1$. If $f(k+1) = f(k) + 1$,
then

$$f(k + 1 + f(k+1)) \quad = \quad k + 1 + f(k+1) - f(f(k + f(k+1)))$$
$$= \quad k + 1 + f(k+1) - f(f(k + f(k)) + 1))$$
$$= \quad k + 1 + f(k+1) - f(k+1) = k + 1.$$

Thus, (30.17) is true for $n = k + 1$ also. The inductive proof is completed.

Testing Questions (A)

1. (THAILAND/2007) Determine if there exists a function $f : \mathbb{Z} \to \mathbb{Z}$ such that
$$f(f(n) - 2n) = 2f(n) + n, \qquad \forall n \in \mathbb{Z}.$$

2. (ESTONIA/TST/2007) Find all continuous functions $f : \mathbb{R} \to \mathbb{R}$ such that
$$f(x + f(y)) = y + f(x + 1) \qquad \text{for any } x, y \in \mathbb{R}. \tag{30.18}$$

3. (IRAN/TST/2008) k is a given natural number. Find all functions $f : \mathbb{N} \to \mathbb{N}$ such that for each $m, n \in \mathbb{N}$ the following holds:
$$f(m) + f(n) \mid (m + n)^k.$$

4. (APMO/2008/Q4) Consider the function $f : \mathbb{N}_0 \to \mathbb{N}_0$, where $\mathbb{N} - 0$ is the set of all non-negative integers, defined by the following conditions :
   (i) $f(0) = 0$;   (ii) $f(2n) = 2f(n)$ and
   (iii) $f(2n + 1) = n + 2f(n)$ for all $n \geq 0$.
   (a) Determine the three sets $L = \{n \mid f(n) < f(n + 1)\}$, $E = \{n \mid f(n) = f(n + 1)\}$, and $G = \{n \mid f(n) > f(n + 1)\}$.
   (b) For each $k \geq 0$, find a formula for $a_k = \max\{f(n) \mid 0 \leq n \leq 2^k\}$ in terms of k.

5. (SSSMO/R2/2008) Find all functions $f : \mathbb{R} \to \mathbb{R}$ so that
   (i) $f(2u) = f(u + v)f(v - u) + f(u - v)f(-u - v)$ for all $u, v \in \mathbb{R}$,
   (ii) $f(u) \geq 0$ for all $u \in \mathbb{R}$.

6. (BMO/2008-2009/R2-Q3) Find all functions f from the real numbers to the real numbers which satisfy
$$f(x^3) + f(y^3) = (x + y)(f(x^2) + f(y^2) - f(xy)) \tag{30.19}$$
for all real numbers x and y.

7. (TURKEY/TST/2009) Find all functions $f : \mathbb{Q}^+ \to \mathbb{Z}$ that satisfy
$$f\left(\frac{1}{x}\right) = f(x) \quad \text{and} \tag{30.20}$$
$$(x + 1)f(x - 1) = xf(x) \tag{30.21}$$
for all rational numbers x that are bigger than 1.

8.   (SLOVENIA/TST/2008-2009) Find all functions $g : \mathbb{R} \to \mathbb{R}$, such that there is a strictly increasing function f which satisfies

$$f(x + y) = f(x)g(y) + f(y), \qquad x, y \in \mathbb{R}. \tag{30.22}$$

9.   (IRAN/TST/2010) Find all non-decreasing functions $f : \mathbb{R}^+ \cup \{0\} \to \mathbb{R}^+ \cup \{0\}$ such that for each $x, y \in \mathbb{R}^+ \cup \{0\}$

$$f\left(\frac{x + f(x)}{2} + y\right) = 2x - f(x) + f(f(y)). \tag{30.23}$$

Testing Questions   (B)

1.   (SMO/2010) Let S be the set of all non-zero real valued functions f defined on the set of all real numbers such that

$$f(x^2 + yf(z)) = xf(x) + zf(y)$$

for all real numbers x, y and z. Find the maximum value of $f(12345)$, where $f \in S$.

2.   (BALKAN MO/2009) Denote by S the set of all positive integers. Find all functions $f : S \to S$ such that

$$f\left(f^2(m) + 2f^2(n)\right) = m^2 + 2n^2 \quad \text{for all } m, n \in S.$$

3.   (GREECE/2009) Find all functions $f : \mathbb{R}^+ \to \mathbb{R}^+$, satisfying

$$f\left(\frac{f(x)}{f(y)}\right) = \frac{1}{y}f(f(x)), \qquad x, y \in \mathbb{R}^+,$$

and that f is strictly monotone on $\mathbb{R}^+$.

4.   (IMO/Shortlist/2008, HUNGARY-ISRAEL MC/2009) We say that a pair (f, g) of functions from $\mathbb{N}$ into $\mathbb{N}$ is a *Spanish Couple* on $\mathbb{N}$, if they satisfy the following conditions:

   (i)   Both functions are strictly increasing, i.e. $f(x) < f(y)$ and $g(x) < g(y)$ for all $x, y \in \mathbb{N}$ with $x < y$;

   (ii)   The inequality $f(g(g(x))) < g(f(x))$ holds for all $x \in \mathbb{N}$.

   Decide whether a Spanish Couple exists?

5.   (IMO/Shortlist/2009) Find all functions f from the set of real numbers into the set of real numbers which satisfy the identity

$$f(xf(x + y)) = f(yf(x)) + x^2 \qquad \text{for all real } x, y.$$

Solutions to Testing Questions

Solutions to Testing Questions

Solutions to Testing Questions 16

Testing Questions (16-A)

1. We prove the general proposition that this can be done for any n-element set, where n is an positive integer, $S_n = \{1, 2, \ldots, n\}$ and integer N with $0 \le N \le 2^n$.

    We induct on n. When $n = 1$, then the only subsets of S_1 are $\emptyset$ and $\{1\}$. If $N = 1$, then let any one element in S be red and the other be black. If $N = 2$, then let the two elements $\emptyset$ and $\{1\}$ be both red.

    Assume that the desired coloring can be done to the subsets of set $S_n = \{1, 2, \ldots, n\}$ and integer N_n with $0 \le N_n \le 2^n$. We now show that there is a desired coloring for set $S_{n+1} = \{1, 2, \ldots, n, n+1\} = S_n \cup \{n+1\}$ and integer N_{n+1} with $0 \le N_{n+1} \le 2^{n+1}$. We consider the following cases:

    (i) $0 \le N_{n+1} \le 2^n$. Applying the induction hypothesis to S_n and $N_n = N_{n+1}$, we get a coloring of all subsets of S_n satisfying conditions (a), (b), (c). All uncolored subsets of S_{n+1} contains the element $n + 1$, and we color all of them blue. It is not hard to see that this coloring of all the subsets of S_{n+1} satisfies conditions (a), (b), (c).

    (ii) $N_{n+1} = 2^n + k$ with $1 \le k \le 2^n$. Applying the induction hypothesis to S_n and $N_n = k$, we get a coloring of all subsets of S_n satisfying conditions (a), (b), (c). All uncolored subsets of S_{n+1} contain the element $n + 1$, and we color all of them red. It is not hard to see that this coloring of all the subsets of S_{n+1} satisfies conditions (a), (b), (c).

    Thus our induction is complete.

2. We use induction on k. Let the statement in the question be P_k. For $k = 1$, then it is enough to let $x_1 = 0, x_2 = 3, y_1 = 1, y_2 = 2$.

    Assume that the propositions $P_l, 1 \le l \le k$ are all true ($k \ge 1$). Then consider P_{k+1}.

127

By the inductive assumption, the set $\{0, 1, 2, \ldots, 2^{k+1} - 1\}$ can be partitioned into two subsets $\{x_1, x_2, \ldots, x_{2^k}\}$ and $\{y_1, y_2, \ldots, y_{2^k}\}$. Then the sets

$$\{x_1, x_2, \ldots, x_{2^k}, 2^{k+1} + y_1, 2^{k+1} + y_2, \ldots, 2^{k+1} + y_{2^k}\}$$

and

$$\{y_1, y_2, \ldots, y_{2^k}, 2^{k+1} + x_1, 2^{k+1} + x_2, \ldots, 2^{k+1} + x_{2^k}\}$$

are disjoint and their union is $\{0, 1, 2, \ldots, 2^{k+2} - 1\}$. Below we prove that

$$\sum_{i=1}^{2^k} x_i^m + \sum_{i=1}^{2^k} (2^{k+1} + y_i)^m = \sum_{i=1}^{2^k} y_i^m + \sum_{i=1}^{2^k} (2^{k+1} + x_i)^m. \qquad (*)$$

In fact,

$$(*) \Leftrightarrow \sum_{i=1}^{2^k} x_i^m + \sum_{i=1}^{2^k} y_i^m + \sum_{i=1}^{m-1} \binom{m}{t} (2^{k+1})^{m-t} \sum_{i=1}^{2^k} y_i^t$$

$$= \sum_{i=1}^{2^k} y_i^m + \sum_{i=1}^{2^k} x_i^m + \sum_{i=1}^{m-1} \binom{m}{t} (2^{k+1})^{m-t} \sum_{i=1}^{2^k} x_i^t$$

$$\Leftrightarrow \sum_{i=1}^{m-1} \binom{m}{t} (2^{k+1})^{m-t} \sum_{i=1}^{2^k} (x_i^t - y_i^t) = 0.$$

The inductive assumption gives that

$$\sum_{i=1}^{2^k} (x_i^t - y_i^t) = 0, \qquad t = 1, 2, \ldots, m - 1,$$

therefore $(*)$ holds, and hence P_{k+1} is true. Thus, the inductive proof is completed.

3. If $a < -2$, then $|f^1(0)| = |a| > 2$, so $a \notin M$.

   If $-2 \le a \le \dfrac{1}{4}$, by definitions, $f^1(0) = f(0) = a$, $f^n(0) = (f^{n-1}(0))^2 + a$, $n = 2, 3, \ldots$. We prove by induction that $|f^n(0)| \le 2$ as follows.

   (i) When $0 \le a \le \dfrac{1}{4}$, it can be obtained that $|f^n(0)| \le \dfrac{1}{2}, n \in \mathbb{N}$.

   In fact, for $n = 1$, $|f^1(0)| = |a| \le \dfrac{1}{2}$. Assume that $|f^n(0)| \le \dfrac{1}{2}$ for $n = k - 1$ ($k \ge 2$), then for $n = k$,

$$|f^k(0)| = |f^{k-1}(0)|^2 + a \le \left(\frac{1}{2}\right)^2 + \frac{1}{4} = \frac{1}{2}.$$

(ii)   When $-2 \leq a < 0$, it can be obtained that $|f^n(0)| \leq |a| \leq 2, n \in \mathbb{N}$. In fact, for $n = 1, |f^1(0)| = |a|$. Assume that $|f^n(0)| \leq |a|$ for $n = k - 1$ ($k \geq 2$), then for $n = k$,

$$a^2 + a \geq (f^{k-1}(0))^2 + a = f^k(0) \geq a = -|a|.$$

Note that $-2 \leq a < 0 \Rightarrow a^2 \leq -2a \Rightarrow a^2 + a \leq -a = |a|$, so $|f^k(0)| \leq |a|$.

Thus, we have proven that $\left[-2, \dfrac{1}{4}\right] \subseteq M$.

(iii)   When $a > \dfrac{1}{4}$, let $a_n = f^n(0), n \in \mathbb{N}$, then

$$a_n \geq a > \frac{1}{4}, \quad n \in \mathbb{N},$$

and $a_{n+1} = f^{n+1}(0) = f(f^n(0)) = f(a_n) = a_n^2 + a$. For any $n \geq 1$,

$$a_{n+1} - a_n = a_n^2 - a_n + a = \left(a_n - \frac{1}{2}\right)^2 + a - \frac{1}{4} \geq a - \frac{1}{4},$$

hence

$$a_{n+1} - a = a_{n+1} - a_1 = \sum_{k=1}^{n}(a_{k+1} - a_k) \geq n\left(a - \frac{1}{4}\right).$$

Hence, for n large enough such that $n > \dfrac{2-a}{a - \frac{1}{4}}$, we have

$$a_{n+1} \geq n\left(a - \frac{1}{4}\right) + a > 2 - a + a = 2 \Rightarrow a \notin M.$$

Combining (i), (ii), (iii), we conclude that $M = \left[-2, \dfrac{1}{4}\right]$.

4.   We prove the conclusion by induction on n.

When $n = 1$, since $\dfrac{1}{x_2} \geq 4(1 - x_2) \Leftrightarrow (2x_2 - 1)^2 \geq 0$,

$$\frac{1}{x_1} + \frac{x_1}{x_2} \geq \frac{1}{x_1} + 4x_1(1 - x_2)$$

$$= 2\left(\frac{1}{2x_1} + 2x_1\right) - 4x_1x_2 \geq 4 - 4x_1x_2 = 4(1 - x_1x_2).$$

Assuming that the given inequality holds for $n = k$, i.e., for any positive numbers $x_1, x_2, \ldots, x_{k+1}$

$$\frac{1}{x_1} + \frac{x_1}{x_2} + \frac{x_1 x_2}{x_3} + \cdots + \frac{x_1 x_2 \cdots x_k}{x_{k+1}} \geq 4(1 - x_1 x_2 \cdots x_{k+1}),$$

then

$$\frac{1}{x_2} + \frac{x_2}{x_3} + \frac{x_2 x_3}{x_4} + \cdots + \frac{x_2 x_3 \cdots x_{k+1}}{x_{k+2}} \geq 4(1 - x_2 x_3 \cdots x_{k+2}),$$

so that for $n = k + 1$,

$$\frac{1}{x_1} + \frac{x_1}{x_2} + \frac{x_1 x_2}{x_3} + \cdots + \frac{x_1 x_2 \cdots x_{k+1}}{x_{k+2}}$$
$$= \frac{1}{x_1} + x_1 \left(\frac{1}{x_2} + \frac{x_2}{x_3} + \cdots + \frac{x_2 x_3 \cdots x_{k+1}}{x_{k+2}} \right)$$
$$\geq \frac{1}{x_1} + 4x_1(1 - x_2 x_3 \cdots x_{k+2}) = 2 \left(\frac{1}{2x_1} + 2x_1 \right) - 4x_1 x_2 \cdots x_{k+2}$$
$$\geq 4 - 4x_1 x_2 \cdots x_{k+2} = 4(1 - x_1 x_2 \cdots x_{k+2}).$$

Hence the inequality is proven for $n = k + 1$. item[5.] $a_1 = 2, a_{n+1} = a_1 a_2 \cdots a_n + 1 \Rightarrow a_n \geq 2$ and $a_1 a_2 \cdots a_n \geq 2^n$ for all $n \in \mathbb{N}$. We first prove the following proposition P_n for $n \in \mathbb{N}$ by induction on n:

$$1 - \left(\frac{1}{a_1} + \frac{1}{a_2} + \cdots + \frac{1}{a_n} \right) = \frac{1}{a_1 a_2 \cdots a_n}. \qquad (*)$$

(i)   When $n = 1$, then $a_1 = 2 \Rightarrow 1 - \frac{1}{a_1} = \frac{1}{a_1}$, $(*)$ is true.

(ii) Assume that $(*)$ is true for $n = k$ $(k \geq 1)$. Since $a_{k+1} = a_1 a_2 \cdots a_k + 1$,

$$1 - \left(\frac{1}{a_1} + \frac{1}{a_2} + \cdots + \frac{1}{a_{k+1}} \right) = \frac{1}{a_1 a_2 \cdots a_k} - \frac{1}{a_{k+1}}$$
$$= \frac{a_{k+1} - a_1 a_2 \cdots a_k}{a_1 a_2 \cdots a_{k+1}} = \frac{1}{a_1 a_2 \cdots a_{k+1}},$$

so $(*)$ holds also for $n = k + 1$.

Combining (i), (ii) and that $a_1 a_2 \cdots a_n \geq \frac{1}{2^n}$, it follows that

$$1 - \left(\frac{1}{a_1} + \frac{1}{a_2} + \cdots + \frac{1}{a_n} \right) = \frac{1}{a_1 a_2 \cdots a_n} \leq \frac{1}{2^n}.$$

Therefore for all $n \in \mathbb{N}$,

$$\frac{1}{a_1} + \frac{1}{a_2} + \cdots + \frac{1}{a_n} \geq 1 - \frac{1}{2^n} = \frac{1}{2} + \frac{1}{4} + \cdots + \frac{1}{2^n}.$$

6. Use induction on n. When $n = 1$, $3^1 > 1^4$ is true.

   Assume that when $n = k$ $(k \geq 1)$, $3^{k^2} > (k!)^4$ holds. Below we show that $3^{(k+1)^2} > [(k+1)!]^4$.

   Since $3^{(k+1)^2} = 3^{k^2} \cdot 3^{2k+1}$ and $[(k+1)!]^4 = (k!)^4(k+1)^4$, it suffices to show that

   $$3^{2k+1} \geq (k+1)^4, \qquad k \in \mathbb{N}.$$

   Use induction again. For $k = 1$, it is clear that $3^3 = 27 > 2^4 = 16$.

   Assume that for $k = j$, $3^{2j+1} > (j+1)^4$ holds. Then for $k = j+1$,

   $$3^{2j+3} \geq (j+2)^4 \Leftrightarrow 3^2 \cdot 3^{2j+1} \geq \frac{(j+2)^4}{(j+1)^4} \cdot (j+1)^4,$$

   and so it suffices to show $3 \geq \dfrac{(j+2)^2}{(j+1)^2}$ for $j \in \mathbb{N}$. This is true since

   $$3 \geq \frac{(j+2)^2}{(j+1)^2} \Leftrightarrow 3j^2 + 6j + 3 \geq j^2 + 4j + 4 \Leftrightarrow 2j^2 + j \geq 1,$$

   and the last inequality is obviously true for $j \in \mathbb{N}$. Thus, the inductive proof is completed.

7. First we prove by induction on t that if $r + s = t$ is odd, where $r, s \in \mathbb{N}_0$, then $ra + sb \in S$.

   When $t = 1$, the conclusion is obvious.

   Assume that the conclusion is true when $t = 2k - 1$ $(k \geq 1)$, then for $t = 2k + 1$, one of r, s is ≥ 2. Say $r \geq 2$. By induction assumption, $(r - 2)a + sb \in S$. By taking $x = (r - 2)a + sb$, $y = z = a$, then the condition (ii) yields

   $$ra + sb = x + y + z \in S,$$

   the conclusion is proven.

   We now return to the original problem. Since $(a, b) = 1$, for any integer $c > 2ab$, there exist integers $r, s \in \mathbb{N}_0$ such that $ra + sb = c$, where r, s are given by

   $$r = r_0 + bt, \quad s = s_0 - at, \quad (t \in \mathbb{Z}).$$

By letting $t = t_1$ and $t_2 = t_1 + 1$ such that $r_1 = r_0 + t_1 b \in [0, b)$ and $r_2 = r_0 + t_2 b \in [b, 2b)$, then

$$s_1 = \frac{c - r_1 a}{b} \in \left(\frac{c - ab}{b}, \frac{c}{b} \right] \Rightarrow s_1 > a, s_2 = \frac{c - r_2 a}{b} > \frac{c - 2ab}{b} > 0$$

$$\Rightarrow (r_2 + s_2) = (r_1 + s_1) + (b - a)$$

$$\Rightarrow (r_1 + s_1) \text{ and } (r_2 + s_2) \text{ have different parities,}$$

so it is possible to select one value from t_1 and t_2 such that $r_i + t_i$ is odd. Then $c = r_i a + s_i b \in S$ is proven as before.

8. In order to apply induction, we generalize the result to be proved so that it reads as follows:

Proposition. If the n-element subsets of a set S with $(n+1)m-1$ elements are partitioned into two classes, then there are at least m pairwise disjoint sets in the same class.

Proof. Fix n and proceed by induction on m. The case of $m = 1$ is trivial.

Assume $m > 1$ and that the proposition is true for $m - 1$. Let P be the partition of the n-element subsets into two classes. If all the n-element subsets belong to the same class, the result is obvious. Otherwise select two n-element subsets A and B from different classes so that their intersection has maximal size. It is easy to see that $|A \cap B| = n - 1$. (If $|A \cap B| = k < n - 1$, then build C from B by replacing some element not in $A \cap B$ with an element of A not already in B. Then $|A \cap C| = k + 1$ and $|B \cap C| = n - 1$ and either A and C or B and C are in different classes.) Removing $A \cup B$ from S, there are $(n + 1)(m - 1) - 1$ elements left. On this set the partition induced by P has, by the inductive hypothesis, $m - 1$ pairwise disjoint sets in the same class. Adding either A or B as appropriate gives m pairwise disjoint sets in the same class.

Remark: The value $n^2 + n - 1$ is sharp. A set S with $n^2 + n - 2$ elements can be split into a set A with $n^2 - 1$ elements and a set B of $n - 1$ elements. Let one class consist of all n-element subsets of A and the other consist of all n-element subsets that intersect B. Then neither class contains n pairwise disjoint sets.

Testing Questions   (16-B)

1. Let P_n be the statement: $\displaystyle\sum_{i=0}^{2n} (-1)^i a_i b_i \geq 0$, where $n = 1, 2, \ldots$.

For $n = 1$, from the given conditions that $a_0 + a_1 \geq 0, a_1 + a_2 \geq 0, a_1 \leq 0, b_0 > 0, b_2 > 0$ and $b_0 + b_1 + b_2 > 0$, it follows that

$$
\begin{aligned}
\sum_{i=0}^{2n} (-1)^i a_i b_i &= a_0 b_0 - a_1 b_1 + a_2 b_2 \\
&= b_0(a_0 + a_1) - a_1(b_0 + b_1 + b_2) + b_2(a_1 + a_2) \geq 0,
\end{aligned}
$$

and the equality holds if and only if $a_0 + a_1 = -a_1 = a_1 + a_2 = 0$, namely, $a_0 = a_1 = a_2 = 0$. Thus, P_1 is proven.

Assume that P_k is true. Then for P_{k+1}, from the conditions $a_{2k+1} + a_{2k+2} \geq 0, b_{2k+2} > 0, a_{2k} + a_{2k+1} \geq 0, a_{2k+1} \leq 0$, it follows that $a_{2k+2} \geq -a_{2k+1}$ and $-a_{2k} \leq a_{2k+1} \leq 0$. Therefore

$$
\begin{aligned}
\sum_{i=0}^{2k+2} (-1)^i a_i b_i &= \sum_{i=0}^{2k-1} (-1)^i a_i b_i + a_{2k} b_{2k} - a_{2k+1} b_{2k+1} + a_{2k+2} b_{2k+2} \\
&\geq \sum_{i=0}^{2k-1} (-1)^i a_i b_i + a_{2k} b_{2k} - a_{2k+1} b_{2k+1} - a_{2k+1} b_{2k+2} \\
&= \sum_{i=0}^{2k-1} (-1)^i a_i b_i + a_{2k} b_{2k} - a_{2k+1}(b_{2k+1} + b_{2k+2}). \qquad (30.1)
\end{aligned}
$$

(i)   When $b_{2k+1} + b_{2k+2} \geq 0$, then $-a_{2k+1}(b_{2k+1} + b_{2k+2}) \geq 0$, so by the induction assumption,

$$
\sum_{i=0}^{2k+2} (-1)^i a_i b_i \geq \sum_{i=0}^{2k} (-1)^i a_i b_i \geq 0.
$$

(ii)   When $b_{2k+1} + b_{2k+2} < 0$, $-a_{2k+1}(b_{2k+1} + b_{2k+2}) \geq a_{2k}(b_{2k+1} + b_{2k+2})$, so

$$\sum_{i=0}^{2k+2} (-1)^i a_i b_i \geq \sum_{i=0}^{2k-1} (-1)^i a_i b_i + a_{2k}(b_{2k} + b_{2k+1} + b_{2k+2}).$$

Letting $c_i = b_i, i = 0, 1, 2, \ldots, 2k-1$ and $c_{2k} = b_{2k} + b_{2k+1} + b_{2k+2}$, then

$$\sum_{i=2p}^{2q} c_i = \sum_{i=2p}^{2q} b_i > 0 \qquad\qquad \text{if } 0 \leq p \leq q \leq k - 1,$$

$$\sum_{i=2p}^{2q} c_i = \sum_{i=2p}^{2k-1} b_i + b_{2k} + b_{2k+1} + b_{2k+2} > 0 \quad \text{if } 0 \leq p \leq q = k.$$

Therefore $c_1, c_2, \ldots, c_{2k}$ satisfy the condition (iii) still, hence, from the induction assumption,

$$\sum_{i=0}^{2k-1} (-1)^i a_i b_i + a_{2k}(b_{2k} + b_{2k+1} + b_{2k+2}) = \sum_{i=0}^{2k} (-1)^i a_i c_i \geq 0.$$

It is clear that $a_0 = a_1 = \cdots = a_{2k+2} = 0 \Rightarrow \sum_{i=0}^{2k+2} (-1)^i a_i b_i = 0$. Conversely, if $\sum_{i=0}^{2k+2} (-1)^i a_i b_i = 0$, then in either case it implies $\sum_{i=0}^{2k} (-1)^i a_i b_i = 0$ or $\sum_{i=0}^{2k} (-1)^i a_i c_i = 0$. Hence, by the induction assumption, $a_0 = a_1 = \cdots = a_{2k} = 0$. Now $a_{2k} + a_{2k+1} \geq 0$ and $a_{2k+1} \leq 0$ together imply that $a_{2k+1} = 0$. $a_{2k+2} b_{2k+2} = 0$ and $b_{2k+2} > 0$ then imply that $a_{2k+2} = 0$, therefore

$$a_0 = a_1 = \cdots = a_{2k+2} = 0$$

is the sufficient and necessary condition for equality to hold.

Thus, P_{k+1} is true and so by induction, P_k is true for all $k \in \mathbb{N}$.

2.   The proof is by induction on n. The base is provided by the $n = 0$ case, where $7^{7^0} + 1 = 7^1 + 1 = 2^3$. To prove the inductive step, it suffices to show that if $x = 7^{2m-1}$ for some positive integer m then $(x^7 + 1)/(x + 1)$

is composite. As a consequence, $x^7 + 1$ has at least two more prime factors than does $x + 1$. To confirm that $(x^7 + 1)/(x + 1)$ is composite, observe that

$$
\begin{aligned}
\frac{x^7 + 1}{x + 1} &= \frac{(x + 1)^7 - ((x + 1)^7 - (x^7 + 1))}{x + 1} \\
&= (x + 1)^6 - \frac{7x(x^5 + 3x^4 + 5x^3 + 5x^2 + 3x + 1)}{x + 1} \\
&= (x + 1)^6 - 7x(x^4 + 2x^3 + 3x^2 + 2x + 1) \\
&= (x + 1)^6 - 7^{2m}(x^2 + x + 1)^2 \\
&= \{(x + 1)^3 - 7^m(x^2 + x + 1)\}\{(x + 1)^3 + 7^m(x^2 + x + 1)\}.
\end{aligned}
$$

It remains to show that each factor exceeds 1. It suffices to check the smaller one. $\sqrt{7x} \le x$ gives

$$
\begin{aligned}
(x + 1)^3 - 7^m(x^2 + x + 1) &= (x + 1)^3 - \sqrt{7x}(x^2 + x + 1) \\
&\ge x^3 + 3x^2 + 3x + 1 - x(x^2 + x + 1) \\
&= 2x^2 + 2x + 1 \ge 113 > 1.
\end{aligned}
$$

Hence $(x^7 + 1)/(x + 1)$ is composite and the proof is complete.

3. (i) For $n = 1, 2$ and 3, we have $a_1 > \dfrac{1}{12} > \dfrac{19}{243}$ and

$$
a_2 > \sqrt{3 \times \frac{19}{243} + 1} = \frac{10}{9}, \quad a_3 > \sqrt{4 \times \frac{10}{9} + 1} = \frac{7}{3},
$$

so $a_n > n - \dfrac{2}{n}$ is true for $n = 1, 2, 3$. For $n \ge 3$ we prove the proposition by induction on n. Let P_n be the proposition: $a_n > n - \dfrac{2}{n}$.

Assume P_k is true ($k \ge 3$), then $a_{k+1} > \sqrt{(k + 2)\left(k - \dfrac{2}{k}\right) + 1}$ implies

P_{k+1} is true if $\sqrt{(k + 2)\left(k - \dfrac{2}{k}\right) + 1} > (k + 1) - \dfrac{2}{k + 1}$. Since

$$
\sqrt{(k + 2)\left(k - \frac{2}{k}\right) + 1} > (k + 1) - \frac{2}{k + 1} \Leftrightarrow \frac{1}{2} > \frac{1}{k} + \frac{1}{(k + 1)^2}
$$

and the last inequality is obvious for $k \ge 3$, P_{k+1} is proven.

(ii) When $a_1 < 1$, it follows by induction that $a_n < n$ and $b_n < 0$ for $n \in \mathbb{N}$. Below we prove that $b_n < b_{n+1}$, i.e., $\dfrac{a_n - n}{2n} < \dfrac{a_{n+1} - (n+1)}{n+1}$.

$$\frac{a_{n+1} - (n+1)}{n+1} = \frac{a_{n+1}^2 - (n+1)^2}{(n+1)(a_{n+1} + n + 1)} = \frac{(n+2)a_n + 1 - (n+1)^2}{(n+1)(a_{n+1} + n + 1)}$$
$$= \frac{(n+2)(a_n - n)}{(n+1)(a_{n+1} + n + 1)},$$

so

$$b_n + 1 < b_{n+1} \Leftrightarrow \frac{1}{2n} > \frac{n+2}{(n+1)(a_{n+1} + n + 1)}$$
$$\Leftrightarrow (n+1)a_{n+1} > 2n(n+2) - (n+1)^2 \Leftrightarrow a_{n+1} > (n+1) - \frac{2}{n+1},$$

and the last is true from the result of (i).

When $a_1 > 1$, then similarly we have $b_n > b_{n+1} > 0$ for $n \in \mathbb{N}$. Thus, $\{b_n\}$ is a monotone and bounded sequence, so it is convergent.

4. Let h_i also denote the student with height h_i. We prove that for $1 \le i < j \le n$, h_j can switch with h_i at most $j - i - 1$ times. We proceed by induction on $j - i$, the base case $j - i = 1$ being evident because h_i is not allowed to switch with h_{i-1}.

For the inductive step, note that h_i, h_{j-1}, h_j can be positioned on the circle either in this order or in the order h_i, h_j, h_{j-1}. Since h_{j-1} and h_j cannot switch, the only way to change the relative order of these three students is for h_i to switch with either h_{j-1} or h_j. Consequently, any two switches of h_i with h_j must be separated by a switch of h_i with h_{j-1}. Since there are at most $j - i - 2$ of the latter, there are at most $j - i - 1$ of the former.

The total number of switches is thus at most

$$\sum_{i=1}^{n-1} \sum_{j=i+1}^{n} (j - i - 1) = \sum_{i=1}^{n-1} \sum_{j=0}^{n-i-1} j = \sum_{i=1}^{n-1} \binom{n-i}{2}$$
$$= \sum_{i=1}^{n-1} \left(\binom{n-i+1}{3} - \binom{n-i}{3} \right) = \binom{n}{3}.$$

5. Yes. There exists such a sequence of moves.

Denote by $(a_1, a_2, \ldots, a_n) \to (a_1', a_2', \ldots, a_n')$ the following: if some consecutive boxes contain $a_1, \ldots, a_n$ coins, then it is possible to perform a series of allowed moves such that the boxes contain $a_1', \ldots, a_n'$ coins respectively, while the contents of the other boxes remain unchanged.

Let $A = 2010^{2010^{2010}}$. Our goal is to show that

$$(1, 1, 1, 1, 1, 1) \to (0, 0, 0, 0, 0, A).$$

First we prove two auxiliary observations.

Lemma 1. $(a, 0, 0) \to (0, 2^k, 0)$ for every $a \geq 1$.

Proof. We prove by induction that $(a, 0, 0) \to (a - k, 2^a, 0)$ for every $1 \leq k \leq a$. For $k = 1$, apply Type 1 to the first box:

$$(a, 0, 0) \to (a - 1, 2, 0) = (a - 1, 2^1, 0).$$

Now assume that $k < a$ and the statement holds for some $k < a$. Starting from $(a-k, 2^k, 0)$, apply Type 1 to the middle box 2^k times, until it becomes empty. Then apply Type 2 to the first box:

$$\begin{aligned} (a - k, 2^k, 0) &\to (a - k, 2^k - 1, 2) \to \cdots \to (a - k, 0, 2^{k+1}) \\ &\to (a - k - 1, 2^{k+1}, 0). \end{aligned}$$

Hence,

$$(a, 0, 0) \to (a - k, 2^k, 0) \to (a - k - 1, 2^{k+1}, 0).$$

$\square$

Lemma 2. For every positive integer n, let $P_n = 2^{2^{\cdot^{\cdot^{2}}}}$ (e.g. $P_3 = 2^{2^2} = 16$). Then $(a, 0, 0, 0) \to (0, P_a, 0, 0)$ for every $a \geq 1$.

Proof. Similarly to Lemma 1, we prove that $(a, 0, 0, 0) \to (a-k, P_k, 0, 0)$ for every $1 \leq k \leq a$.

For $k = 1$, apply Type 1 to the first box:

$$(a, 0, 0, 0) \to (a - 1, 2, 0, 0) = (a - 1, P_1, 0, 0).$$

Now assume that the lemma holds for some $k < a$. Starting from $(a - k, P_k, 0, 0)$, apply Lemma 1, then apply Type 1 to the first box:

$$\begin{aligned} (a - k, P_k, 0, 0) &\to (a - k, 0, 2^{P_k}, 0) = (a - k, 0, P_{k+1}, 0) \\ &\to (a - k - 1, P_{k+1}, 0, 0). \end{aligned}$$

Therefore,

$$(a, 0, 0, 0) \to (a - k, P_k, 0, 0) \to (a - k - 1, P_{k+1}, 0, 0).$$

$\square$

Now we prove the statement of the problem.

First apply Type 1 to box 5, then apply Type 2 to boxes B_4, B_3, B_2 and B_1 in this order. Then apply Lemma 2 twice:

$$(1, 1, 1, 1, 1, 1)$$
$$\rightarrow (1, 1, 1, 1, 0, 3) \rightarrow (1, 1, 0, 3, 0, 0) \rightarrow (1, 0, 3, 0, 0, 0)$$
$$\rightarrow (0, 3, 0, 0, 0, 0) \rightarrow (0, 0, P_3, 0, 0, 0) = (0, 0, 16, 0, 0, 0)$$
$$\rightarrow (0, 0, 0, P_{16}, 0, 0).$$

We already have more than A coins in box B_4, since

$$A \leq 2010^{2010^{2010^{2010}}} < (2^{11})^{2010^{2010}} = 2^{11 \cdot 2010^{2010}} < 2^{2010^{2011}}$$

$$< 2^{(2^{11})^{2011}} = 2^{2^{11 \cdot 2011}} < 2^{2^{2^{15}}} < P_{16}.$$

To decrease the number of coins in box B_4, apply Type 2 to this stack repeatedly until its size decreases to $A/4$. (In every step, we remove a coin from B_4 and exchange the empty boxes B_5 and B_6.)

Finally, apply Type 1 repeatedly to empty boxes B_4 and B_5:

$$(0, 0, 0, A/4, 0, 0) \rightarrow \cdots \rightarrow (0, 0, 0, 0, A/2, 0) \rightarrow \cdots \rightarrow (0, 0, 0, 0, 0, A).$$

Solutions to Testing Questions   17

Testing Questions   (17-A)

1. Since $S_5 = 5a_3$ and $S_9 = 9a_5$, so $5a_3 = 9a_5$, i.e., $a_3 : a_5 = 9 : 5$. The answer is (A).

2. Let d be the common difference of $\{a_n\}$. Then

$$3a_8 = 5a_{13} \Rightarrow 3(a_1 + 7d) = 5(a_1 + 12d) \Rightarrow d = -\frac{2}{39}a_1 < 0,$$

so $\{a_n\}$ is a monotonously decreasing A.P.. Since

$$a_n > 0 \Rightarrow a_1 + (n-1)\left(-\frac{2}{39}a_1\right) > 0 \Rightarrow n < 20.5,$$

S_{20} is the largest partial sum among those listed. The answer is (C).

3. For $n = 1$, $a_1 = 1 + 3 + 4 = 8$, and for $n \geq 2$,

$$a_n = S_n - S_{n-1} = n^2 + 3n + 4 - (n-1)^2 - 3(n-1) - 4 = 2n + 2,$$

therefore $a_{2k-1} = 2(2k - 1) + 2 = 4k$ for $k \geq 2$, hence

$$a_1 + a_3 + a_5 + \cdots + a_{21} = 8 + 4(2 + 3 + \cdots + 11) = 268.$$

4. Since all terms of the sequence are positive, $d > 0$.

    Suppose that $c^r = a_i, c^s = a_j$, where a_i, a_j are two terms of the sequence, c is an integer $> 1, r < s, i < j$. Let $t = s - r$, then $c^s - c^r = c^r(c^t - 1)$. Therefore $c^r(c^t - 1) = a_j - a_i$ is a multiple of d. For c^{r+kt}, we have

    $$c^{r+kt} - c^r = c^r(c^{kt} - 1) = c^r(c^t - 1)(c^{(k-1)t} + \cdots + 1)$$

    which is a multiple of d, therefore the terms of the form c^{r+kt} are terms in the sequence. Thus, there is a geometric progression $\{b_n\}$ with $b_n = b_0 q^n$, where $b_0 = c^r, q = c^t = c^{s-r}$.

5. Let $a = x(y - z)$, then $aq = y(z - x), aq^2 = z(y - x)$. Therefore

    $$a + aq = x(y - z) + y(z - x) = z(y - x) = aq^2 \Rightarrow 1 + q = q^2,$$

    so $q = \dfrac{1 \pm \sqrt{5}}{2}$.

6. Write $a_n = a \cdot r^n, \quad n \geq 0$. First of all, we show that r is an integer. $r = a_n/a_{n-1}$ is a rational number. Let $r = \frac{\alpha}{\beta}$ with $\alpha, \beta \in \mathbb{N}, (\alpha, \beta) = 1$. It suffices to show $\beta = 1$. Otherwise, there exists a prime p which divides β. Then, for sufficiently large n, $a_n = a \cdot \alpha^n/\beta^n$ is not an integer.

    If $4 \mid a_k$ for some $k \in \mathbb{N}$, then $4 \mid a_n$ for $n \geq k$. Therefore $4 \nmid a_0$ and $4 \nmid a_1$. It's clear that $r \neq 1$ (otherwise, all $a_n = 2004$).

    Suppose that $a_k = 2004$, i.e. $a_k = ar^k = 2004 = 2^2 \cdot 3 \cdot 167$. Since $k \geq 2$, if $k > 2$, then $r = 1$, it's impossible. Therefore $k = 2$ and $r = 2$, $a = 3 \times 167 = 501$.

    Thus, $a_n = 501 \cdot 2^n$.

7. When $n = 1$, we have $a_0 a_1 + a_1 a_0 = a_1^2$, so $a_1 = 2a_0$. When $n = 2$, then

    $$a_2^2 = 2a_0 a_2 + 2a_1^2 = 2a_0 a_2 + 8a_0^2 \Rightarrow (a_2 - 4a_0)(a_2 + 2a_0) = 0.$$

    Since $a_2 + 2a_0 > 0$, so $a_2 = 4a_0 = 2a_1$. Below by induction on n we prove that

    $$a_n = 2^n a_0, \quad n = 0, 1, 2, \ldots.$$

    For $n = 0, 1, 2$, the conclusion is true by the above proofs. Assume that the conclusion is true for all $n \leq i$, then for $n = i + 1$,

    $$\begin{aligned} a_{i+1}^2 &= \sum_{k=0}^{i+1} \binom{i+1}{k} a_k a_{i+1-k} = 2a_0 a_{i+1} + 2^{i+1} a_0^2 \sum_{k=1}^{i} \binom{i+1}{k} \\ &= 2a_0 a_{i+1} + 2^{i+1} a_0^2 (2^{i+1} - 2), \end{aligned}$$

which yields $(a_{i+1} - 2^{i+1}a_0)[a_{i+1} + (2^{i+1} - 2)a_0] = 0$, so $a_{i+1} = 2^{i+1}a_0$.

Thus, the conclusion is true for $n = i+1$. The inductive proof is completed, and it is proven that $\{a_n\}$ is a G.P. of initial term a_0 and common ratio 2.

Testing Questions   (17-B)

1. (i) $(1-p)S_n = p - pa_n$ and $(1-p)S_{n+1} = p - pa_{n+1}$ yields

$$(1-p)a_{n+1} = -pa_{n+1} + pa_n \Rightarrow a_{n+1} = pa_n.$$

Let $n = 1$, then $(1-p)a_1 = p - pa_1$ gives $a_1 = p$. Therefore $\{a_n\}$ is a G.P. such that its initial term and common ratio are both p, hence $a_n = p^n$. Then

$$S_n = \frac{p(p^n - 1)}{p - 1},$$
$$1 + \binom{n}{1}a_1 + \binom{n}{2}a_2 + \cdots + \binom{n}{n}a_n = 1 + \binom{n}{1}p + \cdots \binom{n}{n}p^n = (1 + p)^n.$$

Hence

$$f(n) = \frac{1 + \binom{n}{1}a_1 + \binom{n}{2}a_2 + \cdots + \binom{n}{n}a_n}{2^n S_n} = \frac{p-1}{p} \cdot \frac{(p+1)^n}{2^n(p^n - 1)},$$
$$f(n+1) = \frac{p-1}{p} \cdot \frac{(p+1)^{n+1}}{2^{n+1}(p^{n+1} - 1)},$$
$$\frac{p+1}{2p}f(n) = \frac{p-1}{p} \cdot \frac{(p+1)^{n+1}}{2^{n+1}(p^{n+1} - p)},$$
$$\because p > 1, \;\; \therefore f(n+1) < \frac{p+1}{2p}f(n), \quad \text{for } n \in \mathbb{N}.$$

(ii) The given inequality is clear for $n = 1$ since $f(1) = \dfrac{p+1}{2p}$. For $n \geq 2$, from (i),

$$f(n) < \frac{p+1}{2p}f(n-1) < \cdots < \left(\frac{p+1}{2p}\right)^{n-1} \cdot f(1) = \left(\frac{p+1}{2p}\right)^n.$$

Thus, $\displaystyle\sum_{k=1}^{2n-1} f(k) \leq \sum_{k=1}^{2n-1}\left(\frac{p+1}{2p}\right)^k = \frac{p+1}{p-1} \cdot \left[1 - \left(\frac{p+1}{2p}\right)^{2n-1}\right],$
where the equality holds if and only if $n = 1$.

On the other hand, if $n \geq 2$ and $k = 1, 2, \ldots, 2n-1$,

$$
\begin{aligned}
f(k) + f(2n-k) &= \frac{p-1}{p}\left[\frac{(p+1)^k}{2^k(p^k-1)} + \frac{(p+1)^{2n-k}}{2^{2n-k}(p^{2n-k}-1)}\right] \\
&\geq \frac{p-1}{p}\cdot 2\sqrt{\frac{(p+1)^k}{2^k(p^k-1)} \cdot \frac{(p+1)^{2n-k}}{2^{2n-k}(p^{2n-k}-1)}} \\
&= \frac{p-1}{p}\cdot\frac{2(p+1)^n}{2^n}\sqrt{\frac{1}{(p^k-1)(p^{2n-k}-1)}} \\
&= \frac{p-1}{p}\cdot\frac{2(p+1)^n}{2^n}\sqrt{\frac{1}{(p^{2n}-p^k-p^{2n-k}+1}}).
\end{aligned}
$$

Since $p^k + p^{2n-k} \geq 2p^n \Rightarrow p^{2n} - p^k - p^{2n-k} + 1 \leq p^{2n} - 2p^n + 1 = (p^n-1)^2$,

$$
f(k) + f(2n-k) \geq \frac{p-1}{p}\cdot\frac{2(p+1)^n}{2^n(p^n-1)} = 2f(n),
$$

where the equality holds if and only if $k = n$. Thus,

$$
\sum_{k=1}^{2n-1} f(k) = \frac{1}{2}\sum_{k=1}^{2n-1}[f(k) + f(2n-k)] \geq \sum_{k=1}^{2n-1} f(n) = (2n-1)f(n).
$$

2. (i) When $n = 1, a_1 = 5S_1 + 1 = 5a_1 + 1 \Rightarrow a_1 = -\dfrac{1}{4}$.

   For $n \geq 1$,

   $$
   a_{n+1} - a_n = 5S_{n+1} + 1 - 5S_n - 1 = 5a_{n+1} \Rightarrow \frac{a_{n+1}}{a_n} = -\frac{1}{4},
   $$

   therefore $\{a_n\}$ is a G.P. with $a_1 = -\dfrac{1}{4}$ and common ratio $r = -\dfrac{1}{4}$. Thus,

   $$
   a_n = (-1)^n\frac{1}{4^n}, \qquad b_n = \frac{4 + (-1)^n\frac{1}{4^n}}{1 + (-1)^{n+1}\frac{1}{4^n}} = 4 + \frac{5}{(-4)^n - 1}.
   $$

   (ii) Based on the result of (i),

   $$
   \begin{aligned}
   b_{2k-1} + b_{2k} &= 8 + \frac{5}{(-4)^{2k-1} - 1} + \frac{5}{(-4)^{2k} - 1} \\
   &= 8 + \frac{5}{16^k - 1} - \frac{20}{16^k + 4} \\
   &= 8 - \frac{15\cdot 16^k - 40}{(16^k-1)(16^k+4)} < 8.
   \end{aligned}
   $$

For $n = 2m, m \in \mathbb{N}$,

$$R_n = (b_1 + b_2) + (b_3 + b_4) + \cdots + (b_{2m-1} + b_{2m}) < 8m = 4n.$$

For $n = 2m - 1, m \in \mathbb{N}$,

$$\begin{aligned} R_n &= (b_1 + b_2) + (b_3 + b_4) + \cdots + (b_{2m-3} + b_{2m-2}) + b_{2m-1} \\ &< 8(m - 1) + 4 = 4n. \end{aligned}$$

Thus, there is no $k \in \mathbb{N}$ such that $R_k \geq 4k$.

(iii)   Based on $b_n = 4 + \dfrac{5}{(-4)^n - 1}$, it follows that

$$\begin{aligned} c_n &= b_{2n} - b_{2n-1} = \frac{5}{16^n - 1} + \frac{20}{16^n + 4} = \frac{25 \cdot 16^n}{(16^n - 1)(16^n + 4)} \\ &< \frac{25 \cdot 16^n}{(16^n)^2} = \frac{25}{16^n}. \end{aligned}$$

Since $b_1 = 3, b_2 = \dfrac{13}{3} \Rightarrow c_1 = b_2 - b_1 = \dfrac{13}{3} - 3 = \dfrac{4}{3}$, so $T_1 < \dfrac{3}{2}$.

For $n \geq 2$,

$$\begin{aligned} T_n &< \frac{4}{3} + 25 \left(\frac{1}{16^2} + \frac{1}{16^3} + \cdots + \frac{1}{16^n} \right) < \frac{4}{3} + \frac{25 \cdot \frac{1}{16^2}}{1 - \frac{1}{16}} \\ &= \frac{69}{48} < \frac{3}{2}. \end{aligned}$$

3.   For $k = 1$, there is nothing to prove. Henceforth assume $k \geq 2$.

Let $p_1, p_2, \ldots, p_k$ be k distinct primes such that

$$k < p_k < \cdots < p_2 < p_1$$

and let $N = p_1 p_2 \cdots p_k$. By the Chinese Remainder Theorem, there exists a positive integer x satisfying

$$x \equiv -i \pmod{p_i}$$

for all $i = 1, 2, \ldots, k$ and $x > N^2$. Consider the following sequence :

$$\frac{x + 1}{N}, \quad \frac{x + 2}{N}, \quad \ldots, \quad \frac{x + k}{N}.$$

This sequence is obviously an arithmetic sequence of positive rational numbers of length k. For each $i = 1, 2, \ldots, k$, the numerator $x + i$ is divisible by p_i but not by p_j for $j \neq i$. Otherwise, p_j divides $|i - j|$, which is not possible because $p_j > k > |i - j|$. Let

$$a_i := \frac{x + i}{p_i}, \qquad b_i := \frac{N}{p_i} \qquad \text{for all } i = 1, 2, \ldots, k.$$

Then

$$\frac{x + i}{N} = \frac{a_i}{b_i}, \qquad \gcd(a_i, b_i) = 1 \quad \text{for all } i = 1, 2, \ldots, k,$$

and all b_i's are distinct from each other. Moreover, $x > N^2$ implies

$$a_i = \frac{x + i}{p_i} > \frac{N^2}{p_i} > N > \frac{N}{p_j} = b_j \qquad \text{for all } i, j = 1, 2, \ldots, k$$

and hence all a_i's are distinct from b_i's. It only remains to show that all a_i's are distinct from each other. This follows from

$$a_j = \frac{x + j}{p_j} > \frac{x + i}{p_j} > \frac{x + i}{p_i} = a_i \qquad \text{for all } i < j$$

by our choice of $p_1, p_2, \ldots, p_k$. Thus, the arithmetic sequence

$$\frac{a_1}{b_1}, \frac{a_2}{b_2}, \ldots, \frac{a_k}{b_k}$$

of positive rational numbers satisfies the conditions of the problem. $\qquad \square$

Remark. Here is a much easier solution :

For any positive integer $k \geq 2$, consider the sequence

$$\frac{(k!)^2 + 1}{k!}, \quad \frac{(k!)^2 + 2}{k!}, \ldots, \frac{(k!)^2 + k}{k!}.$$

Note that $\gcd(k!, (k!)^2 + i) = i$ for all $i = 1, 2, \ldots, k$. So, taking

$$a_i = \frac{(k!)^2 + i}{i}, \qquad b_i = \frac{k!}{i} \qquad \text{for all } i = 1, 2, \ldots, k,$$

we have $\gcd(a_i, b_i) = 1$ and

$$a_i = \frac{(k!)^2 + i}{i} > a_j = \frac{(k!)^2 + j}{j} > b_i = \frac{k!}{i} > b_j = \frac{k!}{j}$$

for any $1 \leq i < j \leq k$. Therefore this sequence satisfies every condition given in the problem.

4. Suppose that there is an A.P. consisting of 40 distinct positive integers, such that each term is in the form $2^k + 3^l$. Let it be $a, a+d, a+2d, \ldots, a+39d$, where a, d are positive integers. Let

$$m = \lfloor \log_2(a + 39d) \rfloor, \qquad n = \lfloor \log_3(a + 39d) \rfloor.$$

First we show that among $a + 26d, a + 27d, \ldots, a + 39d$ there is at most one term which is not able to be expressed in the form $2^m + 3^l$ or the form $2^k + 3^n$, where k, l are nonnegative integers.

If among them the term $a + hd$ cannot be expressed in $2^m + 3^l$ or $2^k + 3^n$, by assumption, there must be nonnegative integers b and c such that

$$a + hd = 2^b + 3^c.$$

The definition of m and n implies that $b \le m, c \le n$. Since $a + hd$ cannot be expressed as $2^m + 3^l$ or $2^k + 3^n$, so $b \le m - 1$ and $c \le n - 1$.

If $b \le m - 2$, then

$$a + hd \le 2^{m-2} + 3^{n-1} = \frac{1}{4}2^m + \frac{1}{3}3^n \le \frac{7}{12}(a + 39d) < a + 26d,$$

a contradiction.

If $c \le n - 2$, then

$$a + hd \le 2^{m-1} + 3^{n-2} = \frac{1}{2}2^m + \frac{1}{9}3^n \le \frac{11}{18}(a + 39d) < a + 26d,$$

a contradiction. Hence $b = m - 1, c = n - 1$, i.e., there is at most one term among $a + 26d, a + 27d, \ldots, a + 39d$ that it cannot be expressed as $2^m + 3^l$ or $2^k + 3^n$.

Thus, among the 14 numbers, at least 13 of them can be expressed as $2^m + 3^l$ or $2^k + 3^n$. According to the pigeonhole principle, at least 7 numbers belong to the same kind of expression. We have two possible cases as follows.

Case 1: There are seven numbers of the form $2^m + 3^l$. Suppose that they are $2^m + 3^{l_1}, 2^m + 3^{l_2}, \ldots, 2^m + 3^{l_7}$, where $l_1 < l_2 < \cdots < l_7$, then $3^{l_1}, 3^{l_2}, \ldots, 3^{l_7}$ are seven terms of 14 terms in an A.P. with common difference d. However,

$$13d \ge 3^{l_7} - 3^{l_1} \ge \left(3^5 - \frac{1}{3}\right) \cdot 3^{l_2} > 13(3^{l_2} - 3^{l_1}) \ge 13d,$$

a contradiction.

Case 2: There are seven numbers of the form $2^k + 3^n$. Suppose that they are $2^{k_1} + 3^n, 2^{k_2} + 3^n, \ldots, 2^{k_7} + 3^n$, where $k_1 < k_2 < \cdots < k_7$, then $2^{k_1}, 2^{k_2}, \ldots, 2^{k_7}$ are seven terms of 14 terms in an A.P. with common difference d. However,

$$13d \geq 2^{k_7} - 2^{l_1} \geq \left(2^5 - \frac{1}{2}\right) \cdot 2^{k_2} > 13(2^{k_2} - 2^{k_1}) \geq 13d,$$

a contradiction. Thus, the assumption is wrong, and the proposition is true.

Solutions to Testing Questions   18

Testing Questions   (18-A)

1. Let $b_n = \sqrt{1 + 4a_n}, n \geq 1$, then $a_n = \dfrac{b_n^2 - 1}{4}$, and so

$$\frac{b_{n+1}^2 - 1}{4} = 1 + \frac{b_n^2 - 1}{4} + b_n \Rightarrow b_{n+1}^2 = (b_n + 2)^2$$
$$\Rightarrow b_{n+1} = b_n + 2 \Rightarrow b_n = \sqrt{5} + 2(n - 1).$$

Therefore

$$\begin{aligned}
a_n &= \frac{b_n^2 - 1}{4} = \frac{1}{4}[5 + 4\sqrt{5}(n - 1) + 4(n - 1)^2 - 1] \\
&= 1 + \sqrt{5}(n - 1) + (n - 1)^2 = 1 + (n - 1)(n + \sqrt{5} - 1).
\end{aligned}$$

2. (i) Letting $n = 1$ in the given recursive formula yields $(1 - b)a_1 = -ba_1 + 4$, so $a_1 = 4$. When $n \geq 2$, the difference of

$$(1 - b)S_n = -ba_n + 4^n \quad \text{and} \quad (1 - b)S_{n-1} = -ba_{n-1} + 4^{n-1}$$

yields

$$(1 - b)a_n = -b(a_n - a_{n-1}) + 3 \cdot 4^{n-1}, \quad \text{namely } a_n = ba_{n-1} + 3 \cdot 4^{n-1}.$$

(i) If $b = 4$, then $\dfrac{a_n}{4^n} = \dfrac{a_{n-1}}{4^{n-1}} + \dfrac{3}{4}$, so that

$$\frac{a_n}{4^n} = \frac{a_1}{4} + \frac{3}{4}(n - 1) \Rightarrow a_n = (3n + 1)4^{n-1}, \quad n \geq 1.$$

(ii) If $b \neq 4$, then $a_n = ba_{n-1} + 3 \cdot 4^{n-1}$

$$\Rightarrow a_n + \frac{3}{b-4} \cdot 4^n = b\left(a_{n-1} + \frac{3}{b-4} \cdot 4^{n-1}\right)$$

$$\Rightarrow a_n + \frac{3}{b-4} \cdot 4^n = \left(a_1 + \frac{3}{b-4} \cdot 4\right) b^{n-1}$$

$$\Rightarrow a_n = \left(4 + \frac{12}{b-4}\right) b^{n-1} - \frac{3}{b-4} \cdot 4^n \quad \text{for } n \geq 2.$$

Since $a_1 = 4$, so $a_n = \left(4 + \frac{12}{b-4}\right) b^{n-1} - \frac{3}{b-4} \cdot 4^n$ for $n \geq 1$.

Thus,

$$a_n = \begin{cases} (3n+1) \cdot 4^{n-1}, & b = 4, \\ \left(4 + \dfrac{12}{b-4}\right) b^{n-1} - \dfrac{3}{b-4} \cdot 4^n & b \neq 4. \end{cases}$$

(ii) When $b = 4$, then $c_n = \dfrac{3n+1}{4} \to \infty$ as $n \to \infty$, so $b \neq 4$.

When $b \neq 4$, then $c_n = \dfrac{4(b-1)}{b(b-4)} \cdot \left(\dfrac{b}{4}\right)^n - \dfrac{3}{b-4}$. if $b > 4$, then $c_n \to +\infty$ as $n \to +\infty$. So $0 < b < 4$.

(i) When $0 < b < 1$, $\dfrac{4(b-1)}{b(b-4)} > 0, -\dfrac{3}{b-4} > 0$, so c_n is decreasing and positive. Further, $c_1 = 1$, so $b \in (0, 1)$ satisfies the condition.

(ii) $b = 1 \Rightarrow c_n = 1$, so $b = 1$ is allowed.

(iii) When $1 < b < 4$, then $\dfrac{4(b-1)}{b(b-4)} < 0, -\dfrac{3}{b-4} > 0$, so c_n is increasing. Since $c_1 = 1$, so $c_n > 0$. $\lim\limits_{n \to +\infty} c_n = -\dfrac{3}{b-4} \leq 2 \Leftrightarrow 1 < b \leq \dfrac{5}{2}$.

Thus, the allowed range for b is $\left(0, \dfrac{5}{2}\right]$.

3. The given recursive formula gives $a_{n+1} = \left(1 + \dfrac{1}{2^n}\right) a_n$. By induction it is easy to see that $a_n \geq 1$ and $a_{n+1} > a_n$ for all $n \in \mathbb{N}$.

Since $a_{n+1} - a_n = \dfrac{a_n}{2^n} \geq \dfrac{1}{2^n}$, therefore for $n \geq 3$,

$$a_n = a_1 + \sum_{k=1}^{n-1}(a_{k+1} - a_k) > 1 + \sum_{k=1}^{n-1} \frac{1}{2^k} = 2 - \frac{1}{2^{n-1}}.$$

On the other hand, Since $a_{n+1} = \left(1 + \dfrac{1}{2^n}\right) a_n \leq \dfrac{3}{2} a_n$ (the equality holds only when $n = 1$), so for $n \geq 3$

$$a_n = \left(1 + \frac{1}{2^{n-1}}\right)\left(1 + \frac{1}{2^{n-2}}\right)\cdots\left(1 + \frac{1}{2^2}\right)\left(1 + \frac{1}{2}\right)a_1 < \left(\frac{3}{2}\right)^{n-1}.$$

4. Let $b_n = \log_2 a_n, n \geq 1$, then $b_1 = 0, b_2 = 1$ and $b_n = -b_{n-1} + b_{n-2}$ for $n \geq 3$. Let α, β be the real roots of the characteristic equation $x^2 + x - 1 = 0$, then

$$\alpha = \frac{-1 - \sqrt{5}}{2}, \qquad \beta = \frac{-1 + \sqrt{5}}{2},$$

therefore $b_n = A\alpha^{n-1} + B\beta^{n-1}$. Letting $n = 1$ and 2 in this formula leads to

$$A + B = 0, A\alpha + B\beta = 1 \Rightarrow A = \tfrac{1}{\alpha - \beta}, B = -\tfrac{1}{\alpha - \beta}$$

$$\Rightarrow b_n = \frac{\alpha^{n-1} - \beta^{n-1}}{\alpha - \beta} \Rightarrow a_n = 2^{\frac{\alpha^{n-1} - \beta^{n-1}}{\alpha - \beta}}.$$

5. Since $a_{n+2} = 1 - \dfrac{1}{a_{n+1}} = 1 - \dfrac{a_n}{a_n - 1} = \dfrac{-1}{a_n - 1}$, it follows that

$$a_{n+3} = 1 - \frac{1}{a_{n+2}} = 1 + (a_n - 1) = a_n,$$

so $\{a_n\}$ is a periodic sequence with a period 3. Since $a_1 = 2, a_2 = \dfrac{1}{2}, a_3 = -1$, so $P_3 = -1$ and

$$P_{2009} = (P_3)^{669} \cdot P_2 = (-1)^{669} = -1.$$

The answer is (B).

6. (i) By induction it is easy to see that $a_n > 0$ for all $n \geq 1$. Now

$$\frac{a_{n+2}}{a_n} = \frac{a_{n+1}^2 + 1}{a_n^2 + 1} \Rightarrow \frac{a_{n+2}a_{n+1}}{a_{n+1}^2 + 1} = \frac{a_{n+1}a_n}{a_n^2 + 1} \Rightarrow \frac{a_{n+2}}{a_{n+1} + \frac{1}{a_{n+1}}} =$$

$\dfrac{a_{n+1}}{a_n + \frac{1}{a_n}}$. Thus,

$$\frac{a_{n+1}}{a_n + \frac{1}{a_n}} = \frac{a_n}{a_{n-1} + \frac{1}{a_{n-1}}} = \cdots = \frac{a_2}{a_1 + \frac{1}{a_1}} = \frac{2}{1 + \frac{1}{1}} = 1,$$

hence $a_{n+1} = a_n + \dfrac{1}{a_n}, n \geq 1$.

(ii) $a_1 = 1$ and $a_{n+1} = a_n + \frac{1}{a_n}$ imply $a_n \geq 1$ and $0 < \dfrac{1}{a_n^2} \leq 1, n \geq 1$.
For $n \geq 2$,

$$a_n^2 = \left(a_{n-1} + \frac{1}{a_{n-1}}\right)^2 = a_{n-1}^2 + \frac{1}{a_{n-1}^2} + 2$$

implies $2 < a_k^2 - a_{k-1}^2 \leq 3, k \geq 2$. Hence, by adding them up for k from 2 to n yields

$$2(n-1) < a_n^2 - a_1^2 \leq 3(n-1) \Rightarrow 2n - 1 < a_n^2 \leq 3n - 2$$
$$\Rightarrow 4015 < a_{2008}^2 \leq 6022 \Rightarrow 63^2 < a_{2008}^2 < 78^2$$

since $63^2 = 3969 < 4015$ and $78^2 = 6084 > 6022$. Thus, $63 < a_{2008} < 78$.

7.   (i)   We use induction. The given recursive formula gives $a_{n+1} + 1 = a_n(a_n + 1)$.

For $n = 1, a_1 = 3 \Rightarrow a_1 \equiv 3 \pmod 4$.

Assume that $a_n \equiv 3 \pmod 4$ $(n \geq 1)$, i.e., $a_n = 4k + 3$ for some $k \in \mathbb{Z}$. Then

$$a_{n+1} = a_n(a_n + 1) - 1 = 4(4k + 3)(k + 1) - 1 \equiv 3 \pmod 4.$$

The inductive proof is completed.

(ii)   By using $a_{n+1} + 1 = a_n(a_n + 1)$ repeatedly, it follows that

$$a_{n+1} + 1 = a_n(a_n + 1) = a_n a_{n-1}(a_{n-1} + 1) = \cdots = 4a_n a_{n-1} \cdots a_2 a_1.$$

Therefore $a_m \mid (a_n + 1)$ for $m < n$. Let $(a_m, a_n) = d$. Then $d \mid 1$, so $d = 1$.

8.   (i)   By solving the equation $t^2 - 4t + 4 = 0$, it is obtained that

$$\alpha = \beta = 2,$$

therefore $a_n = (A + Bn)2^n$. Letting $n = 1$ and 2 respectively yields

$$2(A + B) = 1 \quad \text{and} \quad 2(A + 2B) = 3 \Rightarrow A = -\frac{1}{2}, B = 1,$$

therefore $a_n = (2n - 1)2^{n-1}, n \in \mathbb{N}$.

(ii) $S_n = \sum_{k=1}^{n}(2k - 1)2^{k-1} = \sum_{k=1}^{n} k2^k - \sum_{k=1}^{n} 2^{k-1} = T - G$, where

$$T = \sum_{k=1}^{n} k \cdot 2^k \Rightarrow 2T = \sum_{k=1}^{n} k \cdot 2^{k+1} = \sum_{k=2}^{n+1}(k - 1)2^k$$

$$\Rightarrow -T = \sum_{k=1}^{n} 2^k - n \cdot 2^{n+1} = 2(2^n - 1) - n \cdot 2^{n+1} = (1 - n)2^{n+1} - 2$$

$$\Rightarrow T = (n - 1)2^{n+1} + 2.$$

$$G = \sum_{k=1}^{n} 2^{k-1} = 2^n - 1,$$

$$\therefore S_n = T - G = (n - 1)2^{n+1} + 2 - 2^n + 1 = (2n - 3) \cdot 2^n + 3.$$

Testing Questions   (18-B)

1. $a_n + 2 = \dfrac{2(2 + a_{n-1})}{1 + a_{n-1}}$ and $a_n - 1 = \dfrac{1 - a_{n-1}}{1 + a_{n-1}}$ lead to

$$\frac{a_n + 2}{a_n - 1} = (-2) \cdot \frac{a_{n-1} + 2}{a_{n-1} - 1}, \quad n \geq 2.$$

By repeatedly applying the recursive relation, it follows that

$$\frac{a_n + 2}{a_n - 1} = (-2)^{n-1} \frac{a_1 + 2}{a_1 - 1} = (-2)^n \Rightarrow a_n = \frac{(-2)^n + 2}{(-2)^n - 1}$$

for $n \geq 2$. Since $a_1 = 0$ satisfies it also, $a_n = \dfrac{(-2)^n + 2}{(-2)^n - 1}, n \geq 1$.

2. (i) $a_{n+1} = \dfrac{4a_n}{2a_n + 1} \Rightarrow \dfrac{1}{a_{n+1}} = \dfrac{1}{2} + \dfrac{1}{4a_n} \Rightarrow b_{n+1} = \dfrac{1}{4}b_n$, where $b_n = \dfrac{1}{a_n} - \dfrac{2}{3}, n \geq 1$. Thus, $b_1 = \dfrac{11}{12} - \dfrac{2}{3} = \dfrac{1}{4}$ and $\{b_n\}$ is a G.P. with initial and common ratio both being $\dfrac{1}{4}$. Therefore

$$b_n = \frac{1}{4^n} \Rightarrow \frac{1}{a_n} = \frac{2}{3} + \frac{1}{4^n} \Rightarrow a_n = \frac{3 \cdot 4^n}{2 \cdot 4^n + 3}.$$

Since $a_n > 0$, for any $x > 0$,

$$\frac{3}{2+x} - \frac{3}{(2+x)^2}\left(\frac{3}{4^n} - x\right) = \frac{3}{2+x} - \frac{3}{(2+x)^2}\left(\frac{3}{4^n} + \frac{6}{3} - \frac{6}{3} - x\right)$$

$$= \frac{3}{2+x} - \frac{9}{(2+x)^2}\left[\frac{1}{4^n} + \frac{2}{3} - \left(\frac{2}{3} + \frac{x}{3}\right)\right]$$

$$= \frac{3}{2+x} - \frac{9}{(2+x)^2}\left(\frac{1}{a_n} - \frac{x+2}{3}\right) = -\frac{1}{a_n}\cdot\frac{9}{(x+2)^2} + \frac{6}{x+2}$$

$$= -\frac{1}{a_n}\left(\frac{3}{x+2} - a_n\right)^2 + a_n \leq a_n.$$

(ii) Based on the resulting inequality, for any $x > 0$

$$a_1 + a_2 + \cdots + a_n \geq \sum_{k=1}^{n}\left[\frac{3}{2+x} - \frac{3}{(2+x)^2}\left(\frac{3}{4^k} - x\right)\right]$$

$$= \frac{3n}{2+x} - \frac{3}{(2+x)^2}\left(\sum_{k=1}^{n}\frac{3}{4^k} - nx\right).$$

Taking $x = \dfrac{1}{n}\displaystyle\sum_{k=1}^{n}\dfrac{3}{4^k} = \dfrac{\frac{3}{4}(1 - \frac{1}{4^n})}{n(1 - \frac{1}{4})} = \dfrac{1}{n}\left(1 - \dfrac{1}{4^n}\right)$, we have

$$a_1 + a_2 + \cdots + a_n \geq \frac{3n}{2 + \frac{1}{n}(1 - \frac{1}{4^n})} = \frac{3n^2}{2n + 1 - \frac{1}{4^n}} > \frac{3n^2}{2n + 1}.$$

3. By induction it is clear that $x_n > 0$ for all $n \in \mathbb{N}$. Simplifying the given recurrence,

$$x_n = \frac{\sqrt{x_{n-1}^2 + 4x_{n-1}} + x_{n-1}}{2} \Rightarrow (2x_n - x_{n-1})^2 = x_{n-1}^2 + 4x_{n-1}$$

$$\Rightarrow x_n^2 - x_n x_{n-1} = x_{n-1} \Rightarrow \frac{1}{x_{n-1}} - \frac{1}{x_n} = \frac{1}{x_n^2}.$$

Then $y_n = \displaystyle\sum_{i=1}^{n}\frac{1}{x_i^2} = \frac{1}{x_1^2} + \sum_{i=2}^{n}\left(\frac{1}{x_{i-1}} - \frac{1}{x_i}\right) = 4 + \left(2 - \frac{1}{x_n}\right) = 6 - \frac{1}{x_n}.$

Since $x_1 = \dfrac{1}{2}$ and

$$x_n = \frac{\sqrt{x_{n-1}^2 + 4x_{n-1}} + x_{n-1}}{2} > \frac{\sqrt{x_{n-1}^2} + x_{n-1}}{2} = x_{n-1},$$

so $x_{n-1}^2 + 4x_{n-1} > x_{n-1}^2 + 2x_{n-1} + 1 = (x_{n-1} + 1)^2$, therefore $x_n > x_{n-1} + \dfrac{1}{2}$. Thus, $\displaystyle\lim_{n\to+\infty}\frac{1}{x_n} = 0$ and $\displaystyle\lim_{n\to+\infty} y_n = 6.$

4. Considering $1 + (2006)(2008)(1) = 2007^2$, let $a = 1, b = 2008$. Below we prove the proposition P_n by induction:

$$1 + 2006x_{n+1}x_n = (x_{n+1} - x_n)^2, \qquad n = 1, 2, \ldots.$$

When $n = 1$, $1 + 2006x_2x_1 = 1 + 2006 \cdot 2008 = 2007^2 = (x_2 - x_1)^2$, so P_1 is true.

Assume that P_k is true ($k \geq 1$), i.e., $1 + 2006x_{k+1}x_k = (x_{k+1} - x_k)^2$, then for $n = k + 1$,

$$
\begin{aligned}
&1 + 2006x_{k+2}x_{k+1} \\
&= 1 + 2006x_{k+1}(2008x_{k+1} - x_k) \\
&= 2007^2 x_{k+1}^2 - x_{k+1}^2 - 2006x_{k+1}x_k + 1 \\
&= 2007^2 x_{k+1}^2 - 5012x_{k+1}x_k + 2006x_{k+1}x_k + 1 - x_{k+1}^2 \\
&= 2007^2 x_{k+1}^2 - 5012x_{k+1}x_k + (x_{k+1} - x_k)^2 - x_{k+1}^2 \\
&= 2007^2 x_{k+1}^2 - 5014x_{k+1}1x_k + x_k^2 \\
&= (2007x_{k+1} - x_k)^2 = (x_{k+2} - x_{k+1})^2
\end{aligned}
$$

hence P_{k+1} is also true, and so the inductive proof is completed.

5. (i) By induction, it's easy to see that each a_i is odd for $i \geq 0$. Let $(a_k, a_n) = m$ for some $k < n$. Below we show that $m = 1$.

$$a_n = 2 + a_0a_1 \cdots a_k \cdots a_{n-1} \Rightarrow m \mid 2.$$

Since a_k, a_n are both odd, so m is odd, therefore $m = 1$.

(ii) Since $a_0a_1 \cdots a_{n-2} = a_{n-1} - 2$, so $a_n - 2 = (a_{n-1} - 2)a_{n-1}$, therefore

$$a_n - 1 = 1 + (a_{n-1} - 2)a_{n-1} = a_{n-1}^2 - 2a_{n-1} + 1 = (a_{n-1} - 1)^2.$$

Thus,

$$a_n - 1 = (a_{n-1} - 1)^2 = (a_{n-2} - 1)^{2^2} = \cdots = (a_0 - 1)^{2^n} = 2^{2^n},$$

namely $a_n = 2^{2^n} + 1$. Hence $a_{2007} = 2^{2^{2007}} + 1$.

Solutions to Testing Questions   19

Testing Questions   (19-A)

1. Since $\dfrac{k}{(k+1)!} = \dfrac{1}{k!} - \dfrac{1}{(k+1)!}$ for any natural number k, therefore

$$S_n = (1 - \frac{1}{2!}) + (\frac{1}{2!} - \frac{1}{3!}) + \cdots + (\frac{1}{n!} - \frac{1}{(n+1)!})$$
$$= 1 - \frac{1}{(n+1)!},$$

therefore

$$\frac{1 - s_{2001}}{1 - s_{2002}} = \frac{\frac{1}{2002!}}{\frac{1}{2003!}} = 2003.$$

2. For $n \geq 2$,

$$a_{n+1} - a_n = \sqrt{\frac{3+a_n}{2}} - \sqrt{\frac{3+a_{n-1}}{2}} = \frac{a_n - a_{n-1}}{2\left(\sqrt{\frac{3+a_n}{2}} + \sqrt{\frac{3+a_{n-1}}{2}}\right)},$$

so $a_2 - a_1, a_3 - a_2, \ldots, a_{n+1} - a_n$ have a same sign, i.e. $\{a_n\}$ is monotone.
Since $a_1 = 4 \Rightarrow a_2 = \sqrt{\frac{7}{2}} < a_1$, so $\{a_n\}$ is a decreasing sequence, therefore

$$S_n = (a_1 - a_2) + (a_2 - a_3) + \cdots + (a_n - a_{n+1}) = a_1 - a_{n+1}.$$

Since $a_{n+2} < a_{n+1} \Rightarrow \sqrt{\frac{3+a_{n+1}}{2}} < a_{n+1} \Rightarrow a_{n+1} > \frac{3}{2}$, so

$$S_n < 4 - \frac{3}{2} = \frac{5}{2}.$$

3. (1)   The two roots of the given equation are $x_1 = 3k, x_2 = 2^k$, therefore

$$S_{2n} = \sum_{k=1}^{n}(3k + 2^k) = 3\sum_{k=1}^{n}k + \sum_{k=1}^{n}2^k = \frac{3n(n+1)}{2} + 2(2^n - 1).$$

(2) $a_1 = 3, a_2 = 2, a_3 = 6, a_4 = 4 \Rightarrow T_1 = \frac{1}{6}, T_2 = \frac{1}{6} + \frac{1}{24} = \frac{5}{24}.$

When $n \geq 3$,

$$
\begin{aligned}
T_n &= \frac{1}{6} + \frac{1}{6 \cdot 2^2} - \frac{1}{a_5 a_6} + \cdots + \frac{(-1)^{f(n+1)}}{a_{2n-1} a_{2n}} \\
&\geq \frac{1}{6} + \frac{1}{24} - \left(\frac{1}{a_5 a_6} + \cdots + \frac{1}{a_{2n-1} a_{2n}} \right) \\
&\geq \frac{1}{6} + \frac{1}{24} - \frac{1}{24} \left(\frac{1}{2} + \frac{1}{4} + \cdots + \frac{1}{2^{n-2}} \right) > \frac{1}{6}
\end{aligned}
$$

and

$$
\begin{aligned}
T_n &= \frac{5}{24} - \frac{1}{3^2 \cdot 2^3} - \frac{1}{a_7 a_8} + \cdots + \frac{(-1)^{f(n+1)}}{a_{2n-1} a_{2n}} \\
&\leq \frac{5}{24} - \frac{1}{9 \cdot 2^3} + \frac{1}{9} \left(\frac{1}{2^4} + \cdots + \frac{1}{2^n} \right) \\
&< \frac{5}{24} - \frac{1}{9 \cdot 2^3} + \frac{1}{9} \cdot \frac{1}{2^3} = \frac{5}{24}.
\end{aligned}
$$

Thus, $\dfrac{1}{6} \leq T_n \leq \dfrac{5}{24}, n \in \mathbb{N}$.

4. Let $S_n = a_1 + a_2 + a_3 + \cdots + a_n$ for $n = 1, 2, 3, \cdots$. For any $n \geq 1$, $\displaystyle\sum_{i=1}^{n} a_{i+2} =$
$\displaystyle\sum_{i=1}^{n} (a_{i+1} - a_i) = a_{n+1} - a_1$, therefore $S_{n+2} = a_{n+1} + a_2$. For $n = 1999$, we have

$$S_{2001} = a_{2000} + 1001.$$

For finding the value of a_{2000}, we note that $a_{n+2} = a_{n+1} - a_n = (a_n - a_{n-1}) - a_n = -a_{n-1}$ for all $n \geq 1$, i.e. $a_n = -a_{n-3} = a_{n-6}$ for all $n = 7, 8, 9, \cdots$, therefore

$$a_{2000} = a_{2+6 \times 333} = a_2 = 1001,$$

and hence

$$S_{2001} = 1001 + 1001 = 2002.$$

5. a is even implies that A must be odd. Let $A = (2k + 1)^2$, then

$$A = (2k + 1)^2 = 4k^2 + 4k + 1 = 4k(k + 1) + 1 = 8p + 1$$

for some positive integer p. Hence

$$8p = A - 1 = a^n + a^{n-1} + \cdots + a \Rightarrow a(a^{n-1} + a^{n-2} + \cdots + 1) = 8p$$
$$\Rightarrow 8 \mid a(a^{n-1} + a^{n-2} + \cdots + 1).$$

Since $a^{n-1} + a^{n-2} + \cdots + 1$ is odd, so $8 \mid a$, the conclusion is proven.

6. $S_n = \dfrac{1}{2}\left(a_n + \dfrac{1}{a_n}\right) = \dfrac{1}{2}\left(S_n - S_{n-1} + \dfrac{1}{S_n - S_{n-1}}\right)$

$\Rightarrow S_n^2 = S_{n-1}^2 + 1$. Since $S_1 = a_1 = 1$, so $S_n^2 = n$, $S_n = \sqrt{n}, n \in \mathbb{N}$.

Since

$$\sqrt{n} + \sqrt{n-1} < 2\sqrt{n} < \sqrt{n+1} + \sqrt{n}$$

$$\Rightarrow \frac{1}{\sqrt{n+1} + \sqrt{n}} < \frac{1}{2\sqrt{n}} < \frac{1}{\sqrt{n} + \sqrt{n-1}}$$

$$\Rightarrow \sqrt{n+1} - \sqrt{n} < \frac{1}{2\sqrt{n}} < \sqrt{n} - \sqrt{n-1}$$

$$\Rightarrow \sqrt{101} - 1 < \frac{1}{2}\sum_{k=1}^{100}\frac{1}{S_k} < \sum_{k=2}^{100}(\sqrt{n} - \sqrt{n-1}) + \frac{1}{2}$$

$$\Rightarrow 18 < \sum_{k=1}^{100}\frac{1}{S_k} < 2(9 + \frac{1}{2}) = 19 \Rightarrow \left\lfloor\sum_{k=1}^{100}\frac{1}{S_k}\right\rfloor = 18.$$

7. From the given assumptions we have

$$a_0 + a_1 + \cdots + a_n = 21 \quad \text{and} \quad a_0 + a_1(25) + \cdots + a_n(25)^n = 78357.$$

Therefore $0 \le a_i \le 21$ for $0 \le i \le n$ and a_0 is the remainder of 78357 when it is divided by 25, i.e., $a_0 = 7$. Similarly, from

$$a_1 + \cdots + a_n = 14 \quad \text{and} \quad a_1 + \cdots + a_n(25)^{n-1} = 3134,$$

a_1 is the remainder of 3134 when it is divided by 25, therefore $a_1 = 9$. Next, from

$$a_2 + a_3 + \cdots + a_n = 5 \quad \text{and} \quad a_2 + a_3(25) + \cdots + a_n(25)^{n-2} = 125$$

we have $a_2 = 0$. Finally, from

$$a_3 + \cdots + a_n = 5 \quad \text{and} \quad a_3 + a_4(25) + \cdots + a_n(25)^{n-3} = 5$$

we obtain $a_3 = 5$ and $a_k = 0$ for all $k > 3$. Thus

$$f(x) = 7 + 9x + 5x^3$$

and $f(10) = 5097$.

8. Since $\dfrac{1}{k(n+1-k)} = \dfrac{1}{n+1}\left(\dfrac{1}{k} + \dfrac{1}{n+1-k}\right)$, so $a_n = \dfrac{2}{n+1}\sum_{k=1}^{n}\dfrac{1}{k}$.

Hence, for any positive integer $n \geq 2$,

$$
\begin{aligned}
\frac{1}{2}(a_n - a_{n+1}) &= \frac{1}{n+1} \sum_{k=1}^{n} \frac{1}{k} - \frac{1}{n+2} \sum_{k=1}^{n+1} \frac{1}{k} \\
&= \left(\frac{1}{n+1} - \frac{1}{n+2} \right) \sum_{k=1}^{n} \frac{1}{k} - \frac{1}{(n+1)(n+2)} \\
&= \frac{1}{(n+1)(n+2)} \left(\sum_{k=1}^{n} \frac{1}{k} - 1 \right) > 0,
\end{aligned}
$$

namely $a_{n+1} < a_n$, as desired.

9. We prove the conclusion by induction on n. For $n = 1$, we have $f(1) = 2$.
    For $k \geq 1$ we have $\binom{n+1+k}{k} = \binom{n+k}{k-1} + \binom{n+k}{k}$, so

$$
\begin{aligned}
&f(n + 1) \\
&= \sum_{k=0}^{n+1} \binom{n+1+k}{k} 2^{-k} = 1 + \sum_{k=1}^{n+1} \binom{n+k}{k-1} 2^{-k} + \sum_{k=1}^{n+1} \binom{n+k}{k} 2^{-k} \\
&= \frac{1}{2} \sum_{i=0}^{n} \binom{n+i+1}{i} 2^{-i} + \binom{2n+1}{n+1} 2^{-n-1} + f(n) \\
&= \frac{1}{2} f(n+1) + f(n),
\end{aligned}
$$

that is, $f(n + 1) = 2f(n) = 2^{n+1}$.

10 The sequence $\{a_n\}_{n \geq 1}$ is

$$21,\ 90,\ 11,\ 1,\ 12,\ 13,\ 25,\ 38,\ 63,\ 1,\ 64,\ 65,\ 29,\ 94,\ 23,\ 17,\ 40,\ \cdots$$

and the sequence of the remainders of a_n^2 when it is divided by 8 is

$$1,\ 4,\ 1,\ 1,\ 0,\ 1,\ 1,\ 4,\ 1,\ 1,\ 0,\ 1,\ 1,\ 4, 1, 1, 0, \cdots .$$

Thus, the sequence is periodic and 6 is its period. From $1 + 4 + 1 + 1 + 0 + 1 \equiv 0 \pmod 8$ and $2005 = 6 \times 334 + 1$, we find that the remainder of $a_1^2 + a_2^2 + \cdots + a_{2005}^2$ when it is divided by 8 is equal to that of a_1^2, i.e. 1.

Testing Questions   (19-B)

1. Such a sequence does not exist. It suffices to show that $\displaystyle\sum_{i=2}^{2^n} \frac{1}{a_i} > \frac{n}{4}$.

    By the Cauchy-Schwartz inequality, $\displaystyle\left(\sum_{i=2^k+1}^{2^{k+1}} a_i\right) \cdot \left(\sum_{i=2^k+1}^{2^{k+1}} \frac{1}{a_i}\right) \geq 2^{2k}$.

    Therefore

    $$\sum_{i=2^k+1}^{2^{k+1}} \frac{1}{a_i} \geq \frac{2^{2k}}{\displaystyle\sum_{i=2^k+1}^{} a_i} > \frac{2^{2k}}{\displaystyle\sum_{i=1}^{} a_i} \geq \frac{2^{2k}}{2^{2k+2}} = \frac{1}{4}.$$

    Hence

    $$\sum_{i=2}^{2^n} \frac{1}{a_i} = \sum_{k=0}^{n-1} \left(\sum_{i=2^k+1}^{2^{k+1}} \frac{1}{a_i}\right) > \frac{n}{4}.$$

    Thus, condition (b) is not satisfied provided n is big enough.

2. The left hand of the given equation is an integer, with the right hand side must be as well. Let $x = \dfrac{n}{44}$ and $n = 44m + r$, where $n, m, r \in \mathbb{Z}$ and $0 \leq r \leq 43$. Then the given equation becomes

    $$\sum_{k=1}^{9} \left\lfloor k\left(m + \frac{r}{44}\right)\right\rfloor = 44m + r \Leftrightarrow \sum_{k=1}^{9} km + \sum_{k=1}^{9} \left\lfloor \frac{kr}{44}\right\rfloor = 44m + r$$

    $$\Leftrightarrow m = r - \sum_{k=1}^{9} \left\lfloor \frac{kr}{44}\right\rfloor.$$

    Thus, the m is uniquely determined by r. There are a total of 44 possible values for r, so the given equation has 44 real solution for x. Let S be the sum of all these solutions. Letting m_r denote the value of m corresponding to r,

    $$44S = \sum_{r=0}^{43}(44m_r + r) = \sum_{r=0}^{43}\left[44\left(r - \sum_{k=1}^{9}\left\lfloor\frac{kr}{44}\right\rfloor\right) + r\right]$$
    $$= 45\sum_{r=0}^{43} r - 44\sum_{r=0}^{43}\sum_{k=1}^{9}\left\lfloor\frac{kr}{44}\right\rfloor = \frac{45 \cdot 43 \cdot 44}{2} - 44\sum_{r=0}^{43}\sum_{k=1}^{9}\left\lfloor\frac{kr}{44}\right\rfloor.$$

Hence

$$S = \frac{45 \cdot 43}{2} - \sum_{r=0}^{43}\sum_{k=1}^{9}\left\lfloor\frac{kr}{44}\right\rfloor = \frac{1935}{2} - \sum_{r=0}^{43}\sum_{k=1}^{9}\left\lfloor\frac{kr}{44}\right\rfloor.$$

Let $T = \sum_{r=0}^{43}\sum_{k=1}^{9}\left\lfloor\frac{kr}{44}\right\rfloor = \sum_{r=1}^{43}\sum_{k=1}^{9}\left\lfloor\frac{kr}{44}\right\rfloor = \sum_{k=1}^{9}\sum_{r=1}^{43}\left\lfloor\frac{kr}{44}\right\rfloor$. Then $2T =$

$$\sum_{k=1}^{9}\sum_{r=1}^{43}\left(\left\lfloor\frac{kr}{44}\right\rfloor + \left\lfloor\frac{k(44-r)}{44}\right\rfloor\right) = \sum_{k=1}^{9}\sum_{r=1}^{43}\left(\left\lfloor\frac{kr}{44}\right\rfloor + \left\lfloor k - \frac{kr}{44}\right\rfloor\right).$$

Since

$$\left\lfloor\frac{kr}{44}\right\rfloor + \left\lfloor k - \frac{kr}{44}\right\rfloor = \begin{cases} k, & \text{if } \frac{kr}{44} \text{ is an integer;} \\ k-1, & \text{if } \frac{kr}{44} \text{ is not an integer,} \end{cases}$$

and $1 \le k \le 9, 1 \le r \le 43$, so $\frac{kr}{44}$ is an integer only when (k,r) are one of

$$(4,11),(8,11),(2,22),(4,22),(6,22),(8,22),(4,33),(8,33).$$

Therefore

$$2T - 8 = \sum_{k=1}^{9}\sum_{r=1}^{43}(k-1) = \sum_{k=1}^{9}43(k-1) = 43\cdot 36 = 1548,$$

i.e., $T = \frac{1548+8}{2} = 778$. Thus, $S = \frac{1935}{2} - 778 = \frac{379}{2}$.

3. (i) Since $\left(k+\frac{3}{2}\right)^2 < k^2 + 3k + 3 < (k+2)^2$, it follows that

$$\sum_{k=a}^{b}\left(k+\frac{3}{2}\right) < \sum_{k=a}^{b}\sqrt{k^2+3k+3} < \sum_{k=a}^{b}(k+2).$$

By taking the arithmetic average of each term, it is obtained that

$$\frac{a+b}{2} + \frac{3}{2} < M(a,b) < \frac{a+b}{2} + 2,$$

i.e., $\frac{a+b+3}{2} < M(a,b) < \frac{a+b+4}{2}$, so $K(a,b) = \left\lfloor\frac{a+b+3}{2}\right\rfloor$.

(ii)　Since $k + 1 = \lfloor k + \dfrac{3}{2} \rfloor = \lfloor \sqrt{k^2 + 3k + 3} \rfloor$,

$$\sum_{k=a}^{b} (k + 1) = \sum_{k=a}^{b} \lfloor \sqrt{k^2 + 3k + 3} \rfloor$$

which implies that

$$N(a, b) = \frac{1}{b - a + 1} \sum_{k=a}^{b} (k + 1) = \frac{a + b}{2} + 1 = \frac{a + b + 2}{2}.$$

4.　*Necessity:*　Suppose that there exists $\{x_n\}$ satisfying the conditions (i), (ii) and (iii). Note that the equality in (iii) can be changed to the form

$$x_n - x_{n-1} = \sum_{k=1}^{2008} a_k (x_{n+k} - x_{n+k-1}), \quad n \in \mathbb{N},$$

where $x_0 = 0$. By adding up above first n equalities, considering $x_0 = 0$, we obtain

$$x_n = \sum_{k=1}^{2008} a_k \sum_{m=1}^{n} (x_{m+k} - x_{m+k-1}) = \sum_{k=1}^{2008} a_k (x_{n+k} - x_k)$$

Letting $n \to \infty$ and $b = \lim_{n \to \infty} x_n$,

$$\begin{aligned} b &= a_1 (b - x_1) + a_2 (b - x_2) + \cdots + a_{2008} (b - x_{2008}) \\ &= b \cdot \sum_{k=1}^{2008} a_k - (a_1 x_1 + a_2 x_2 + \cdots + a_{2008} x_{2008}) < b \cdot \sum_{k=1}^{2008} a_k, \end{aligned}$$

therefore $\displaystyle\sum_{k=1}^{2008} a_k > 1$.

Sufficiency:　Suppose that $\displaystyle\sum_{k=1}^{2008} a_k > 1$. Define the polynomial f by

$$f(s) = -1 + \sum_{k=1}^{2008} a_k s^k, \quad s \in [0, 1],$$

then f is increasing on $[0, 1]$, and $f(0) = -1 < 0$, $f(1) = -1 + \displaystyle\sum_{k=1}^{2008} a_k > 0$. Hence f has a unique root s_0 on $(0, 1)$, i.e., $f(s_0) = 0$ with $0 < s_0 < 1$.

Define the sequence $\{x_n\}$ by $x_n = \displaystyle\sum_{k=1}^{n} s_0^k$ for $n = 1, 2, \ldots$. Then $\{x_n\}$ satisfies the condition (i), and

$$x_n = \sum_{k=1}^{n} s_0^k = \frac{s_0 - s_0^{n+1}}{1 - s_0}.$$

Since $0 < s_0 < 1$, so $\lim_{n\to\infty} s_0^{n+1} = 0$ and hence $\lim_{n\to\infty} x_n = \dfrac{s_0}{1 - s_0}$, namely $\{x_n\}$ satisfies the condition (2). Finally, since $f(s_0) = 0 \Rightarrow \displaystyle\sum_{k=1}^{2008} a_k s_0^k = 1$,

$$
\begin{aligned}
x_n - x_{n-1} &= s_0^n = \left(\sum_{k=1}^{2008} a_k s_0^k\right) s_0^n = \sum_{k=1}^{2008} a_k s_0^{n+k} \\
&= \sum_{k=1}^{2008} a_k (x_{n+k} - x_{n+k-1}).
\end{aligned}
$$

Thus, $\{x_n\}$ satisfies the condition (3) as well.

5. Below we prove the following general conclusion: Let $n \geq 4$, $X_n = (x_1, x_2, \cdots, x_n)$ be a permutation of the first n natural numbers $\{1, 2, 3, \cdots, n\}$, and A be the set of all such X_n. Let

$$f(X_n) = x_1 + 2x_2 + 3x_3 + \cdots + nx_n$$

and $M_n = \{f(X_n) | X_n \in A\}$, then $|M_n| = \dfrac{n^3 - n + 6}{6}$.

By induction we first prove that $M_n =$

$$\left\{ \frac{n(n+1)(n+2)}{6}, \frac{n(n+1)(n+2)}{6} + 1, \cdots, \frac{n(n+1)(2n+1)}{6} \right\}.$$

When $n = 4$, the rearrangement inequality indicates that the minimum element in M_4 is $f(\{4, 3, 2, 1\}) = 20$ and the maximum element in M_4 is $f(\{1, 2, 3, 4\}) = 30$. Besides,

$$
\begin{array}{lll}
f(\{3, 4, 2, 1\}) = 21, & f(\{3, 4, 1, 2\}) = 22, & f(\{4, 2, 1, 3\}) = 23, \\
f(\{4, 1, 2, 3\}) = 24, & f(\{2, 4, 1, 3\}) = 25, & f(\{1, 4, 3, 2\}) = 26, \\
f(\{1, 4, 2, 3\}) = 27, & f(\{2, 1, 4, 3\}) = 28, & f(\{1, 2, 4, 3\}) = 29.
\end{array}
$$

Therefore $M_4 = \{20, 21, \cdots, 30\} \Rightarrow |M_4| = 11 = \dfrac{4^3 - 4 + 6}{6}$, so the conclusion is proven for $n = 4$.

Assume that the conclusion is true for $n - 1$ ($n \geq 5$), then for the n, since we can get an element X_n any element X_{n-1} by taking $x_n = n$, and for such X_n, $f(X_n) = n^2 + \sum_{k=1}^{n-1} k x_k$. By the inductive assumption, $f(X_n)$ can take on the value of any integer in the interval

$$\left[n^2 + \frac{(n-1)n(n+1)}{6}, n^2 + \frac{(n-1)n(2n-1)}{6} \right]$$
$$= \left[\frac{n(n^2 + 6n - 1)}{6}, \frac{n(n+1)(2n+1)}{6} \right].$$

Next consider X_n with $x_n = 1$, then

$$\sum_{k=1}^{n} k x_k = n + \sum_{k=1}^{n-1} k x_k = n + \sum_{k=1}^{n-1} k(x_k - 1) + \frac{n(n-1)}{2}$$
$$= \frac{n(n+1)}{2} + \sum_{k=1}^{n-1} k(x_k - 1).$$

The inductive assumption then indicates that such $f(X_n)$ takes all integral values of the interval

$$\left[\frac{n(n+1)}{2} + \frac{(n-1)n(n+1)}{6}, \frac{n(n+1)}{2} + \frac{(n-1)n(2n-1)}{6} \right]$$
$$= \left[\frac{n(n+1)(n+2)}{6}, \frac{2n(n^2+2)}{6} \right].$$

Since $\dfrac{2n(n^2+2)}{6} - \dfrac{n(n^2+6n-1)}{6} = \dfrac{n(n^2-6n+5)}{6} = \dfrac{n(n-1)(n-5)}{6}$

≥ 0, so $f(X_n)$ takes all integer values of the interval

$$\left[\frac{n(n+1)(n+2)}{6}, \frac{n(n+1)(2n+1)}{6} \right],$$

so the conclusion is also true for n. The inductive proof is completed.

In conclusion, $|M| = \dfrac{n(n+1)(2n+1)}{6} - \dfrac{n(n+1)(n+2)}{6} + 1 = \dfrac{n^3 - n + 6}{6}$.

Returning to the original problem, for $n = 9$, we have $|M_9| = 121$.

Solutions to Testing Questions   20

Testing Questions   (20-A)

1. The conclusion is obvious for $n = 1$. For $n \geq 2$, let $n = p_1^{\alpha_1} p_2^{\alpha_2} \cdots p_k^{\alpha_k}$, where $\alpha_1, \alpha_2, \ldots, \alpha_k$ are non-negative integers. Since

$$\tau(n) = (\alpha_1 + 1)(\alpha_2 + 1) \cdots (\alpha_k + 1) \geq 2^k,$$

and

$$\varphi(n) = n \left(1 - \frac{1}{p_1}\right) \left(1 - \frac{1}{p_2}\right) \cdots \left(1 - \frac{1}{p_k}\right) \geq n \left(1 - \frac{1}{2}\right)^k = \frac{n}{2^k},$$

$$\therefore \varphi(n) \cdot \tau(n) \geq \frac{n}{2^k} \cdot 2^k = n.$$

2. Write $m = 2^\alpha M_1, n = 2M_2$, where $\alpha \geq 1$, $(2, M_1) = 1$, $(2, M_2) = 1$ and $(M_1, M_2) = 1$. Then
$$\varphi(mn) = \varphi(2^{\alpha+1} M_1 M_2) = \varphi(2^{\alpha+1})\varphi(M_1)\varphi(M_2) = 2^\alpha \varphi(M_1)\varphi(M_2)$$
$$= 2 \cdot [2^{\alpha-1}\varphi(M_1)] \cdot \varphi(M_2) = 2\varphi(m)\varphi(n).$$

3. It is clear that $Q(n) \equiv n \pmod 9$, so

$$2005^{2005} \equiv (9 \times 222 + 7)^{2005} \equiv 7^{2005} \equiv 7^{6 \times 334 + 1} \pmod 9.$$

From Euler's Theorem and $\varphi(9) = 6$,

$$7^{\varphi(9)} \equiv 7^6 \equiv 1 \Rightarrow 2005^{2005} \equiv 7 \pmod 9.$$
$$\therefore Q(Q(Q(2005^{2005}))) \equiv Q(Q(2005^{2005})) \equiv Q(2005^{2005}) \equiv 2005^{2005}$$
$$\equiv 7 \pmod 9.$$

On the other hand, $2005^{2005} < (10^4)^{2005} = 10^{8020}$ implies that 2005^{2005} has at most 8020 digits, therefore $Q(2005^{2005}) \leq 9 \times 8020 = 72180$, i.e., $Q(2005^{2005})$ has at most 5 digits, so $Q(Q(2005^{2005})) \leq 9 \times 5 = 45$. Then

$$Q(Q(2005^{2005})) \leq 45 \Rightarrow Q(Q(Q(2005^{2005}))) \leq 3 + 9 = 12$$
$$\Rightarrow Q(Q(Q(2005^{2005}))) = 7$$

since $Q(Q(Q(2005^{2005}))) \equiv 7 \pmod 9$.

4. Suppose that f is a non-constant solution. We may assume that the first coefficient is positive, then there exists a positive integer N such tht $f(n) \geq 2$ when $n \geq N$.

Take any positive integer $n \geq N$. If p is a prime factor of $f(n)$, then $f(n) \mid (2^n - 1)$ yields $2^n \equiv 1 \pmod{p}$. Since $f(n+p) \equiv f(n) \equiv 0 \pmod{p}$ and $f(n+p) \mid (2^{n+p} - 1)$, so $2^{n+p} \equiv 1 \pmod{p}$, therefore $2^p \equiv 1 \pmod{p}$. However, Fermat's Little Theorem gives $2^p \equiv 2 \pmod{p}$, so $1 \equiv 2 \pmod{p}$, a contradiction! Thus, f must be a constant polynomial. Let $f(x) = a$ for all real x.

$f(1) \mid 2^1 - 1 \Rightarrow a \mid 1 \to a = \pm 1$. Thus, $f(x) = 1$ identically or $f(x) = -1$ identically.

5. Take a prime factor p of the number $a_1 + 2a_2 + \cdots + ma_m$. By the Fermat's Little Theorem, $k^p \equiv k \pmod{p}$ for each of positive integer k in $\{1, 2, \ldots, m\}$, therefore for any positive integer n,

$$a_1 \cdot 1^{p^n} + a_2 \cdot 2^{p^n} + \cdots + a_n \cdot m^{p^n} \equiv a_1 + 2a_2 + \cdots + a_n m \equiv 0 \pmod{p},$$

hence all the numbers $a_1 \cdot 1^{p^n} + a_2 \cdot 2^{p^n} + \cdots + a_n \cdot m^{p^n}$ are composite.

6. Write $s_n = 2^{\varphi(n)} + 3^{\varphi(n)} + \cdots + n^{\varphi(n)}$. If n has a prime factor p such that $p^2 \mid n$, let $n = p^2 m$, then

$$1 + s_n \equiv 0^{\varphi(n)} + 1^{\varphi(n)} + 2^{\varphi(n)} + 3^{\varphi(n)} + \cdots + (n-1)^{\varphi(n)}$$

$$\equiv \sum_{j=0}^{mp-1} \sum_{k=0}^{p-1} (jp+k)^{\varphi(n)} \equiv \sum_{j=0}^{mp-1} \sum_{k=0}^{p-1} k^{\varphi(n)}$$

$$\equiv mp \sum_{k=0}^{p-1} k^{\varphi(n)} \equiv 0 \pmod{p},$$

namely $p \mid (1 + s_n)$. However $p \mid n \Rightarrow p \mid s_n$, a contradiction. Thus, $n = p_1 p_2 \cdots p_k$. Without loss of generality we can assume that $p_1 < p_2 < \cdots < p_k$. Then

$$\varphi(n) = (p_1 - 1)(p_2 - 1) \cdots (p_k - 1),$$

which implies that $(p_i - 1) \mid \varphi(n)$ for $i = 1, 2, \ldots, k$. By Fermat's Little Theorem,

$$x^{\varphi(n)} \equiv \begin{cases} 1 \pmod{p_i}, & \text{if } (x, p_i) = 1, \\ 0 \pmod{p_i}, & \text{if } (x, p_i) > 1. \end{cases}$$

Among the $n - 1$ numbers $2, 3, \ldots, n$ there are $\dfrac{n}{p_i}$ numbers not relatively prime to p_i, so $s_n \equiv n - \dfrac{n}{p_i} - 1 \pmod{p_i}$. Since $p_i \mid n$, so $p_i \mid s_n$ and hence $p_i \mid 1 + \frac{n}{p_i}$. Thus,

$$p_1 p_2 \cdots p_k \left| \left(\frac{n}{p_1} + \frac{n}{p_2} + \cdots + \frac{n}{p_k} + 1 \right), \right.$$

namely

$$n \left| n \left(\frac{1}{p_1} + \frac{1}{p_2} + \cdots + \frac{1}{p_1 p_2 \cdots p_k} \right), \right.$$

therefore $\dfrac{1}{p_1} + \dfrac{1}{p_2} + \cdots + \dfrac{1}{p_1 p_2 \cdots p_k}$ is an integer.

7. From Fermat's Little Theorem,

$$2^p \equiv 2 \pmod{p} \Rightarrow m = 2^p - 1 \equiv 1 \pmod{p} \Rightarrow p \mid (m-1),$$

therefore $(2^p - 1) \mid (2^{m-1} - 1)$, and hence $m \mid (2^{m-1} - 1)$. On the other hand,

$$6 \mid (p-1) \Rightarrow 63 = (2^6 - 1) \mid (2^{p-1} - 1) \Rightarrow 7 \mid (2^p - 2) \Rightarrow 7 \mid (m-1).$$

Thus, $127 = (2^7 - 1) \mid (2^{m-1} - 1)$. Now it suffices to show that $(127, m) = 1$. It's enough to show that $127 \nmid m$ since 127 is a prime number.

Since $p > 7$, write $p = 7s + n$, where $0 < n < 7, s \geq 1$. Then

$$127 = (2^7 - 1) \mid (2^{7s} - 1) \Rightarrow 127 \mid (2^{7s+n} - 2^n) = (2^p - 2^n).$$

If $127 \mid m$, then $127 \mid (2^n - 1)$ which contradicts $0 < 2^n - 1 < 127$, hence $(127, m) = 1$.

8. The general term of the given A.P. is $a_{i+1} = 18 + 19i, i = 0, 1, 2, 3, \ldots$. We show that there are infinitely many i such that a_{i+1} consists of only the digit 1, i.e.

$$18 + 19i = \frac{10^k - 1}{9} \quad \text{or} \quad 10^k = 163 + 171i.$$

Since $10^k = 163 + 19 \cdot 9i \Rightarrow 10^k \equiv 11 \pmod{19}$, we observe the remainders of 10^k modulo 19 as k changes, which are listed in the following table:

k	1	2	3	4	5	6
Remainder	10	5	12	6	3	11

therefore $k = 6$ is the minimum k satisfying the necessary condition that $10^k \equiv 11 \pmod{19}$. In fact,

$$\frac{10^6 - 1}{9} = 111111 = 18 + 19 \cdot 5847 \quad \text{or} \quad 10^6 = 163 + 171 \cdot 5847$$

indicates that 111111 is really in the given A.P.. For getting other k satisfying the necessary condition, by Fermat's Little Theorem, $10^{18} \equiv 1 \pmod{19}$, so $10^{18t} \equiv 1 \pmod{19}$, therefore

$$10^{6+18t} \equiv 11 \cdot 1 \equiv 11 \pmod{19}, \quad \text{for all } t \in \mathbb{N}.$$

Below we show that $\dfrac{10^{6+18t} - 1}{9}$ is in the given A.P. for any $t \in \mathbb{N}$.

It suffices to show that $10^{6+18t} = 163 + 171m$ for some $m \in \mathbb{N}$. Since $9 \mid (10^{18t} - 1)$ and $19 \mid (10^{18t} - 1)$, so $171 \mid (10^{18t} - 1)$, namely $10^{18t} = 1 + 171n, n \in \mathbb{N}$. Then

$$10^{6+18t} = (163 + 171 \cdot 5847)(1 + 171n) = 163 + 171m$$

for some $m \in \mathbb{N}$. Thus, $\dfrac{10^{6+18t} - 1}{9}$ is in the given A.P. for any $t \in \mathbb{N}$.

(Note: It is explained that $10^k \equiv 11 \pmod{19}$ is actually a sufficient condition for $\frac{10^k - 1}{9}$ is in the given A.P. in the next lecture.)

9. First of all we show the following lemma.

Lemma: If $x \in \mathbb{Z}$, each odd prime factor p of $x^2 + 1$ must be of the form $4k + 1$.

Proof. If $p \mid (x^2 + 1)$, then $(p, x) = 1$. Then

$$x^2 + 1 \equiv 0 \Rightarrow x^2 \equiv -1 \Rightarrow x^{p-1} \equiv (-1)^{\frac{p-1}{2}} \pmod{p}.$$

Fermat's Little Theorem yields $x^{p-1} \equiv 1 \pmod{p}$, so $\dfrac{p-1}{2} = 2k$ for some $k \in \mathbb{N}$. Hence $p = 4k + 1$.

Now return to the original problem. The given equation is equivalent to

$$x^{2010} + 1 = 4y^{2009} + 4y^{2008} + 2007y + 2007,$$
$$\therefore x^{2010} + 1 = (4y^{2008} + 2007)(y + 1).$$

$4y^{2008} + 2007 \equiv 3 \pmod{4}$ implies that $y^{2008} + 2007$ must have prime factor with form $4k + 3$, but the lemma shows that $(x^{1005})^2 + 1$ has no such prime factor, a contradiction. Thus, the given equation has no required solution.

Testing Questions   (20-B)

1. First of all, for prime number p and integer n with $p \nmid n$, Fermat's Little Theorem yields
$$n^{p-1} \equiv 1 \quad (\text{mod } p). \qquad (30.2)$$

If $p > 103$, then (30.2) holds for each $n \le 103$, so $\displaystyle\sum_{n=1}^{103} n^{p-1} \equiv 103$.
However it is impossible to have $103 \equiv 0$ (mod p) for some $p > 103$. Therefore $p \le 103$, and so there exist positive integer q and non-negative integer $r < p$ such that $103 = pq + r$. Thus, in the numbers from 1 to 103, the number of multiples of p is
$$q = \frac{103 - r}{p},$$

which gives
$$\sum_{n=1}^{103} n^{p-1} \equiv 103 - q \equiv pq + r - q \equiv r - q \quad (\text{mod } p).$$

Since $\displaystyle\sum_{n=1}^{103} n^{p-1} \equiv 0$, so
$$r \equiv q \quad (\text{mod } p). \qquad (30.3)$$

(i)   When $p > q$, then (30.3) implies that $r = q$, so $103 = (p+1)r$. Since 103 is prime, so $p = 102, r = 1$, but this contracts the fact that p is prime.

(ii) When $p \le q$, then $103 = pq + r \ge p^2$. Thus, p can only be one of $3, 5, 7$. By checking each of them, it is found that only $p = 3$ satisfies the given conditions.

Thus, $p = 3$.

2. It is clear that $b \ge 2$. Suppose that $\dfrac{b^n - 1}{b - 1} = p^l$ (where p is a prime, $n \ge 2, l \ge 1$).
When $n = xy$, where $x, y > 1$, then
$$\frac{b^{xy} - 1}{b - 1} = \frac{b^{xy} - 1}{b^y - 1} \cdot \frac{b^y - 1}{b - 1} = (1 + b^y + \cdots + b^{y(x-1)})\frac{b^y - 1}{b - 1}.$$

Since $\dfrac{b^{xy} - 1}{b - 1} = p^l$ and $x > 1, y > 1$, so each factor of the right hand side is a power of p. Thus, $p \mid (b^y - 1)$, namely $b^y \equiv 1 \pmod{p} \Rightarrow 1 + b^y + \cdots + b^{y(x-1)} \equiv x \pmod{p}$. Thus, $p \mid x$.

Since x is an arbitrary factor of n, so the above analysis implies that $n = p^m$ $(m \in \mathbb{N})$. Thus,

$$\frac{b^{p^m} - 1}{b - 1} = \frac{b^{p^m} - 1}{b^{p^{m-1}} - 1} \cdots \cdots \frac{b^{p^2} - 1}{b^p - 1} \cdot \frac{b^p - 1}{b - 1},$$

where each factor is a power p and greater than 1, therefore $p \mid (b^p - 1)$, namely $b^p \equiv 1 \pmod{p}$.

On the other hand, Fermat's Little Theorem gives $b^p \equiv b \pmod{p}$, so $b \equiv 1 \pmod{p}$ or $p \mid (b - 1)$.

Since $p \left| \dfrac{b^p - 1}{b - 1} \right.$, , so $p^2 \mid (b^p - 1)$, namely $b^p \equiv 1 \pmod{p^2}$.

Suppose that $m \geq 2$. Consider

$$\frac{b^{p^2} - 1}{b^p - 1} = 1 + b^p + \cdots + b^{p(p-1)}. \qquad (*)$$

The right hand side of $(*)$ has remainder $p \pmod{p^2}$, and it must be greater than p, so the power of p must be divisible by p^2, a contradiction. Thus, $m = 1$ and $n = p$.

3. We prove the conclusion by induction on k. For $k = 1$, $n = 1$ satisfies the requirement.

    Assume that the conclusion is true for $k = t$ $(t \geq 1)$. Then for $k = t + 1$, since there exists positive integer n_0 such that $n_0^{n_0} \equiv m \pmod{2^t}$, n_0 must be odd. If $n_0^{n_0} \equiv m \pmod{2^{t+1}}$, then n_0 satisfies the requirement for $k = t + 1$.

    When $n_0^{n_0} \not\equiv m \pmod{2^{t+1}}$, since $n_0^{n_0} = m + s \cdot 2^t$, so s must be odd, i.e., $s = 2l + 1$, hence $n_0^{n_0} \equiv m + 2^t \pmod{2^{t+1}}$. Below we show that $n = n_0 + 2^t$ satisfies the requirement for $k = t + 1$.

    Since n is odd, so $(n, 2^{t+1}) = 1$ and since $\varphi(2^{t+1}) = 2^t$, by Euler's theorem, $n^{2^t} \equiv 1 \pmod{2^{t+1}}$, therefore $n^n \equiv n^{n_0 + 2^t} \equiv n^{n_0} \cdot n^{2^t} \equiv n^{n_0} \pmod{2^{t+1}}$. By the Binomial expansion,

$$n^{n_0} = (n_0 + 2^t)^{n_0} = \sum_{i=0}^{n_0} \binom{n_0}{i} 2^{it} n_0^{n_0 - i} \equiv n_0^{n_0} + 2^t n_0^{n_0} \pmod{2^{t+1}}$$

$$\equiv m + 2^t + 2^t n_0^{n_0} \equiv m + 2^t (n_0^{n_0} + 1) \equiv m \pmod{2^{t+1}}.$$

    Thus, the conclusion is true also for $k = t + 1$.

4. Since $2009 = 223 \times 9 + 2$, so the desired number has at least 224 digits. Write it as $x = \overline{c_{223}c_{222}\cdots c_1 c_0}$. It is obvious that $c_{223} \geq 2$.

If $c_{223} = 2$, then $c_{222} = c_{221} = \cdots = c_1 = c_0 = 9$. Note that $2009 = 49 \times 41$, so

$$x = 3 \times 10^{223} - 1 \equiv 30 - 1 \equiv 1 \pmod 7 \Rightarrow 2009 \nmid x.$$

If $c_{223} = 3$, then only one c_i is 8 and others are all 9. Therefore

$$x = 3 \underbrace{99\ldots98}_{222-i} \underbrace{99\ldots9}_{i} = 4 \times 10^{223} - 10^i - 1.$$

Based on the fact $10^5 \equiv 1 \pmod{41}$, it follows that

$$10^{5k} \equiv 1, \quad 10^{5k+1} \equiv 10, \quad 10^{5k+2} \equiv 18 \pmod{41}$$
$$10^{5k+3} \equiv 16, \quad 10^{5k+4} \equiv 37 \pmod{41},$$
$$\therefore x \equiv 22 - 10^i \not\equiv 0 \pmod{41} \Rightarrow 2009 \nmid x.$$

If $c_{223} = 4$, then among $c_{222}, c_{221}, \ldots, c_1, c_0$ two are 8 and the rest are 9, or one is 7 and rest are 9. Thus

$$x = 5 \times 10^{223} - 10^i - 10^j - 1 \equiv 38 - (10^i + 10^j) \pmod{41},$$

where i and j may be equal. To obtain $10^i + 10^j \equiv 38 \pmod{41}$, it is necessary that $(i, j) \equiv (0, 4)$ or $(4, 0) \pmod 5$, so $i \neq j$ and $i, j \leq 220$. To let x be minimum we can take $j = 220$ and $i \equiv 0 \pmod 5$. Below it suffices to choose i such that $49 \mid x$.

By Euler's Theorem, $1 \equiv 10^{\varphi(49)} \equiv 10^{42} \pmod{49}$, and for the factors of 42, if $k = 1, 2, 3, 6, 7, 14, 21$, then $10^k \equiv 10, 2, 20, 8, 31, 30, 48 \pmod{49}$, therefore there is no positive integer $k \leq 41$ such that $10^k \equiv 1 \pmod{49}$. Thus,

$$x = 5 \times 10^{223} - 10^{220} - 10^i - 1 \equiv 5 \times 10^{13} - 10^{10} - 10^i - 1$$
$$\equiv 31 - 10^i \pmod{49},$$

hence $49 \mid x$ if and only if $10^i \equiv 31 \pmod{49}$, namely $i \equiv 7 \pmod{42}$. Since $i \equiv 4 \pmod 5$, by solving the system, it is obtained that

$$i \equiv 49 \pmod{210}.$$

Thus, the unique solution for i is $i = 49$, and the corresponding value of x is

$$4998 \underbrace{99\ldots98}_{170} \underbrace{99\ldots9}_{49}.$$

5. The substitution $n = p$, a prime, yields $p \equiv (f(p))^p \equiv 0 \pmod{f(p)}$, so p must be divisible by $f(p)$. Hence, for each prime p, $f(p) = 1$ or $f(p) = p$.

Let $S = \{p : p \text{ is prime and } f(p) = p\}$. If S is infinite, then $f(n)^p \equiv n \pmod{p}$ for infinitely many primes p. By Fermat's little theorem, $n \equiv f(n)^p \equiv f(n) \pmod{p}$, so that $f(n) - n$ is a multiple of p for infinitely many primes p. This can happen only if $f(n) = n$ for all values of n, and it can be verified that this is a solution.

If S is empty, then $f(p) = 1$ for all primes p, and any function satisfying this condition is a solution.

Now suppose that S is finite and non-empty. Let q be the largest prime in S. Suppose that $q \geq 3$. Then for any prime p exceeding q,

$$p \equiv f(p)^q \equiv 1^q \pmod{f(q)} \Rightarrow p \equiv 1 \pmod{q}.$$

However, this is not true. Let Q be the product of all the odd primes up to q. Then all the prime factors of $Q + 2$ must exceed q. Let p be any prime factor of $Q + 2$, then $f(p) = 1$, so $p \equiv f(p)^q \pmod{f(q)} \Rightarrow p \equiv 1 \pmod{q}$, so $Q + 2 \equiv 1 \pmod{q}$. However this contradicts $Q + 2 \equiv 2 \pmod{q}$.

The only remaining case is that $S = \{2\}$. Then $f(2) = 2$ and $f(p) = 1$ for every odd prime p. Since $f(n)^2 \equiv n \pmod{2}$, $f(n)$ and n must have the same parity. Conversely, any function f for which $f(n) \equiv n \pmod{2}$ for all n, $f(2) = 2$ and $f(p) = 1$ for all odd primes p satisfies the condition.

Therefore the only solutions are (i) $f(n) = n$ for all $n \in \mathbb{N}$; (ii) any function f with $f(p) = 1$ for all primes p; (iii) any function for which $f(2) = 2$, $f(p) = 1$ for primes p exceeding 2 and $f(n)$ and n have the same parity.

6. We prove thr statement by contradiction. Suppose that $n^7 + 7 = x^2$ holds for some pair (n, x) of positive integers. Then

    (i) n must be odd. Otherwise, it follows that $x^2 \equiv 3 \pmod{4}$.

    (ii) $n \equiv 1 \pmod{4}$, since n is odd implies that $4 \mid n^7 + 7$, so $n \equiv 1 \pmod{4}$.

(iii) $x^2 = n^7 + 7$ yields

$$x^2 + 11^2 = n^7 + 128$$
$$= (n + 2)(n^6 - 2n^5 + 4n^4 - 8n^3 + 16n^2 - 32n + 64) \quad (30.4)$$

If $11 \nmid x$, then each prime factor p of $x^2 + 11^2$ is odd and $p \equiv 1$ (mod 4), since if $p = 4k + 3$ then by Fermat's Little Theorem,

$$x^2 \equiv -11^2 \Rightarrow x^{p-1} \equiv -11^{p-1} \equiv -1 \quad (\text{mod } p),$$

a contradiction.

(30.4) yields $(n + 2) \mid x^2 + 11^2$. But $n + 2 \equiv 3$ (mod 4) implies that $x^2 + 11^2$ has at least one prime factor with remainder 3 modulo 4, a contradiction.

If x is a multiple of 11, let $x = 11y$, then (30.4) becomes

$$121(y^2 + 1) = (n + 2)(n^6 - 2n^5 + 4n^4 - 8n^3 + 16n^2 - 32n + 64).$$

By substituting n with remainder $0, \pm 1, \pm 2, \pm 3, \pm 4, \pm 5$ (mod 11) and doing a direct calculation, it is found that $n^6 - 2n^5 + 4n^4 - 8n^3 + 16n^2 - 32n + 64$ is not a multiple of 11, so $121 \mid (n + 2)$, and this indicates that

$$y^2 + 1 = \frac{n + 2}{121}(n^6 - 2n^5 + 4n^4 - 8n^3 + 16n^2 - 32n + 64). \quad (30.5)$$

By similar reasoning, it can be proven that each prime factor of $y^2 + 1$ must have remainder 1 modulo 4, so its odd divisors must also have that property. However, $\dfrac{n + 2}{121} \equiv 3$ (mod 4), so (30.5) cannot hold.

Thus, $n^7 + 7$ is not a perfect square in all the possible cases. (Given by **Xiaosheng Mu**)

Solutions to Testing Questions 21

Testing Questions (21-A)

1. There are infinitely many primes, so for any $n \in \mathbb{N}$ it is always possible to pick out n distinct primes $p_1, p_2, \cdots, p_n$. Then $(p_i^2, p_j^2) = 1$ for any $i \neq j$ with $1 \leq i < j \leq n$. By the Chinese Remainder Theorem, the system

$$x \equiv -1 \ (\text{mod } p_1^2), \quad x \equiv -2 \ (\text{mod } p_2^2), \quad \cdots, \quad x \equiv -n \ (\text{mod } p_n^2),$$

has infinitely many solutions. Let $x_0 > 0$ be a positive solution, then the n consecutive numbers $x_0 + 1, x_0 + 2, \cdots, x_0 + n$ are divisible by $p_1^2, p_2^2, \cdots, p_n^2$ respectively.

2. $x_1 = 5$ satisfies the equation $x \equiv 1 \pmod{4}$. Let $x_2 = 5 + 4k$ where $k \in \mathbb{N}$. Then $x_2 = 5 + 12 = 17$ satisfies the first two equations.

    Let $x_3 = 17 + 20k$ where $k \in \mathbb{N}$. From

    $$17 + 20k \equiv 3 - k \equiv 4 \pmod{7},$$

    we find $k = 6$, i.e. $x_3 = 137$ is a solution of the system. Thus, the solutions are given by the the congruence class satisfying the equation

    $$x \equiv 137 \pmod{140}.$$

3. It is obvious that $p \neq q$. We may assume that $p < q$.

    When $p = 2$, since $q^q + 5 \equiv 5 \equiv 0 \pmod{q}$, q can only be 5. By checking, $(2, 5)$ is a solution.

    When p, q are both odd primes, $q \mid (p^p + 1) \Rightarrow q \mid (p^{p-1} - p^{p-2} + \cdots - p + 1)$ and $p^{2p} \equiv 1 \pmod{q}$.

    On the other hand, Fermat's Little Theorem gives $p^{q-1} \equiv 1 \pmod{q}$. If $(2p, q - 1) = 2$, then $\mathrm{ord}_q p = 2$ so that $p^2 \equiv 1 \pmod{q}$ and hence $p \equiv 1 \pmod{q}$ or $p \equiv -1 \pmod{q}$. Then

    $$0 \equiv p^{p-1} - p^{p-2} + \cdots - p + 1 \equiv 1 \ \text{ or } \ 1 + p \pmod{q},$$

    a contradiction.

    If $(2p, q - 1) = 2p$, i.e., $q \equiv 1 \pmod{p}$, then $0 \equiv p^p + q^q + 1 \equiv 2 \pmod{p}$, a contradiction also. Thus, the solutions are $(2, 5)$ and $(5, 2)$.

4. Let $p_1, p_2, \cdots, p_n$ and $q_1, q_2, \cdots, q_n$ be $2n$ distinct prime numbers. Let x be any positive solution of the system

    $$x \equiv -1 \pmod{p_1 q_1}, \quad x \equiv -2 \pmod{p_2 q_2}, \quad \cdots, \quad x \equiv -n \pmod{p_n q_n},$$

    then each of the n consecutive positive integers $x + 1, x + 2, \cdots, x + n$ has at least two different prime factors, satisfying the requirement of the problem.

5. If all the primes not greater than k are prime factors of m, the conclusion is clearly true. If this is not the case, let N be the product of all primes which are not greater than k, and are relatively prime to m. Then $N > 1$. For positive integer r with $(r, m) = 1$, the system of equations

    $$x \equiv r \pmod{m}, \quad x \equiv 1 \pmod{N}$$

has positive solutions, and for any such solution, each of its prime factor is relatively prime to N, so it must be greater than k. By taking each number of a reduced residue system modulo m as r to get the corresponding solution x, the group of $\varphi(m)$ solutions form a required reduced residue system modulo m.

6. For positive integer $n > 1$,

$$
\begin{aligned}
F_{n-1}^{2^{n+1}} &= (2^{2^{n-1}} + 1)^{2^{n+1}} = (2^{2^n} + 2^{2^{n-1}+1} + 1)^{2^n} \equiv (2^{2^{n-1}+1})^{2^n} \\
&= (2^{2^n})^{2^{n-1}+1} \equiv (-1)^{2^{n-1}+1} \equiv -1 \pmod{F_n}.
\end{aligned}
$$

Therefore $F_{n-1}^{2^{n+1}} + 1 \equiv 0 \pmod q$, and it yields $F_{n-1}^{2^{n+2}} \equiv 1 \pmod q$. By Fermat's Little Theorem, $2^{n+2} \mid q - 1$, namely $q = 2^{n+2}k + 1$ for some positive integer k.

7. Assume that $p \le q$.

When $p = 2$, then $q = 2$ satisfies the condition, $(2, 2)$ is a solution. If $q > 2$, then q is odd and satisfies $q \mid (1 + 2^{q-2})$, so

$$
2^{q-2} \equiv -1 \pmod q.
$$

By Fermat's Little Theorem, $1 \equiv 2^{q-2} \cdot 2 \equiv -2 \pmod q$, so $3 \equiv 0 \pmod q$, therefore $q = 3$. By checking, $(2, 3)$ is a solution also.

When $p > 2$, $2^{q-1} \equiv 1 \pmod q$ and $q \mid (2^p + 2^q)$ imply that $0 \equiv 2^p + 2^q \equiv 2^p + 2 \pmod q$. Therefore

$$
2^{p-1} + 1 \equiv 0 \pmod q \Rightarrow 2^{p-1} \equiv -1 \pmod q \Rightarrow 2^{2(p-1)} \equiv 1 \pmod q.
$$

Let $k = \mathrm{ord}_q 2$, then $k \mid 2(p-1)$ but $k \nmid p-1$, so $k = 2^{s+1}m$, $p-1 = 2^s t$, where $s, t \in \mathbb{N}$, t is odd and $m \mid t$. Since $k \mid q - 1$, so in the prime factorization of $p-1$, the power of 2 is less than that of $q-1$. By exchanging p and q and applying the same reasoning, the power of 2 in $p - 1$ then is greater than that of $q - 1$, a contradiction. Thus, there must be no solution in case of $q \ge p > 2$.

Thus, the solutions for pairs (p, q) are $(2, 2)$, $(2, 3)$ and $(3, 2)$.

8. We prove by contradiction.

Solution 1. Suppose that there is a positive integer $n > 1$ such that $n \mid 2^n - 1$, then n is odd. Let

$$
n_0 = \min\{n \in \mathbb{N} : \quad n > 1, \ n \mid (2^n - 1)\}.
$$

Let $k = \mathrm{ord}_{n_0} 2$. $2^{n_0} \equiv 1 \pmod{n_0}$ implies $k \mid n_0$, hence $1 \leq k \leq n_0$. If $k = 1$ then $2 - 1 \equiv 0 \pmod{n_0}$, i.e. $n_0 = 1$, a contradiction, therefore $1 < k$. Since n_0 is odd and $k \mid \varphi(n_0) \leq n_0 - 1$, therefore $1 < k < n_0$.

However, $2^k \equiv 1 \pmod{n_0}$ implies $n_0 \mid (2^k - 1)$, and hence implies $k \mid (2^k - 1)$ also, which contradicts the definition of n_0.

Solution 2. Suppose that there is a positive integer $n > 1$ such that $n \mid 2^n - 1$, then n is odd. Suppose that p is an odd prime factor of n . Let $k = \mathrm{ord}_p 2$, then

$$n \mid 2^n - 1 \Rightarrow p \mid (2^n - 1) \Rightarrow 2^n \equiv 1 \pmod{p},$$

therefore $k \mid n$. By Fermat's Little Theorem, $2^{p-1} \equiv 1 \pmod{p}$, so $k \mid p - 1$, and hence $k \mid (n, p - 1)$.

Now we take p be the minimum prime factor of n, then $(n, p - 1) = 1$, hence $k \mid 1$ i.e. $k = 1$. However, it implies $2 \equiv 1 \pmod{p}$ i.e. $p = 1$, a contradiction.

Testing Questions   (21-B)

1. Let p be a prime factor of m with $p^\alpha \| m, \alpha \geq 1$. Then there is an infinite subset A_1 of A such that all elements in A_1 are not divisible by p. By the pigeonhole principle, there exists an infinite subset A_2 of A_1 such that $x \equiv a \pmod{mn}$ for any element $x \in A_2$, where a is an integer not divisible by p.

The condition $(m, n) = 1$ yields $\left(p^\alpha, \dfrac{mn}{p^\alpha} \right) = 1$. By the Chinese Remainder Theorem the system

$$\begin{cases} x \equiv a^{-1} \pmod{p^\alpha}, \\ x \equiv 0 \pmod{\dfrac{mn}{p^\alpha}} \end{cases} \tag{30.6}$$

has infinitely many solutions. Let x be anyone of them, and denote by B_p the set consisting of the first x elements of A_2, and by S_p the sum of all elements in B_p, then $S_p \equiv ax \pmod{mn}$, and from (30.6),

$$S_p \equiv ax \equiv 1 \pmod{p^\alpha}, \quad S_p \equiv 0 \pmod{\dfrac{mn}{p^\alpha}}.$$

Let $m = p_1^{\alpha_1} \cdots p_k^{\alpha_k}$, and suppose that corresponding to each p_i ($1 \leq i \leq k-1$), the finite subset B_{p_i} has selected, where $B_{p_i} \subset A \backslash \{B_{p_1} \cup \cdots \cup B_{i-1}\}$, such that the elements in B_{p_i} and S_{p_i} satisfy

$$S_{p_i} \equiv 1 \pmod{p_i^{\alpha_i}}, \quad S_{p_i} \equiv 0 \pmod{\frac{mn}{p_i^{\alpha_i}}}. \tag{30.7}$$

Now define the set $B = \bigcup_{i=1}^{k} B_{p_i}$ and the sum S of elements of B given by

$S = \sum_{i=1}^{k} S_{p_i}$. Then (30.7) yields

$$S \equiv 1 \pmod{p_i^{\alpha_i}} \text{ for } 1 \le i \le k \quad \text{and} \quad S \equiv 0 \pmod{n},$$

hence B satisfies the requirements in question.

2. $2550 = 2 \cdot 3 \cdot 5^2 \cdot 17$, we first consider the remainder of $p_k^{p_k^4-1}$ modulo 2550 when $p_k \ne 2, 3, 5, 17$.

For $p_k > 5$, Fermat's Little Theorem gives $5 \mid (p_k^4 - 1)$. Since

$$p_k^4 - 1 = (p_k - 1)(p_k + 1)(p_k^2 + 1)$$

and one between the two consecutive even numbers $(p_k - 1)$ and $(p_k + 1)$ must be divisible by 4, so $16 \mid (p_k^4 - 1)$.

It is easy to see that

$$\varphi(2) = 1, \varphi(3) = 2, \varphi(5) = 4, \varphi(5^2) = 5 \cdot 4 = 20, \varphi(17) = 16.$$

By Euler's Theorem, $p_k^{p_k^4-1} \equiv 1 \pmod{p}$ for $p = 2, 3, 5^2, 17$, so

$$p_k^{p_k^4-1} \equiv 1 \pmod{2550}.$$

Below we consider the cases: $p_k = 2, 3, 5, 17$. Since $2 = p_1$, so it is not needed to consider $p_k = 2$ for the sum $\sum_{k=2}^{2550} p_k^{p_k^4-1}$.

Since $p_2 = 3$, $p_3 = 5$, $p_7 = 17$, for convenience write $A = p_2^{p_2^4-1}$, $B = p_3^{p_3^4-1}$, $C = p_7^{p_7^4-1}$. Then simple calculations yield the following table of remainders:

	mod 2	mod 3	mod 5^2	mod 17
A	1	0	1	1
B	1	1	0	1
C	1	1	1	0
$A + B + C$	1	2	2	2

Therefore $A + B + C = (3 \cdot 5^2 \cdot 17)k + 2$ and $k = 2m + 1$ for some $k, m \in \mathbb{N}$. Thus,

$$A + B + C = (2 \cdot 3 \cdot 5^2 \cdot 17)m + 1277.$$

Hence $\displaystyle\sum_{k=2}^{2550} p_k^{p_k^4 - 1} \equiv 1277 + (2550 - 4) \equiv 1273 \pmod{2550}$.

3. We first prove the following lemma.

Lemma: Let p be a given odd prime number, $u > 1$ be an integer with $p \nmid u$. Let $d = \text{ord}_p u$ and $p^v \| (u^d - 1)$. Then $p^{t+v} \| (u^{dmp^t} - 1)$, where m is a positive integer with $(p, m) = 1$ and t is an arbitrary non-negative integer.

Proof of Lemma: We use induction on t. When $t = 0$, the definition of v gives $u^d = 1 + p^v k$ (where $p \nmid k$). The Binomial expansion yields

$$u^{md} = (1 + p^v k)^m = 1 + p^v km + p^{2v} k^2 \binom{m}{2} + \cdots$$

$$= 1 + p^v \left(km + p^v k^2 \binom{m}{2} + \cdots \right) = 1 + p^v k_1,$$

where k_1 is an integer with $(k_1, p) = 1$, so the lemma is true for $t = 0$.

Assume that the lemma is true for t, i.e., $u^{dmp^t} = 1 + p^{t+v} k_t$, where $p \nmid k_t$. Then the binomial expansion yields

$$u^{dmp^{t+1}} = (1 + p^{t+v} k_t)^p = 1 + p^{t+1+v} (k_t + \binom{p}{2} p^{v+t-1} k_t^2 + \cdots)$$

$$= 1 + p^{t+1+v} k_{t+1},$$

where $p \nmid k_{t+1}$ (Note that p is an odd prime, so $p \mid \binom{p}{2}$). Thus, the lemma is proven.

Now we return to the original problem. The conclusion is clearly true for $u = 1$. For $u > 1$, the given equation can be written in the form

$$n! = u^r (u^s - 1), \qquad r, s \in \mathbb{N}. \tag{30.8}$$

Take an odd prime p with $p \nmid u$. It suffices to consider the case $n > p$. If $p^\alpha \| n!$, then $\alpha \geq 1$. (30.8) and $p \nmid u$ implies that $p^\alpha \| (u^s - 1)$, and in particular, $p \mid (u^s - 1)$. Let $d = \text{ord}_p u$, then $d \mid s$. Let $s = dmp^t$, where $t \geq 0$, $p \nmid m$. Then $p^\alpha \mid (u^s - 1)$ and the Lemma implies that $\alpha = t + v$, namely $t = \alpha - v$, where $p^v \| (u^d - 1)$. Hence

$$u^s - 1 = u^{dmp^{\alpha-v}} - 1. \tag{30.9}$$

It is well known that

$$\alpha = \sum_{i=1}^{\infty} \left\lfloor \frac{n}{p^i} \right\rfloor \geq \left\lfloor \frac{n}{p} \right\rfloor > an, \tag{30.10}$$

where a is a positive constant depending p only. Write $b = u^{dp^{-v}}$.

d, p, u, v are all fixed positive integers, so that b is a constant greater than 1, therefore (30.9) and (30.10) together yields

$$u^s - 1 \geq u^{dp^{\alpha-v}} - 1 > b^{p^{an}} - 1. \tag{30.11}$$

It is easy to see that for n big enough,

$$b^{p^{an}} - 1 > n^n - 1 \tag{30.12}$$

(i.e., $b^{p^{an}} > n^n$, namely $p^{an} > n \log_b n$). Thus, (30.9), (30.11), (30.12) implies that $u^s - 1 > n!$ and hence $u^r (u^s - 1) > n!$ for big n. Thus, the equation (30.8) has no solution for n big enough. On the other hand, for each given n, the equation (30.8) has only finitely many solutions, so the conclusion is proven.

4. As is known by us (for example, see Example 5), let p be a prime factor of F_n, then $p = 2^{n+1}x + 1$, where x is a positive integer. Let

$$F_n = p_1^{\alpha_1} p_2^{\alpha_2} \cdots p_s^{\alpha_s}, \tag{30.13}$$

then $p_i = 2^{n+1}x_i + 1, i = 1, 2, \ldots, s$. By (30.13) and the binomial expansion,

$$2^{2^n} + 1 \geq (2^{n+1} + 1)^{(\alpha_1 + \cdots + \alpha_s)} > 2^{(n+1)(\alpha_1 + \cdots + \alpha_s)} + 1,$$

hence

$$\sum_{i=1}^{s} \alpha_i < \frac{2^n}{n+1}. \tag{30.14}$$

On the other hand, by the binomial expansion,

$$p_i^{\alpha_i} = (2^{n+1}x_i + 1)^{\alpha_i} \equiv 1 + 2^{n+1}\alpha_i x_i \pmod{2^{2n+2}}.$$

For $n \geq 3$, since $2^n \geq 2n + 2$ holds, by taking mod 2^{2n+2} to (30.13) it follows that

$$1 \equiv 2^{2^n} + 1 = \prod_{i=1}^{s} p_i^{\alpha_i} \equiv \prod_{i=1}^{s}(1 + 2^{n+1}\alpha_i x_i)$$

$$\equiv 1 + 2^{n+1} \sum_{i=1}^{s} \alpha_i x_i \pmod{2^{2n+2}},$$

which implies that $2^{n+1} \sum_{i=1}^{s} \alpha_i x_i \equiv 0 \pmod{2^{2n+2}}$, namely $\sum_{i=1}^{s} \alpha_i x_i \equiv 0$

$\pmod{2^{n+1}}$, hence $\sum_{i=1}^{s} \alpha_i x_i \geq 2^{n+1}$. Thus, there must be some x_j such

that

$$x_j \sum_{i=1}^{s} \alpha_i \geq 2^{n+1}. \qquad (30.15)$$

Combining (30.13), (30.14), (30.15), it is obtained that $2^{n+1} \leq x_j \sum_{i=1}^{s} \alpha_i <$

$x_j \cdot \dfrac{2^n}{n+1}$, so $x_j > 2(n+1)$, hence

$$p_j = 2^{n+1} x_j + 1 > 2^{n+1}(2n+2) = 2^{n+2}(n+1),$$

as desired.

5. Let p be a prime number, a a positive integer, and $V_p(a)$ denotes the component of p in the prime factorization of a, i.e., $p^\alpha \| a \Rightarrow V_p(a) = \alpha$.

    Lemma: Let p be a given odd prime number, $u > 1$ be an integer with $p \nmid u$. Let $d = \mathrm{ord}_p u$ and $p^v \| (u^d - 1)$. Then $p^{t+v} \| (u^{dmp^t} - 1)$, where m is a positive integer with $(p, m) = 1$ and t is an arbitrary non-negative integer.

    Proof of Lemma: We use induction on t. When $t = 0$, the definition of v gives $u^d = 1 + p^v k$ (where $p \nmid k$). The binomial expansion yields

    $$
    \begin{aligned}
    u^{md} = (1 + p^v k)^m &= 1 + p^v k m + p^{2v} k^2 \binom{m}{2} + \cdots \\
    &= 1 + p^v \left(k m + p^v k^2 \binom{m}{2} + \cdots \right) = 1 + p^v k_1,
    \end{aligned}
    $$

    where k_1 is an integer with $(k_1, p) = 1$, so the lemma is true for $t = 0$.

    Assume that the lemma is true for t, i.e., $u^{dmp^t} = 1 + p^{t+v} k_t$, where $p \nmid k_t$. Then the binomial expansion yields

    $$
    \begin{aligned}
    u^{dmp^{t+1}} &= (1 + p^{t+v} k_t)^p = 1 + p^{t+1+v}(k_t + \binom{p}{2} p^{v+t-1} k_t^2 + \cdots) \\
    &= 1 + p^{t+1+v} k_{t+1},
    \end{aligned}
    $$

    where $p \nmid k_{t+1}$ (Note that p is odd prime, so $p \mid \binom{p}{2}$). Thus, the lemma is proven.

Now we return to the original problem. Suppose $b > 1$. Write the given equation in the form

$$a^b - 1 = b(1 + b + \cdots + b^{n-1}) \Rightarrow b \mid a^b - 1.$$

Let p be the minimum prime factor of b, then $\text{ord}_p a \mid b$. Since $\text{ord}_p a \leq p - 1 < p$, so $\text{ord}_p a = 1$. If p is odd, let $p^v \| (a - 1)$, $p^t \| b$. Then the Lemma gives $p^{v+t} \| (a^b - 1)$, i.e., $p^{v+t} \| b(1 + b + \cdots + b^{n-1})$.

On the other hand, $(b, 1 + b + \cdots + b^{n-1}) = 1$, so $p^t \| b(1 + b + \cdots + b^{n-1})$, i.e., $t = v + t$, or $v = 0$. However, this contradicts the fact that $p \mid (a - 1)$. Thus, p is not odd.

If $p = 2$, let $2^v \| \dfrac{a^2 - 1}{2}$, $2^t \| b$. b is even implies a is odd, so $v > 0$. Besides, $\dfrac{a^2 - 1}{2} = \dfrac{1}{2}(a + 1)(a - 1)$ and $(a + 1)$, $(a - 1)$ are both even but one of them is not a multiple of 4, and the index of 2 in other factor is v. Hence one of $a - 1$ and $a + 1$ can be expressed in the form $2^v m$ (where $2 \nmid m$).

Write $b = 2^t k$, where $2 \nmid k$, then

$$a^b - 1 = (2^v m \pm 1)^{2^t k} - 1 = (2^{2v} m^2 \pm 2^{v+1} m + 1)^{2^{t-1} k} - 1$$
$$= (2^{v+1} m_1 + 1)^{2^{t-1} k} - 1 = \cdots = (2^{v+t} m_t + 1)^k - 1$$

$$= 2^{(v+t)k} m_t^k + \binom{k}{1} 2^{(v+t)(k-1)} m_t^{k-1} + \cdots + \binom{k}{k-1} 2^{v+t} m_t,$$

where $m_1, \cdots, m_t$ are all odd. Since $\binom{k}{k-1} = k$ is odd, so $2^{v+t} \| (a^b - 1)$. Since $1 + b + \cdots + b^{n-1}$ is odd, so $2^t \| b(1 + b + \cdots + b^{n-1})$, which yields $v = 0$ again, a contradiction. Thus, we have proven that b has no prime factor p, i.e., $b = 1$, and the solutions for (a, b, n) are $(a, 1, a - 1)$, $a > 1$.

Solutions to Testing Questions   22

Testing Questions   (22-A)

1. It is clear that $p \neq q$ and we can assume that $p < q$.

    If $p = 2$, then $q \mid (q^q + 5) \Rightarrow q = 5$. it's easy to verify that $(2, 5)$ is a solution.

When p, q are both odd, then

$$p^p + 1 \equiv 0 \pmod{q} \Rightarrow q \mid (p^{p-1} - p^{p-2} + \cdots - p + 1),\ p^{2p} \equiv 1 \pmod{q}.$$

On the other hand, Fermat's Little Theorem gives $p^{q-1} \equiv 1 \pmod{q}$. If $(2p, q - 1) = 2$, then $p^2 \equiv 1 \pmod{q}$ so that $p \equiv 1 \pmod{q}$ or $p \equiv -1 \pmod{q}$, hence

$$0 \equiv p^{p-1} - p^{p-2} + \cdots - p + 1 \equiv 1 \text{ or } p \pmod{q},$$

a contradiction. If $(2p, q - 1) = 2p$, namely $q \equiv 1 \pmod{p}$, then

$$0 \equiv p^p + q^q + 1 \equiv 2 \pmod{p},$$

which also leads to a contradiction. Thus, the only solutions are $(2, 5)$ and $(5, 2)$.

2.   Without loss of generality we assume that $a \leq b \leq c$. Then

$$abc = 2009(a + b + c) \leq 6027c,$$

namely $ab \leq 6027$, so the number of pairs (a, b) satisfying $a \leq b \leq c$ for at least one $c \in \mathbb{N}$ and the equation

$$abc = 2009(a + b + c)$$

is finite.

On the other hand, for any given pair (a, b), $c(ab - 2009) = 2009(a + b)$ implies that c is determined uniquely by (a, b). Thus, the number of c is finite, so the number of (a, b, c) with $a \leq b \leq c$ which satisfies the given equation is finite. Thus, the conclusion is proven.

3.   Let the pair $\{x, y\}$ satisfy the equation

$$x^2 - 2xy + 126y^2 - 2009 = 0. \tag{$*$}$$

If we consider $(*)$ as a quadratic equation in x, then the value of its discriminant

$$\Delta = 4y^2 - 4(126y^2 - 2009) = 500(4^2 - y^2) + 36$$

is a perfect square. $4^2 - y^2 \geq 0$ implies that $|y|$ can only be $0, 1, 2, 3, 4$. By checking, $|y| = 4$ is the unique possible value. Then, from $(*)$,

$$y = 4 \Rightarrow x^2 - 8x + 7 = 0 \Rightarrow \{x, y\} = \{1, 4\} \text{ or } \{7, 4\};$$
$$y = 4 \Rightarrow x^2 + 8x + 7 = 0 \Rightarrow \{x, y\} = \{-1, -4\} \text{ or } \{-7, -4\}.$$

Thus, the solutions are $\{1, 4\}$, $\{7, 4\}$, $\{-1, -4\}$, $\{-7, -4\}$.

4. If $n < m$, then $n![m(m-1)\cdots(n+1)+1] = m^n$. However

$$(m(m-1)\cdots(n+1)+1, m) = 1 \Rightarrow m^n \nmid (m!+n!) \Rightarrow \text{no solution.}$$

Therefore $m \le n$.

If $m > 2$, then $[(m-2)!](m-1)m[1+(m+1)\cdots n] = m^n$. Since $(m-1, m) = 1$, so there is no factor $(m-1)$ in m^n, hence there is no solution in this case. Thus, $m = 1$ or 2.

But if $m = 1$, then $1 + n! = 1$ has no solution for n.

When $m = 2$, the given equation becomes $2! + n! = 2^n$. For $n \ge 4$,

$$n! = 1 \cdot 2 \cdot 3 \cdots n > 2 \cdot 2 \cdot 2 \cdots 2 = 2^n,$$

therefore $n = 2$ or 3. When $n = 2$, then $2! + 2! = 2^2$, so $(2, 2)$ is a solution. When $n = 3$, then $2! + 3! = 2^3$, so $(2, 3)$ is also a solution. Thus, $(m, n) = (2, 2)$ or $(2, 3)$.

5. Assume that $x \ge y \ge z$. If $z > \dfrac{1}{\sqrt[3]{34} - 3}$, then $3z + 1 < \sqrt[3]{34}z$, hence similarly

$$3x + 1 < \sqrt[3]{34}x \quad \text{and} \quad 3y + 1 < \sqrt[3]{34}y.$$

Multiplying them up yields $(3x+1)(3y+1)(3z+1) < 34xyz$ which contradicts the given equation. Therefore

$$z \le \frac{1}{\sqrt[3]{34} - 3} < 5.$$

$z = 1 \Rightarrow 4(3x+1)(3y+1) = 34xy \Rightarrow xy + 6(x+y) + 2 = 0$, no positive integer solution.

$z = 2 \Rightarrow 7(3x+1)(3y+1) = 68xy \Rightarrow 5xy - 21(x+y) - 7 = 0$
$\Rightarrow (5x-21)(5y-21) = 476 = 2^2 \times 7 \times 17 = 119 \times 4 = 34 \times 14$, therefore $(x, y) = (28, 5)$ or $(11, 7)$.

$z = 3 \Rightarrow 10(3x+1)(3y+10) = 102xy \Rightarrow 6xy - 15(x+y) - 5 = 0$. By taking modulo 3, then we have $-5 \equiv 0 \pmod 3$, so there is no solution in this case.

$z = 4 \Rightarrow 13(3x+1)(3y+1) = 136xy \Rightarrow 19xy - 39(x+y) - 13 = 0$
$\Rightarrow (19x-39)(19y-39) = 1768 = 2^3 \times 13 \times 17$. After discussing all the possible cases we find no solution.

Thus, $(28, 5, 2)$, $(11, 7, 2)$ and all their permutations are the solutions.

6. The given equation is equivalent to $(2y + x)^2 = 4x^3 - 27x^2 + 44x - 12$.

$$
\begin{aligned}
4x^3 - 27x^2 + 44x - 12 &= (x - 2)(4x^2 - 19x + 6) \\
&= (x - 2)[(x - 2)(4x - 11) - 16],
\end{aligned}
$$

implies that $x = 2$, $y = -1$ is a solution.

When $x \neq 2$, since $(2y + x)^2$ is a perfect square, let $x - 2 = ks^2$, where s is a positive integer, then $k \in \{-2, -1, 1, 2\}$. In fact, if $p^{2m+1} \| x - 2$, then $p \mid (x - 2)(x - 11) - 16$, so $p \mid 16$, i.e. $p = 2$.

When $k = \pm 2$, then $4x^2 - 19x + 6 = \pm 2n^2$, where n is a positive integer, i.e. $(8x - 19)^2 - 265 = \pm 32n^2$. Since

$$\pm 32n^2 \equiv 0, \pm 2 \quad (\text{mod } 5), \ (8x - 19)^2 \equiv 0, \pm 1 \quad (\text{mod } 5) \ \text{and} \ 25 \nmid 265,$$

we obtain a contradiction.

When $k = 1$, then $4x^2 - 19x + 6 = n^2$ for some positive integer n, i.e. $265 = (8x - 19)^2 - 16n^2 = (8x - 19 - 4n)(8x - 19 + 4n)$. From $265 = 1 \times 265 = 5 \times 53 = (-265) \times (-1) = (-53) \times (-5)$, in these four cases, only $x = 6$ has $x - 2 = s^2$ being a perfect square, so $n = 6$. Therefore $y = 3$ or $y = -9$.

When $k = -1$, then $4x^2 - 19x + 6 = -n^2$ for some positive integer n, so that $265 = (8x - 19)^2 + 16n^2$. From $16n^2 \leq 265$ we obtain $n \leq 4$. for $n = 1, 2$, the equation $265 = (8x - 19)^2 + 16n^2$ has no integer solution, for $n = 3$, $x = 1$ and hence $y = 1$ or -2 are solutions; for $n = 4$, $x = 2$ is a solution, a contradiction.

Thus, the solutions are $(2, -1)$, $(6, 3)$, $(6, -9)$, $(1, 1)$, $(1, -2)$.

7. We have $n^2 \equiv 1 \pmod 3$. Thus $n = 3k + 1$ or $3k + 2$ for some nonnegative integer k.

   (i)   When $n = 3k + 1$, then given equation gives $2^m = 3k^2 + 2k = k(3k + 2)$. Hence k and $3k + 2$ are both powers of 2. It is clear that $k = 2$ is a solution but $k = 1$ is not.
         If $k = 2^p$ where $p \geq 2$, then $3k + 2 = 2(3 \cdot 2^{p-1} + 1)$ is not a power of 2, hence there are no solutions in this case.
         $k = 2 \Rightarrow n = 7, m = 4$, so $n = 7, m = 4$ is a solution.

(ii) When $n = 3k + 2$, then the given equation yields $2^m = 3k^2 + 4k + 1 = (3k + 1)(k + 1)$, so both $k + 1$ and $3k + 1$ are powers of 2. $k = 1$ is an acceptable solution, so $n = 5, m = 3$ is a solution also. When $k = 0, 2^m = 1$ gives $m = 0$, so there is no solution in this case.

For $k > 1$, since $4(k + 1) > 33k + 1 > 2(k + 1)$, if $k + 1 = 2^p$ for some positive integer $p > 1$, then

$$2^{p+2} > 3k + 1 > 2^{p+1},$$

i.e., $3k + 1$ must be not a power of 2. Hence there is no solution in this case.

In summary, there are two solutions for (n, m): $(7, 4)$ and $(5, 3)$.

8. Note that if (x, t, y, s, v, r) is a solution, then

$$(xa^{kn}, ta^{km}, ya^{kn}, sa^{km}, va^{kn}, ra^{km}), \quad \text{where } a \in \mathbb{N}$$

is also a solution. Hence it suffices to find a fundamental solution. Note that $2 + 3 = 5$ implies that

$$2^{mn+1} \times 3^{mn} \times 5^{mn} + 2^{mn} \times 3^{mn+1} \times 5^{mn} = 2^{mn} \times 3^{mn} \times 5^{mn+1}.$$

Since $(m, n) = 1$, there exists an integer $i \in [1, n]$ such that $mi \equiv -1$ (mod n), so $mi + 1 = nd$, where $d \in \mathbb{N}$. Thus,

$$mn + 1 = m(n - i) + mi + 1 = m(n - i) + nd,$$

so

$$(2^n \times 3^{n-i} \times 5^n)^m \times (2^d)^n + (2^n \times 3^{n-i} \times 5^n)^m \times (3^d)^n$$
$$= (2^n \times 3^n \times 5^{n-i})^m \times (5^d)^n,$$

i.e., $(2^n \times 3^n \times 5^n, 2^d, 2^n \times 3^{n-i} \times 5^n, 3^d, 2^n \times 3^n \times 5^{n-i}, 5^d)$ is a fundamental solution. Thus, there are infinitely many solutions.

Testing Questions (22-B)

1. First of all we show that if x is a solution then x is a positive integer.

Let $4x^5 - 7 = m^2, 4x^{13} - 7 = n^2$ $(m, n \in \mathbb{N}_0)$. Then $x^5 = \dfrac{m^2 + 7}{4}, x^{13} = \dfrac{n^2 + 7}{4}$. It is clear that $x > 0$, so $x = \dfrac{x^{40}}{x^{39}} = \left(\dfrac{m^2 + 7}{4}\right)^8 / \left(\dfrac{n^2 + 7}{4}\right)^3 \in$

Q. Let $x = \dfrac{p}{q}$ with $p, q \in \mathbb{N}$, $(p, q) = 1$, then $\dfrac{4p^5}{q^5} = m^2 + 7 \in \mathbb{N}$ implies that $q = 1$ and hence $x \in \mathbb{N}$. $4x^5 = m^2 + 7 \geq 7$ implies $x \geq 2$.

(i) $x = 2$ yields $4x^5 - 7 = 11^2$ and $4x^{13} - 7 = 181^2$, so $x = 2$ is a solution.

(ii) When $x \geq 3$, then

$$(mn)^2 = (4x^5 - 7)(4x^{13} - 7) = (4x^9)^2 - 7 \cdot 4x^{13} - 7 \cdot 4x^5 + 49$$

$$< (4x^9)^2 - 7 \cdot 4x^{13} + \frac{49}{4}x^8 = \left(4x^9 - \frac{7}{2}x^4\right)^2. \qquad (30.16)$$

Below we verify

$$(mn)^2 = (4x^5 - 7)(4x^{13} - 7) > \left(4x^9 - \frac{7}{2}x^4 - 1\right)^2,$$

i.e., $8x^9 - \dfrac{49}{4}x^8 - 28x^5 - 7x^4 + 48 > 0$. In fact,

$$8x^9 - \frac{49}{4}x^8 - 28x^5 - 7x^4 + 48 > 24x^8 - 13x^8 - 28x^5 - 7x^4 + 48$$
$$\geq 99x^6 - 28x^5 - 7x^4 + 48 > 0.$$

hence $\left(4x^9 - \dfrac{7}{2}x^4 - 1\right)^2 < (mn)^2 < \left(4x^9 - \dfrac{7}{2}x^4\right)^2$, i.e.,

$$4x^9 - \frac{7}{2}x^4 - 1 < mn < 4x^9 - \frac{7}{2}x^4.$$

Hence x must be odd so that it is possible to solve $mn = 4x^9 - \dfrac{7}{2}x^4 - \dfrac{1}{2}$. By substituting it into (30.16), it is obtained that

$$\left(4x^9 - \frac{7}{2}x^4 - \frac{1}{2}\right)^2 = (4x^{13} - 7)(4x^5 - 7),$$

namely $4x^9 - \frac{49}{4}x^8 - 28x^5 - \frac{7}{2}x^4 + 49 - \frac{1}{4} = 0$. When multiplying both sides by 4 and then taking modulo 16, we have

$$-x^8 + 2x^4 + 3 \equiv 0 \pmod{16} \quad \text{or} \quad (x^4 + 1)(x^4 - 3) \equiv 0 \pmod{16}.$$

However, this contradicts $x^4 \equiv 1 \pmod 8$. Thus, there is no solution when $x \geq 3$.

In conclusion, $x = 2$ is the unique solution.

2. We prove the conclusion by contradiction. Suppose that a, b are not relatively prime to each other, let the prime p be a common factor of a and b, and let $V_p(n)$ denote the highest component of p in n.

Let (x_1, y_1) and (x_2, y_2) be two distinct solutions of the given equation with $x_1 > x_2 > 1$. Then

$$a(ba^{x_1-1} - a^{x_1-1} + 1) = b(ab^{y_1-1} - b^{y_1-1} + 1) \Rightarrow V_p(a) = V_p(b).$$

By $\left(\dfrac{a^{x_1} - 1}{a - 1} = \dfrac{b^{y_1} - 1}{b - 1}\right) - \left(\dfrac{a^{x_2} - 1}{a - 1} = \dfrac{b^{y_2} - 1}{b - 1}\right)$, we have

$$a^{x_2} \cdot \frac{a^{x_1-x_2} - 1}{a - 1} = b^{y_2} \cdot \frac{b^{y_1-y_2} - 1}{b - 1}. \tag{30.17}$$

Since $\left(n, \dfrac{n^l - 1}{n - 1}\right) = 1$ for any positive integer $n > 1, l \in \mathbb{N}$, (30.17) implies $x_2 V_p(a) = y_2 V_p(b)$, hence $x_2 = y_2$. It is then easy to see that

$$\frac{a^{x_2} - 1}{a - 1} = \frac{b^{x_2} - 1}{b - 1} \Rightarrow a = b$$

which contradicts $a > b$. Thus, a and b must be relatively prime.

3. Let $k = 2$. For any odd positive integer $2l + 1$, let $n_1 = 5^l, n_2 = 2 \cdot 5^l$, then $n_1^2 + n_2^2 = 5^{2l+1}$. For any even nonnegative integer $2l$, let $n_1 = 3 \cdot 5^l, n_2 = 4 \cdot 5^l$, then $n_1^2 + n_2^2 = 5^{2l+2}$. Thus, there must be two positive integers n_1, n_2 such that $n_1^2 + n_2^2 = 5^t$ for $t \in \mathbb{N}$.

Next, consider $k = 3$. Let $n_1 = 3a_1, n_2 = 4a_1, n_3 = 5a_2$, then $n_1^2 + n_2^2 + n_3^2 = 5^2(a_1^2 + a_2^2)$. Since for any nonnegative integer m, there exist positive integers a_1, a_2 such that $a_1^2 + a_2^2 = 5^{1+m}$, so for any nonnegative integer m, there must be three positive integers n_1, n_2, n_3 such that $n_1^2 + n_2^2 + n_3^2 = 5^{3+m}$.

For $k \geq 3$ we use mathematical induction on k. Assume that for all nonnegative integers m and all integers l with $2 \leq l \leq k$ the equation

$$a_1^2 + a_2^2 + \cdots + a_l^2 = 5^{l+m}$$

has a positive integer solution $a_1, a_2, \ldots, a_l$. Below we show that for all nonnegative integers m, the equation

$$n_1^2 + n_2^2 + \cdots + n_{k+1}^2 = 5^{k+1+m}$$

has also a positive integer solution. It is obvious that for $2 \leq k - 1 < k$

$$5^{k+1+m} = 5^{k+m} + 4 \cdot 5^{k+m}.$$

By the induction assumption, for any nonnegative integer i, there exist positive integers $a_1, a_2, \ldots, a_{k-1}$ satisfying

$$a_1^2 + a_2^2 + \cdots + a_{k-1}^2 = 5^{k-1+i}.$$

Hence there exist positive integers $a_1, a_2, \ldots, a_{k-1}$ satisfying

$$a_1^2 + a_2^2 + \cdots + a_{k-1}^2 = 5^{k+m}.$$

We have proven above that there exist positive integers b_1, b_2 satisfying $b_1^2 + b_2^2 = 5^{k+m}$. Hence for any nonnegative integer m

$$
\begin{aligned}
5^{k+1+m} &= 5^{k+m} + 4 \cdot 5^{k+m} \\
&= a_1^2 + a_2^2 + \cdots + a_{k-1}^2 + 4(b_1^2 + b_2^2) \\
&= a_1^2 + a_2^2 + \cdots + a_{k-1}^2 + (2b_1)^2 + (2b_2)^2.
\end{aligned}
$$

4. It is obvious that $a \geq 0$ and $b \geq 0$. Suppose that $3^a + 7^b = n^2$, then n is a positive even integer, and so

$$n^2 \equiv (-1)^a + (-1)^b \equiv 0 \pmod 4.$$

It follows that (i) a is odd and b is even or (ii) a is even and b is odd.

Case (i): Let $b = 2c$. Then $3^a = (n - 7^c)(n + 7^c)$. It cannot be the case that 3 divides both $n - 7^c$ and $n + 7^c$. However each of these is a power of 3. It follows that $n - 7^c = 1$, and therefore $3^a = 2 \cdot 7^c + 1$. If $c = 0$, then $a = 1$, and we obtain the solution $a = 1, b = 0$.

So suppose that $c \geq 1$. Then $3^a \equiv 1 \pmod 7$. This is impossible, since $\mathrm{ord}_7 3 = 6$, the value of a such that $3^a \equiv 1 \pmod 7$ must be a multiple of 6, so must be even, contradicting the fact that a is odd.

Case (ii): Let $a = 2c$. Then $7^b = (n - 3^c)(n + 3^c)$. Thus each of $n - 3^c$ and $n + 3^c$ is a power of 7. But 7 cannot divide both of these, so it follows that $n - 3^c = 1$, and therefore $7^b = 2 \cdot 3^c + 1$. If $c = 1$, then $b = 1$, and we obtain the solution $a = 2, b = 1$.

Now we assume that $c > 1$. Then $7^b \equiv 1 \pmod 9$. Since $\mathrm{ord}_9 7 = 3$, the integer b must be a multiple of 3. Let $b = 3d$, then $(7^3)^d = 2 \cdot 3^c + 1$. Since $7^3 \equiv 1 \pmod{19}$, so $2 \cdot 3^c \equiv 0 \pmod{19}$, a contradiction.

Thus, the solutions are $(1, 0)$ and $(2, 1)$.

5. When $x \neq 0, y \neq 0$, let $(x, y) = d$ so that $x = x_0 d, y = y_0 d$ with $(x_0, y_0) = 1$. Then

$$y_0^3 d^3 = 8x_0^6 d^6 + 2x_0^3 y_0 d^4 - y_0^2 d^2 \Rightarrow y_0^3 d = 8x_0^6 d^4 + 2x_0^3 y_0 d^2 - y_0^2$$
$$\Rightarrow d^2 \mid (y_0^3 d + y_0^2) \Rightarrow d^2 \mid y_0^2(y_0 d + 1) \Rightarrow d^2 \mid y_0^2 \Rightarrow d \mid y_0.$$

Let $y_0 = dy_1$, then

$$y_1^3 d^4 = 8x_0^6 d^4 + 2x_0^3 y_1 d^3 - y_1^2 d^2 \Rightarrow y_1^3 d^2 = 8x_0^6 d^2 + 2x_0^3 y_1 d - y_1^2.$$

Let $(d, y_1) = e, d = ed_0, y_1 = ey_2$. The last equality implies that $d \mid y_1^2$, so $ed_0 \mid e^2 y_2^2 \Rightarrow d_0 \mid e y_2^2$. Since $(d_0, y_2) = 1$, so $d_0 \mid e$. Let $e = kd_0$. Then

$$(kd_0^2)^2 (kd_0 y_2)^3 = 8(kd_0^2)^2 x_0^6 + 2kd_0^2 x_0^3 \cdot kd_0 y_2 - (kd_0 y_2)^2$$
$$\Rightarrow k^3 d_0^5 y_2^3 = 8d_0^2 x_0^6 + 2d_0 x_0^3 y_2 - y_2^2 \Rightarrow d_0 \mid y_2^2.$$

But $(d_0, y_2) = 1$, so $d_0 = 1$, hence $d = e = k, y_1 = dy_2$. Then

$$d^3 y_2^3 = 8x_0^6 + 2x_0^3 y_2 - y_2^2. \tag{30.18}$$

Thus, $y_2 \mid 8x_0^6$. Then $(x_0, y_2) = 1$ implies that $y_2 \mid 8$.

(i)   If $y_2 = 1$, then (30.18) implies that $d^3 = 8x_0^6 + 2x_0^3 - 1$. Since

$$(2x_0^2 + 1)^3 = 8x_0^6 + 12x_0^4 + 6x_0^2 + 1 > 8x_0^6 + 2x_0^3 + 1, \quad \text{and}$$
$$(2x_0^2 - 1)^3 = 8x_0^6 - 12x_0^4 + 6x_0^2 - 1 < 8x_0^6 + 2x_0^3 - 1,$$
$$\Rightarrow d = 2x_0^2 \Rightarrow 2x_0^3 = 1 \Rightarrow x_0 \notin \mathbb{Z},$$

a contradiction. A similar contradiction can be obtained for $y_2 = -1$, so there are no solutions when $y_2 = \pm 1$.

(ii) If $y_2 = 2$, (30.18) yields $8d^3 = 8x_0^6 + 4x_0^3 - 4$, i.e., $(2d)^3 = 8x_0^6 + 4x_0^3 - 4$, so $8x_0^6 + 4x_0^3 - 4$ is a cube of an even number. If $x_0 > 0$, then $(2x_0^2)^3 \leq 8x_0^6 + 4x_0^3 - 4 < (2x_0^2 + 1)^3$, so $8x_0^6 + 4x_0^3 - 4 = 8x_0^6$, i.e., $x_0 = 1$. Then

$$d = x_0 = 1 \Rightarrow x = 1 \Rightarrow y^3 = 8 + 2y - y^2 \Rightarrow y = 2.$$

Thus, $(1, 2)$ is a solution.
If $x_0 = -1$, then $8d^3 = 0 \Rightarrow d = 0$, which is impossible. If $x_0 \leq -2$, then

$$(2x_0^2 - 2)^3 < 8x_0^6 + 4x_0^3 - 4 < (2x_0^2)^3 \Rightarrow x_0 \mid -3 \Rightarrow \text{no solution.}$$

If $y_2 = -2$, then $-8d^3 = 8x_0^6 - 4x_0^3 - 4$. $d > 0$ implies $2x_0^6 - x_0^3 - 1 < 0$, then x_0 must be 0 and d has no solution.

(iii) If $y_2 = \pm 4$, (30.18) yields

$$(dy_2)^3 = 8x_0^6 \pm 8x_0^3 - 16 < 8x_0^6 + 12x_0^4 + 6x_0^2 + 1 = (2x_0^2 + 1)^3.$$

If $dy_2 = 2x_0^2$, then $8x_0^3 = \pm 16$, but this has no integer solution for x_0.

When $|x_0| \geq 2$, then $(dy_2)^3 = 8x_0^6 \pm 8x_0^3 - 16 > 8x_0^6 - 12x_0^4 + 6x_0^2 - 1 = (2x_0^2 - 1)^3$, hence there is no solution for x_0.

When $x_0 = \pm 1$, then $(dy_2)^3 = 0$ or -16, hence there is no integer solution for y_2.

(iv) When $y_2 = \pm 8$, Then $(x_0, y_2) = 1 \Rightarrow 2 \nmid x_0 \Rightarrow 16 \nmid 8x_0^6$. However

$$8x_0^6 = (dy_2)^3 - 2x_0^3 y_2 + y_2^2 \Rightarrow 16 \mid 8x_0^6,$$

a contradiction.

Finally, If $x = 0$ then $y^3 = -y^2 \Rightarrow y = 0$ or -1, so $(0, 0)$ and $(0, -1)$ are solutions. If $y = 0$, then $8x^6 = 0$ implies $x = 0$ only. Thus, the solutions for (x, y) are

$$(0, 0), \quad (0, -1), \quad (1, 2).$$

Solutions to Testing Questions   23

Testing Questions   (23-A)

1. The question is equivalent to finding all Pythagorean triples (x, y, z) of the equation $x^2 + y^2 = z^2$ with $z = x + 1$ or $z = y + 1$. Without loss of generality we only discuss the case where $z = y + 1$. Then

$$x^2 = (y + 1)^2 - y^2 = 2y + 1,$$

hence x is odd. Let $x = 2m + 1$ with $m \in \mathbb{N}_0$, then

$$y = 2m(m + 1), \qquad z = 2m^2 + 2m + 1,$$

where $m \in \mathbb{N}$. Thus, the required Pythagorean triples (x, y, z) are $(2m + 1, 2m(m + 1), 2m^2 + 2m + 1)$ and $(2m(m + 1), 2m + 1, 2m^2 + 2m + 1)$, where $m \in \mathbb{N}$.

2. If y is odd, then $x^2 + 2y^2 \equiv 2$ or $3 \pmod 4$ and $z^2 \equiv 0$ or $1 \pmod 4$, a contradiction. Therefore y is even. As y is even, x and z must both be odd.

From $(x, y) = 1$ we have $(x, z) = 1$. Thus, one of $(z + x)/2$ and $(z - x)/2$ is odd and

$$2y^2 = (z - x)(z + x).$$

When $\dfrac{z + x}{2}$ is odd, then

$$\left(\frac{z + x}{2}, z - x \right) = \left(\frac{z + x}{2}, z \right) = (z + x, z) = (x, z) = 1.$$

When $\dfrac{z - x}{2}$ is odd, then

$$\left(z + x, \frac{z - x}{2} \right) = \left(\frac{z + x}{2}, z \right) = (z + x, z) = (x, z) = 1.$$

Thus, when $y^2 = \dfrac{z + x}{2}(z - x)$ and $(z + x)/2$ is odd, we have $n, m \in \mathbb{N}$ where m is odd with $(n, m) = 1$, such that

$$z - x = 4n^2 \qquad \text{and} \qquad \frac{z + x}{2} = m^2.$$

Therefore $x = m^2 - 2n^2$ and $z = m^2 + 2n^2$.

When $y^2 = (z + x)\left(\dfrac{z - x}{2} \right)$ and $(z - x)/2$ is odd, we have $n, m \in \mathbb{N}$ where m is odd with $(n, m) = 1$, such that

$$z + x = 4n^2 \qquad \text{and} \qquad \frac{z - x}{2} = m^2.$$

Therefore $x = 2n^2 - m^2$ and $z = m^2 + 2n^2$. Thus, $x = |m^2 - 2n^2|$, $z = m^2 + 2n^2$.

3.   Suppose that x_0, y_0, z_0 is a positive integer solution of the equation $x^4 + y^4 = z^4$. Then x_0, y_0, z_0^2 is a positive integer solution of the equation $x^4 + y^4 = z^2$, which contradicts the fact that the equation $x^4 + y^4 = z^2$ has no positive integer solution (cf. Example 1).

4.   Take $t \in \mathbb{N}$ odd and $t \geq n$. Let $a = 2^{n-1}t^n$, $b = (2t - 1)^n - a$, then

$$b = (2t)^n \left[\left(1 - \frac{1}{2t} \right)^n - \frac{1}{2} \right] > (2t)^n \left[1 - \frac{n}{2t} - \frac{1}{2} \right] \geq 0,$$

i.e. b is a positive integer. It is obvious that a is even and b is odd. If p is an odd prime factor of a, then $p \mid t$, hence $b \equiv (-1)^n \pmod{p}$ which means that p does not divide b, therefore $(a, b) = 1$. It's clear that $a > b$.

For the pair (a, b), the primitive Pythagorean triple $a^2 - b^2, 2ab, a^2 + b^2$ has the sum

$$a^2 - b^2 + 2ab + a^2 + b^2 = 2a(a + b) = (2t)^n (2t - 1)^n = [2t(2t - 1)]^n$$

which is an nth power of $2t(2t - 1)$.

5. Answer (A). Let the two legs of the right triangle be a and b, and the hypotenuse be c. Then

$$a = k(p^2 - q^2), \qquad b = 2kpq, \qquad c = k(p^2 + q^2),$$

where $k, p, q \in \mathbb{N}$ with $(p, q) = 1$ and p, q are one odd and one even. Since $ab = 6(a + b + c)$,

$$2k^2 pq(p^2 - q^2) = 12k(p^2 + pq),$$
$$kq(p - q) = 6.$$

When $k = 1$, then $(p, q) = (5, 2), (7, 6)$.

When $k = 2$, then $q(p - q) = 3$, and $(p, q) = (4, 1), (4, 3)$.

When $k = 3$, then $q(p - q) = 2$, and $(p, q) = (3, 2)$.

When $k = 6$, then $q(p - q) = 1$, and $(p, q) = (2, 1)$.

Thus, $(k, p, q) = (1, 5, 2), (1, 7, 6), (2, 4, 1), (2, 4, 3), (3, 3, 2), (6, 2, 1)$.

6. When consider the value of y_1 by checking $1, 2, 3, \cdots$, we find that the fundamental solution is $x_1 = 3, y_1 = 2$. Therefore, by formula (23.3), all the positive integer solutions (x_n, y_n) are given by

$$x_n = \frac{1}{2}[(3 + 2\sqrt{2})^n + (3 - 2\sqrt{2})^n],$$

$$y_n = \frac{1}{2\sqrt{2}}[(3 + 2\sqrt{2})^n - (3 - 2\sqrt{2})^n],$$

$$n = 1, 2, \cdots.$$

For example, for $n = 1, 2, 3, 4, (x, y) = (3, 2), (17, 12), (99, 70), (577, 408)$ respectively.

7. Suppose that the equation has an integer solution. It is obvious that d has a prime factor of the form $4k + 3$, say q. Taking mod q to the given equation, we obtain

$$x^2 \equiv -1 \pmod{q}.$$

However, by Euler's Criterion (cf. Appendix C),

$$\left(\frac{-1}{q}\right) = (-1)^{(q-1)/2} = -1,$$

i.e., -1 is not a quadratic residue modulo q, we have arrived at a contradiction.

Testing Questions   (23-B)

1. Squaring the first equation and then subtracting four times the second, we obtain
$$x^2 - 6xy + y^2 = (z - u)^2,$$
from which we obtain
$$\left(\frac{x}{y}\right)^2 - 6\left(\frac{x}{y}\right) + 1 = \left(\frac{z-u}{y}\right)^2. \qquad (*)$$

The quadratic $\omega^2 - 6\omega + 1$ takes the value 0 for $\omega = 3 \pm 2\sqrt{2}$, and is positive for $\omega > 3 + 2\sqrt{2}$. Because $x/y \geq 1$ and the right side of $(*)$ is a square, the left side of $(*)$ is positive, so we must have $x/y > 3 + 2\sqrt{2}$. We now show that x/y can be made as close to $3 + 2\sqrt{2}$ as we like, so the desired m is $3 + 2\sqrt{2}$. We prove this by showing that the term $((z-u)/y)^2$ in $(*)$ can be made as small as we like.

To this end, we first find a way to generate solutions of the system. If p is a prime divisor of z and u, then p is a divisor of both x and y. Thus we may assume, without loss of generality, that z and u are relatively prime. If we square both sides of the first equation and subtract twice the second equation, it is obtained that
$$(x - y)^2 = z^2 + u^2.$$

Thus $(z, u, x - y)$ is a primitive Pythagorean triple, and we may assume that u is even. Hence there are relatively prime positive integers a and b, one of them even and the other odd, such that
$$z = a^2 - b^2, \qquad u = 2ab, \qquad \text{and} \qquad x - y = a^2 + b^2.$$

Combining these equations with $x + y = z + u$, we find that
$$x = a^2 + ab \qquad \text{and} \qquad y = ab - b^2.$$

Observe that $z - u = a^2 - b^2 - 2ab = (a-b)^2 - 2b^2$. When $z - u = 1$, we get the Pell equation $1 = (a-b)^2 - 2b^2$, which has solution $a - b = 3$, $b = 2$. By well known facts, this equation has infinitely many positive integer solutions $a - b$ and b, and both of these quantities can be made arbitrarily large. It follows that $y = ab - b^2$ can be made arbitrarily large. Hence the right side of $(*)$ can be made as small as we like, and the corresponding value of x/y can be made as close to $3 + 2\sqrt{2}$ as we like.

2. Consider a special case: let $a_i = i, i = 1, 2, \ldots, n$ andd $b_k = \sum_{i=1}^{k} a_i$. Then

$$b_k = m^2 \quad \Leftrightarrow \quad \frac{k(k+1)}{2} = m^2 \Rightarrow \frac{k}{2} < m$$

$$\Rightarrow \quad b_{k+1} = \frac{(k+1)(k+2)}{2} = m^2 + k + 1 < (m+1)^2,$$

therefore b_{k+1} is not a perfect square. If we interchange a_k and a_{k+1}, then b_{k+1} is not a perfect square and $b_k = m^2 + 1$ is not also. By continuing this operation, i.e. when a term b_k in the sequence $\{b_1, \cdots, b_n\}$ is a perfect square, interchange a_k and a_{k+1}, we can obtain a non-quadratic permutation unless $k = n$, namely $\frac{n(n+1)}{2} = m^2$. Conversely, It is clear that when n satisfies this equation, every permutation of $\{1, \cdots, n\}$ is a quadratic permutation.

Let $x = 2n + 1, y = 2m$, then $x^2 - 2y^2 = 1$, so it becomes a Pell's equation. From the result of the Q6 in TQ(A), $x_k = \frac{1}{2}[(3 + 2\sqrt{2})^k + (3 - 2\sqrt{2})^k]$, and

$$n = \frac{x_k - 1}{2} = \frac{(3 + 2\sqrt{2})^k + (3 - 2\sqrt{2})^k - 2}{4}, \quad k = 1, 2, \cdots.$$

3. Note that for each prime p, the Pell's equation $x^2 - py^2 = 1$ has infinitely many positive integer solutions (x, y). Therefore for any prime p, there exist infinitely many positive integers s, t with $s > 3$ such that

$$s^2 - 1 = pt^2.$$

For each such pair (s, t), let $x = s^2 - 1, y = s + 1, z = s - 1$, then x, y, z are distinct and $s^2 - 1 = pt^2 \neq t$.

If $y = s + 1 = t$, then

$$(s - 1)(s + 1) = pt^2 = p(s + 1)^2 \Rightarrow s - 1 = p(s + 1) > s - 1,$$

a contradiction.

If $z = s - 1 = t$, then

$$(s-1)(s+1) = pt^2 = p(s-1)^2 \Rightarrow s+1 = p(s-1) \geq 2(s-1) \Rightarrow s \leq 3,$$

a contradiction also. Thus, x, y, z, t are distinct. With these values of x, y, z, t,

$$(x^2 + s^2 - 1)(y^2 + s^2 - 1)(z^2 + s^2 - 1)$$
$$= (s^2 - 1)s^2(s + 1) \cdot 2s \cdot 2s(s - 1) = [2s^2(s^2 - 1)]^2.$$

The conclusion is proven.

4. First of all we prove a lemma as follows.

Lemma. If the equation

$$Ax^2 - By^2 = 1 \tag{30.19}$$

has positive integer solutions (where A, B are both not perfect squares). Let its minimum positive integer solution be (x_0, y_0), then the Pell's equation

$$x^2 - ABy^2 = 1 \tag{30.20}$$

has positive integer solutions. If its minimum solution is (a_0, b_0), then (a_0, b_0) satisfies the equations

$$a_0 = Ax_0^2 + By_0^2, \quad b_0 = 2x_0y_0.$$

Proof of Lemma. Since (x_0, y_0) is a solution of (30.19), so $Ax_0^2 - By_0^2 = 1$. Let $u = Ax_0^2 + By_0^2, v = 2x_0y_0$. Then

$$u^2 - ABv^2 = (Ax_0^2 + By_0^2)^2 - AB(2x_0y_0)^2 = (Ax_0^2 - By_0^2)^2 = 1,$$

so (u, v) is a solution of (30.20). Let (a_0, b_0) be the minimum solution of (30.20). If $(u, v) \neq (a_0, b_0)$, then $u > a_0, v > b_0$. Then

$$a_0 - \sqrt{AB}b_0 < (a_0 - \sqrt{AB}b_0)(a_0 + \sqrt{AB}b_0) = a_0^2 - ABb_0^2 = 1$$
$$\Rightarrow (a_0 - \sqrt{AB}b_0)(\sqrt{A}x_0 + \sqrt{B}y_0) < \sqrt{A}x_0 + \sqrt{B}y_0$$
$$\Rightarrow (a_0x_0 - Bb_0y_0)\sqrt{A} + (a_0y_0 - Ab_0x_0)\sqrt{B} < \sqrt{A}x_0 + \sqrt{B}y_0.$$

On the other hand,

$$a_0 + \sqrt{AB}b_0 < u + \sqrt{AB}v = (\sqrt{A}x_0 + \sqrt{B}y_0)^2$$
$$\Rightarrow (a_0x_0 - Bb_0y_0)\sqrt{A} - (a_0y_0 - Ab_0x_0)\sqrt{B}$$
$$= (a_0 + \sqrt{AB}b_0)(\sqrt{A}x_0 - \sqrt{B}y_0)$$
$$< (\sqrt{A}x_0 + \sqrt{B}y_0)^2(\sqrt{A}x_0 - \sqrt{B}y_0) = \sqrt{A}x_0 + \sqrt{B}y_0.$$

Let $s = a_0x_0 - Bb_0y_0, t = a_0y_0 - Ab_0x_0$. The two inequalities above can be written as

$$\sqrt{A}s + \sqrt{B}t < \sqrt{A}x_0 + \sqrt{B}y_0, \tag{30.21}$$
$$\sqrt{A}s - \sqrt{B}t < \sqrt{A}x_0 + \sqrt{B}y_0. \tag{30.22}$$

Now

$$As^2 - Bt^2 = A(a_0 x_0 - Bb_0 y_0)^2 - B(a_0 y_0 - Ab_0 x_0)^2$$
$$= (a_0^2 - ABb_0^2)(Ax_0^2 - By_0^2) = 1.$$

Note that

$$s > 0$$
$$\Leftrightarrow a_0 x_0 > Bb_0 y_0 \Leftrightarrow a_0^2 x_0^2 > B^2 b_0^2 y_0^2 \Leftrightarrow a_0^2 x_0^2 > Bb_0^2 (Ax_0^2 - 1)$$
$$\Leftrightarrow (a_0^2 - ABb_0^2)x_0^2 > -Bb_0^2 \Leftrightarrow x_0^2 > -Bb_0^2,$$

and the last inequality is obvious so $s > 0$. Since

$$t = 0 \quad \Leftrightarrow \quad a_0 y_0 = Ab_0 x_0 \Leftrightarrow a_0^2 y_0^2 = A^2 b_0^2 x_0^2$$
$$\Leftrightarrow (ABb_0^2 + 1)y_0^2 = Ab_0^2 (By_0^2 + 1) \Leftrightarrow y_0^2 = Ab_0^2,$$

so $t \neq 0$ as A is not a perfect square.

If $t > 0$, then (s, t) is a solution of (30.19), so $s \geq x_0, t \geq y_0$, which contradicts the inequality (30.21).

If $t < 0$, then $(s, -t)$ is a solution Of (30.19), so $s \geq x_0, -t \geq y_0$, then it contradicts the inequality (30.22).

In summary, we must have $u = a_0, v = b_0$. The lemma is proven.

We now return to the original problem. Suppose that

$$ax^2 - by^2 = 1 \qquad\qquad (30.23)$$
$$by^2 - ax^2 = 1 \qquad\qquad (30.24)$$

both have positive integer solutions.

Let (m, n) be the minimum solution of the equation $x^2 - aby^2 = 1$, (x_1, y_1) be the minimum solution of the equation (30.23), and (x_2, y_2) be the minimum solution of (30.24). By the lemma it is obtained that

$$\begin{cases} m = ax_1^2 + by_1^2, \\ n = 2x_1 y_1, \end{cases} \quad \text{and} \quad \begin{cases} m = bx_2^2 + ay_2^2, \\ n = 2x_2 y_2. \end{cases}$$

Since $ax_1^2 = by_1^2 + 1, ay_2^2 = bx_2^2 - 1$, so

$$ax_1^2 + by_1^2 = bx_2^2 + ay_2^2 \Leftrightarrow 2by_1^2 + 1 = 2bx_2^2 - 1 \Leftrightarrow b(x_2^2 - y_1^2) = 1.$$

This is a contradiction since $b > 1$.

5. The given equation is equivalent to $y^2 = (x - p)(x + p)x$. Discuss the following two cases separately.

(i) $p \nmid y$. Then $(x, p) = (x - p, p) = (x + p, p) = 1$, therefore $(x - p, x) = (x + p, x) = 1$.

If x is even, then $(x - p, x + p) = (x, x + p) = 1$. Hence $x, x - p, x + p$ are all perfect squares. However $x + p \equiv 3$ or $7 \pmod 8$, a contradiction.

Therefore x is odd, and hence $(x - p, x + p) = 2$.

Let $x = r^2, x - p = 2s^2, x + p = 2t^2$. Then $t^2 - s^2 = p$, i.e. $t = \dfrac{p+1}{2}, s = \dfrac{p-1}{2}$. However

$$r^2 = x = (x - p) + p = \frac{(p-1)^2}{2} + p = \frac{p^2 + 1}{2} = \frac{(8k+3)^2 + 1}{2}$$
$$\equiv 5 \pmod 8,$$

a contradiction. Thus, the equation has no integer solution in the case $p \nmid y$.

(ii) $p \mid y$. Note that $x = 0$ or p or $-p$ if $y = 0$, so $(0, 0), (p, 0), (-p, 0)$ are three solutions of the given equation.

If $y > 0$, then $(x - p, x, x + p) = p$, so $p^2 \mid y$. Canceling p^3 from both sides of the given equation, we obtain

$$pb^2 = (a - 1)a(a + 1),$$

where $a = \dfrac{x}{p}, b = \dfrac{y}{p^2}$. Since $y > 0$, thus $a, a - 1, a + 1$ are all positive, and p divides one of them.

When $p \mid a - 1$, (i) if $a - 1$ is odd, then $a, a + 1$ must be both perfect squares, and forcing a to be 0, a contradiction. (ii) If $a - 1$ is even, let $a - 1 = 2pm, a = 2pm + 1, a + 2 = 2(pm + 1)$. Then $m, 2pm + 1, pm + 1$ are relatively prime pairwise, so they are all perfect squares, which we denoted by r^2, s^2, t^2 respectively. We have $a - 1 = 2pr^2, a = s^2, a + 1 = 2t^2$, so $s^2 = t^2 + pr^2$. Since $(s, t) = 1$, p divides only one of $s + t$ and $s - t$.

First we prove that when $p \mid (s + t)$, the equation

$$pb^2 = a(a - 1)(a + 1) \tag{30.25}$$

has no solution. In fact,

$$a + 1 = 2t^2 \Rightarrow a - 1 = s^2 - 1 \text{ is divisible by } 4 \Rightarrow 2 \mid r \Rightarrow 4 \mid (s^2 - t^2).$$

Now $(s, t) = 1$ gives $\left(\dfrac{s+t}{2p}, \dfrac{s-t}{2}\right) = 1$. As $\left(\dfrac{r}{2}\right)^2 = \left(\dfrac{s+t}{2p}\right)\left(\dfrac{s-t}{2}\right)$

we can let $\dfrac{s+t}{2p} = m^2, \dfrac{s-t}{2} = n^2$ with $(m, n) = 1$. Thus,

$$t = pm^2 - n^2, \tag{30.26}$$
$$s = pm^2 + n^2, \tag{30.27}$$

$$r = 2mn. \tag{30.28}$$

Since $(t, r) = 1$, so m, n have different parities, hence (30.26) yields $t \equiv 3$ or $7 \pmod 8$. From $a + 1 = s^2 + 1 = 2t^2$ we have the Pell's type equation

$$2t^2 - s^2 = 1. \tag{30.29}$$

$t_1 = 1, s_1 = 1$ and $t_2 = 5, s_2 = 7$ are two specific solutions, and we have the recursive formula for the general solution further

$$
\begin{align}
t_{n+1} &= 3t_n + 2s_n, \tag{30.30} \\
s_{n+1} &= 4t_n + 3s_n. \tag{30.31}
\end{align}
$$

Thus,
$$t_{n+2} = 6t_{n+1} - t_n. \tag{30.32}$$

By induction, it is easy to find that for the sequence $\{t_n\}$ and $k \in \mathbb{N}$,

$$t_{4k+1} \equiv 1, t_{4k+2} \equiv 5, t_{4k+3} \equiv 5, t_{4k+4} \equiv 1 \quad (\text{mod } 8).$$

Therefore $t_i \equiv 1$ or $5 \pmod 8$, which contradicts $t \equiv 3$ or $7 \pmod 8$. Thus, we have proven that the equation (30.25) has no solution.

Next, we prove that when $p \mid (s - t)$, the equation (30.25) also has no solution.

In this case we let $\dfrac{t + s}{2} = m^2, \dfrac{s - t}{2p} = q^2$, then $(m, q) = 1$. From (30.30) and (3.31) we obtain

$$
\begin{cases}
t_{n+2} &= 6t_{n+2} - t_n, \\
s_{n+2} &= 6s_{n+1} - s_n.
\end{cases}
$$

Let $d_n = \dfrac{s_n + t_n}{2}$. Then $d_{n+2} = 6d_{n+1} - d_n$ with $d_1 = 1, d_2 = 6$. Solving the recurrence relation yields

$$d_n = \frac{1}{4\sqrt{2}}[(3 + 2\sqrt{2})^n - (3 - 2\sqrt{2})^n].$$

Below we prove that starting from the second term, all the terms of the sequence $\{d_n\}$ are not perfect squares. For this we construct the sequence of integers $\{c_n\}$ defined by

$$c_n = \frac{1}{2}[(3 + 2\sqrt{2})^n + (3 - 2\sqrt{2})^n].$$

Then $c_n^2 - 8d_n^2 = 1$. To prove our claim, it remains to show that the equation $x^2 - 8y^4 = 1$ has no other positive integer solutions besides $x = 3, y = 1$.

We need to prove the following lemma first:

Lemma. The equation $x^4 - 2y^2 = -1$ has no other positive integer solution besides $(x, y) = (1, 1)$.

Proof of Lemma. The given equation is equivalent to $y^4 = x^4 + (y^2 - 1)^2$. From the result of Example 2, the only positive integer solution is $(1, 1)$.

Now we return to the original problem.

Note that $x^2 - 8y^4 = 1 \Leftrightarrow x^2 - 1 = 8y^4 \Leftrightarrow (x + 1)(x - 1) = 8y^4$. Since $(x + 1, x - 1) = 2$, let $y = 2^r uv$, where $(u, v) = 1$ and $2 \nmid uv$. The equality $\dfrac{x-1}{2} \cdot \dfrac{x+1}{2} = 2(2^r uv)^4$ yields

$$\begin{cases} \dfrac{x+1}{2} = 2^{4r+1}v^4, \\ \dfrac{x-1}{2} = u^4, \end{cases} \quad \text{or} \quad \begin{cases} \dfrac{x+1}{2} = u^4, \\ \dfrac{x-1}{2} = 2^{4r+1}v^4. \end{cases}$$

Thus,

$$1 = 2^{4r+1}v^4 - u^4 \tag{30.33}$$

or

$$1 = u^4 - 2^{4r+1}v^4. \tag{30.34}$$

(30.33) is also $u^4 - 2^{4r+1}v^4 = -1$. From above Lemma, its only solution is $u = v = 1, r = 0$, so $x = 3, y = 1$.

(30.34) is also $(u^2)^2 - 2(2^r v)^4 = 1$. Let $u^2 = \alpha, 2^r v = \beta$, (30.34) then becomes

$$\alpha^2 - 2\beta^4 = 1. \tag{30.35}$$

Below we show that (30.35) has no integer solution. Otherwise, α must be odd. Let $\alpha = 2l + 1$, then $2l(l + 1) = \beta^4$. Let $\beta = \lambda_1 \lambda_2$, where $(\lambda_1, \lambda_2) = 1$.

If l is odd, then $(2(l+1), l) = 1$, so $l = \lambda_1^4, 2(l+1) = \lambda_2^4$. Let $\lambda_2 = 2^t \lambda_3$, then $\lambda_1^4 + 1 = 2^{4t-1}\lambda_3^4$, therefore

$$\lambda_1^4 + (1 - 2^{4t-2}\lambda_3^4)^2 = (2^{2t-1}\lambda_3^2)^4.$$

By the result of Example 2, it has no positive integer solution.

If l is even, then $(2l, l + 1) = 1$, and $2l = \lambda_1^4, l + 1 = \lambda_2^4$. Let $\lambda_1 = 2^w \lambda_4$, then $\lambda_2^4 = 1 + 2^{4w-1}\lambda_4^4$, so

$$\lambda_2^4 + (2^{2w-1}\lambda_4^2)^4 = (1 + 2^{4w-2}\lambda_4^4)^2.$$

By the result of Example 1, above equation has no positive integer solution.

Thus, we have proven that starting from the second term, the terms of $\{d_n\}$ are all not perfect squares. Since $d_1 = 1, s_1 = t_1 = 1$, so $a = 1, b = 0$, i.e., there is no non-zero integer solution when $p \mid (s - t)$.

(2) When $p \mid (a + 1)$, (i) if $a + 1$ is odd, similar to (1) (i) a contradiction will be obtained.

(ii) If $a + 1$ is even, let $a + 1 = 2px^2, a = y^2, a - 1 = 2z^2$. a is odd implies that $y = 2l + 1$, so $a = (2l + 1)^2 = 4l(l + 1) + 1$. Therefore $a + 1 = 2[2l(l + 1) + 1]$. Hence px^2 is odd, so x is odd. Since $z^2 + 1 \equiv 1$ or $5 \pmod 8$, but $px^2 \equiv 3 \pmod 8$, which contradicts the equality $z^2 + 1 = px^2$.

(3) When $p \mid a$, if $a \neq 0$ is even, then $a - 1, \dfrac{a}{p}, a + 1$ are relatively prime pairwise, so they are all perfect squares, but $a - 1$ and $a + 1$ cannot be squares at the same time, so a is odd, hence $(a - 1, a + 1) = 2$. Let $a - 1 = 2r^2, a + 1 = 2s^2$. Then $s^2 - r^2 = 1$, namely $s = 1, r = 0$. Thus, $a = 1$ and it contradicts $p \mid a$. Hence there is no solution in this case.

In summary, this problem has exactly three integer solutions

$$(0, 0), \quad (p, 0), \quad (-p, 0).$$

Solutions to Testing Questions   24

Testing Questions   (24-A)

1. (i) For modulo 7,

j	1	2	3
$r \equiv j^2 \pmod 7$	1	-3	2

Thus, $1, 2, -3$, are the quadratic residues modulo 7, and $-1, -2, 3$ are the quadratic non-residues modulo 7.

(ii) For modulo 11,

j	1	2	3	4	5
$r \equiv j^2 \pmod{11}$	1	4	-2	5	3

Thus, $1, -2, 3, 4, 5$ are the quadratic residues modulo 11, and $-1, 2, -3, -4, -5$ are the quadratic non-residues modulo 11.

(iii) For modulo 17,

j	1	2	3	4	5	6	7	8
$r \equiv j^2$ (mod 17)	1	4	-8	-1	8	2	-2	-4

Thus, $\pm 1, \pm 2, \pm 4, \pm 8$ are the quadratic residues modulo 17, and $\pm 3, \pm 5,$ $\pm 6, \pm 7$ are the quadratic non-residues modulo 17.

(iv) For modulo 19,

j	1	2	3	4	5	6	7	8	9
$r \equiv j^2$ (mod 19)	1	4	9	-3	6	-2	-8	7	5

Thus, $1, -2, -3, 4, 5, 6, 7, -8, 9$ are the quadratic residues modulo 19, and $-1, 2, 3, -4, -5, -6, -7, 8, -9$ are the quadratic non-residues modulo 19.

2. Suppose that there are finitely many primes of form $4k + 1$. Denote them by $p_1, p_2, \ldots, p_m$.

Consider the number $p = (2p_1 p_2 \cdots p_m)^2 + 1$. Note that p cannot have any p_i as a prime factor and $p \equiv 1 \pmod{4}$, so p is not a prime. Let p_0 be a prime factor of p, then p_0 is odd and -1 is a quadratic residue modulo p_0, so $p_0 \equiv 1 \pmod{4}$. However, p_0 is different from each of $p_1, p_2, \ldots, p_m$, a contradiction.

Thus, there must be infinitely many primes of form $4k + 1$.

3. When -1 is a quadratic residue modulo p, then, by Euler's Criterion, $(-1)^{\frac{p-1}{2}} = 1$, so $\frac{p-1}{2} = 2k$ for some $k \in \mathbb{N}$, hence $p = 4k + 1$.

Conversely, if $p = 4k + 1$ where $k \in \mathbb{N}$, then $(-1)^{\frac{p-1}{2}} = (-1)^{2k} = 1$, so by Euler's Criterion again, -1 is a quadratic residue modulo p.

4. (i) $(-2)^{33} \equiv -(2^6)^5 \cdot 8 \equiv -(-3)^5 \cdot 8 \equiv 1944 \equiv 1 \pmod{67}$, therefore it has two solutions;

(ii) $2^{33} \equiv -(-2)^{33} \equiv -1 \pmod{67}$, no solution;

(iii) $(-2)^{18} \equiv (2^5)^3 \cdot 2^3 \equiv (-5)^3 \cdot 8 \equiv -1000 \equiv -1 \pmod{37}$, no solution;

(iv) Since $2^{18} = (-2)^{18} \equiv -1 \pmod{37}$, no solution;

(v) Since $221 = 13 \times 17$, and $(-1)^6 \equiv 1 \pmod{13}$, $(-1)^8 \equiv 1 \pmod{17}$, so there are $2 \times 2 = 4$ solutions;

(vi) Since $427 = 7 \times 61$, and $(-1)^3 \equiv -1 \pmod{7}$, $(-1)^{30} \equiv 1 \pmod{61}$, no solution.

5. As shown in the Theorem I, $1^2, 2^2, \cdots, \left(\dfrac{p-1}{2}\right)^2$ are the non-zero quadratic residues modulo p, and -1 is also a quadratic residue modulo p.

When $k > \dfrac{p}{2}$, then $p - k < \dfrac{p}{2}$, i.e. $p - k \le \dfrac{p-1}{2}$, and $\left\{\dfrac{k^2}{p}\right\} =$

$\left\{\dfrac{(p-k)^2}{p}\right\}$, so

$$\sum_{k=1}^{p-1}\left\{\dfrac{k^2}{p}\right\} = 2\sum_{k=1}^{\frac{p-1}{2}}\left\{\dfrac{k^2}{p}\right\}.$$

Since $\left(\dfrac{b}{p}\right) = \left(\dfrac{-1}{p}\right)\left(\dfrac{-b}{p}\right)$, so b is a quadratic residue modulo p if and only if $-b$ (and therefore $p - b$) is a quadratic residue. Therefore with respect to modulo p, the set $S = \left\{1^2, 2^2, \ldots, \left(\dfrac{p-1}{2}\right)^2\right\}$ is a set of the

form $\{a_1, p-a_1, a_2, p-a_2, \ldots, a_{\frac{p-1}{4}}, p-a_{\frac{p-1}{4}}\}$, where $1 \le a_i \le \dfrac{p-1}{2}$.

For any two integers a_i and $p - a_i$ with $1 \le a_i \le \dfrac{p-1}{2}$, since $\dfrac{a_i}{p} +$

$\dfrac{p-a_i}{p} = 1$, so $\left\{\dfrac{a_i}{p}\right\} + \left\{\dfrac{p-a_i}{p}\right\} = 1$. Therefore

$$2\sum_{k=1}^{\frac{p-1}{2}}\left\{\dfrac{k^2}{p}\right\} = 2\cdot\dfrac{p-1}{4} = \dfrac{p-1}{2}.$$

6. Note that if $x^2 \equiv a \pmod{p}$ has a solution x_0 with $0 < x_0 < p$, where $(a, p) = 1$, then $-x_0$ is also a solution, and there is no other solution x' with $0 < |x'| < p$. Otherwise, let x' be another solution. We can assume $0 < x' < p$ (since $-x'$ is also a solution). Note that $x' - x_0 \not\equiv 0 \pmod{p}$ and $x' + x_0 \not\equiv 0 \pmod{p}$,

$$x_0^2 \equiv (x')^2 \equiv a \pmod{p} \Rightarrow (x_0 - x')(x_0 + x') \equiv 0 \pmod{p}$$

which implies $p \mid (x_0 - x')$ or $p \mid (x_0 + x')$, a contradiction.

Let $f(x) = x^3 - x$. Then $f(x)$ is an odd function.

For $p = 2$, there are two desired solutions for (x, y): $(0, 0), (1, 0)$.

For prime p with $p \equiv 3 \pmod 4$, since $\left(\dfrac{-a}{p}\right) = \left(\dfrac{-1}{p}\right)\left(\dfrac{a}{p}\right) = -\left(\dfrac{a}{p}\right)$,

so between $y^2 \equiv f(x)$ and $y^2 \equiv -f(x)$ exactly one has a solution and the other has no solution. If among the equations $y^2 \equiv f(x)$ on $x = 2, 3, \ldots, \dfrac{p-1}{2}$, k equations have solutions, then on $x = \dfrac{p+1}{2}, \dfrac{p+3}{2}, \ldots,$

$p-2$, $\dfrac{p-3}{2} - k$ equations of $y^2 \equiv f(x)$ have solutions. In conclusion, on $x = 2, 3, \ldots, p-2$, among the $p-3$ equations $y^2 \equiv f(x)$, $\dfrac{p-3}{2}$ of them have solutions. Since an equation if has one solutions, then it has exactly two solutions, so there are a total of $2 \cdot \frac{p-3}{2} = p-3$ solutions. Since the equation has one solution on $x = 0, 1$ and $p-1$, so it has a total of p solutions.

For prime p with $p \equiv 1 \pmod 4$, since $\left(\dfrac{-a}{p}\right) = \left(\dfrac{-1}{p}\right)\left(\dfrac{a}{p}\right) = \left(\dfrac{a}{p}\right)$, so $y^2 \equiv f(x)$ and $y^2 \equiv f(-x)$ both have two solutions or both have no solutions. If k of the equations $y^2 \equiv f(x)$ have two solutions each on $x = 2, 3, \ldots, \frac{p-1}{2}$, then k of the equations on $x = \dfrac{p+1}{2}, \dfrac{p+3}{2}, \cdots, p-2$ have two solutions each. Hence, on $x = 2, 3, \ldots, p-2$ the equation $y^2 \equiv f(x) \pmod p$ has a total of $4k$ solutions. Since the equation has one solution on each of $x = 0, 1, p-1$, so $y^2 \equiv f(x) \pmod p$ has a total of $4k + 3$ solutions, i.e., any $p = 4k + 1$ does not satisfy the requirement.

Thus, $p = 2$ or all primes p with $p \equiv 3 \pmod 4$ satisfy the requirement.

7. We first prove the proposition by induction on k: For each positive integer k, there exists a positive integer a_k such that $a_k^2 \equiv -7 \pmod{2^k}$.

By observation, $4 \cdot 2 - 7 = 1^2, 2 \cdot 2^2 - 7 = 1^2$ and $1 \cdot 2^3 - 7 = 1^2$ means that a_1, a_2, a_3 can be 1.

Assume that for some $k \geq 3$ there exists a positive integer a_k such that $a_k^2 \equiv -7 \pmod{2^k}$. Below we consider the remainder of a_k^2 modulo 2^{k+1}.

(i) If $a_k^2 = (2n)(2^k) - 7 \equiv -7 \pmod{2^{k+1}}$, then let $a_{k+1} = a_k$.

(ii) If $a_k^2 = (2n+1)(2^k) - 7 \equiv 2^k - 7 \pmod{2^{k+1}}$, then let $a_{k+1} = a_k + 2^{k-1}$. Since a_k is odd, so

$$a_{k+1}^2 = a_k^2 + 2^{2k-2} + 2^k a_k \equiv a_k^2 + 2^k \equiv -7 \pmod{2^{k+1}}.$$

The induction proof is completed.

As a_k has no upper bound, so it is possible to let $a_k^2 \geq 2^k - 7$, therefore $\{a_k\}$ contains infinitely many distinct numbers. Thus, the conclusion is proven.

Testing Questions   (24-B)

1. First we need to prove two lemmas:

Lemma 1. If p is a prime of form $4k + 3$, then $x^2 + y^2 \equiv 0 \pmod{p}$ $\Leftrightarrow$ $x \equiv y \equiv 0 \pmod{p}$.

Proof of Lemma 1 Euler's Criterion gives $\left(\dfrac{-1}{p}\right) = (-1)^{\frac{p-1}{2}} = -1$,

therefore $\left(\dfrac{-d}{p}\right) = \left(\dfrac{-1}{p}\right)\left(\dfrac{d}{p}\right) = -\left(\dfrac{d}{p}\right)$ for $1 \le d \le p - 1$. Hence between d and $-d$ only one is quadratic residue modulo p. If $xy \not\equiv 0$ $\pmod{p}$, then $x^2 + y^2 \not\equiv 0 \pmod{p}$, therefore $x^2 + y^2 \equiv 0 \pmod{p}$ $\Rightarrow$ $x \equiv y \equiv 0 \pmod{p}$. The inverse is obvious: $x \equiv y \equiv 0 \pmod{p}$ $\Rightarrow$ $x^2 + y^2 \equiv 0 \pmod{p}$.

Lemma 2. If p is a prime of form $4k + 3$, there exist $x_0, y_0 \in \mathbb{N}$ such that $x_0^2 + y_0^2 + 1 \equiv 0 \pmod{p}$.

Proof of Lemma 2 Since $\left(\dfrac{-d}{p}\right) = -\left(\dfrac{d}{p}\right)$ for $1 \le d \le p - 1$, there are $\dfrac{p+1}{2}$ quadratic residue modulo p (including 0). The sets $A = \{0, 1^2, 2^2, \ldots, (p-1)^2\}$ and $B = \{0^2 - 1, -1^2 - 1, \ldots, -(p-1)^2 - 1\}$ each contains $\frac{p+1}{2}$ distinct values modulo p, so by the pigeonhole principle, there must be at least one common value of A and B, namely there exist $x_0^2 \in A$, $-y_0^2 - 1 \in B$ such that

$$x_0^2 \equiv -y_0^2 - 1 \pmod{p} \quad \text{or equivalently,} \quad x_0^2 + y_0^2 + 1 \equiv 0 \pmod{p}.$$

Now we return to the original problem.

By Lemma 2, take $x_0, y_0 \in \mathbb{N}$ such that $x_0^2 + y_0^2 \equiv -1 \pmod{p}$. By Lemma 1, $x_0 y_0 \not\equiv 0 \pmod{p}$. Therefore

$$[(ix_0)^2 + (iy_0)^2]^2 \equiv i^4 \pmod{p}, \quad i = 0, 1, 2, \ldots, p - 1.$$

Let $q_i \equiv i^2 \pmod{p}$, then there are distinct $\dfrac{p+1}{2}$ such q_i. Lemma 1 indicates that there are no $i, j \in \{1, 2, \ldots, p - 1\}$ such that $q_i + q_j \equiv 0$ $\pmod{p}$, so $\{q_0^2, q_1^2, \ldots, q_{p-1}^2\}$ takes $\dfrac{p+1}{2}$ values modulo p.

In summary, the number of different residues mod p is $\dfrac{p+1}{2}$.

2. Based on Theorem I in this lecture for any odd prime p there are $1 + \frac{p-1}{2} = \frac{p+1}{2}$ distinct quadratic residues modulo p.

   If there is integer x such that $x^2 + ax + b \equiv 0 \pmod{167}$, then there is a pair (x, y) of integers which satisfies the given equation. Since

   $$x^2 + ax + b \equiv 0 \pmod{167} \Leftrightarrow 4x^2 + 4ax + 4b \equiv 0 \pmod{167}$$
   $$\Leftrightarrow (2x + a)^2 \equiv a^2 - 4b \pmod{167},$$

so for fixed a, b can take on any value such that $a^2 - 4b$ is a quadratic residue modulo 167. Since the number of allowed distinct values for $a^2 - 4b$ modulo 167 is $\dfrac{167 + 1}{2} = 84$, so each a corresponds to 84 choices for b modulo 167. Since $2004/167 = 12$, so to a given a there are 84×12 choices for b. Since there are 2004 choices for a, the number of the pairs (a, b) is

$$2004 \times 84 \times 12 = 2020032.$$

3. Since $(q^{\frac{p-1}{2}} + 1, q^{\frac{p-1}{2}} - 1) = 2$ for any odd number q, if there exist an integer x and a prime number p satisfying $px^2 = q^{p-1} - 1$, then there exist integers y, z which satisfy one of the following systems.

(i) $q^{\frac{p-1}{2}} - 1 = 2py^2$ and $q^{\frac{p-1}{2}} + 1 = 2z^2$;

(ii) $q^{\frac{p-1}{2}} - 1 = 2y^2$ and $q^{\frac{p-1}{2}} + 1 = 2pz^2$.

(a) Let $q = 7$. Since $7^{\frac{p-1}{2}} - 1 = 2y^2 \Rightarrow 2y^2 \equiv 6 \pmod 7 \Rightarrow y^2 \equiv 3 \pmod 7$, but $\left(\dfrac{3}{7}\right) \equiv 3^{\frac{7-1}{2}} \equiv -1 \pmod 7$, we arrive at a contradiction. Hence (ii) is not satisfied.

For system (i), $6 \mid (7^{\frac{p-1}{2}} - 1) = 2py^2 \Rightarrow 3 \mid py^2$. If $p = 3$, then $\dfrac{7^2 - 1}{3} = 4^2$, so $p = 3$ satisfies the requirement. If $p \neq 3$, then $3 \mid y^2 \Rightarrow 9 \mid (7^{\frac{p-1}{2}} - 1) \Rightarrow 3 \mid \left|\dfrac{p-1}{2}\right|$. Let $k = (p-1)/6$, then $2z^2 = 7^{3k} + 1 = (7^k + 1)(7^{2k} - 7^k + 1)$. Since $\gcd\{(7^k + 1)/2, 7^{2k} - 7^k + 1\} = \gcd\{7^k + 1, 7^{2k} - 7^k + 1\} = 1$, so $7^{2k} - 7^k + 1$ is a perfect square. However

$$(7^k - 1)^2 < 7^{2k} - 7^k + 1 < (7^k)^2$$

indicates that it is impossible. Thus, $p = 3$ is the unique solution for p.

(b) Let $q = 11$. Since $11^{\frac{p-1}{2}} + 1 = 2z^2 \Rightarrow 2z^2 \equiv 12 \pmod{11} \Rightarrow z^2 \equiv 6 \pmod{11}$, but $\left(\dfrac{6}{11}\right) \equiv 6^{\frac{11-1}{2}} \equiv -1 \pmod{11}$, we arrive at a contradiction. Hence (i) is not satisfied, we only have to consider the system (ii).

$$11^{\frac{p-1}{2}} + 1 = 2pz^2 \Rightarrow 11^{\frac{p-1}{2}} \equiv -1 \pmod p$$
$$\Rightarrow 2y^2 = 11^{\frac{p-1}{2}} - 1 \equiv -2 \pmod p$$
$$\Rightarrow y^2 \equiv -1 \pmod p \Rightarrow (-1)^{\frac{p-1}{2}} = 1 \Rightarrow p \equiv 1 \pmod 4.$$

Now $11^{\frac{p-1}{2}} - 1 = 2y^2 \Rightarrow (11^{\frac{p-1}{4}} - 1)(11^{\frac{p-1}{4}} + 1) = 2y^2$, and it implies $11^{\frac{p-1}{4}} + 1 = u^2$ or $11^{\frac{p-1}{4}} + 1 = 2u^2$ for some u. However

$$11^{\frac{p-1}{4}} + 1 = u^2 \Rightarrow ((u-1),(u+1)) \geq 11,$$
$$11^{\frac{p-1}{4}} + 1 = 2u^2 \Rightarrow 2u^2 \equiv 1 \pmod{11} \Rightarrow u^2 \equiv 6 \pmod{11},$$

and both are impossible. Thus, there is no prime p such that $\dfrac{11^{p-1} - 1}{p}$ is a perfect square.

4.   "First we prove the following lemma:

Lemma   For any odd natural number n, the number $3^n + 1$ has no odd prime factor of form $3k + 2$.

Proof of Lemma   Let p be an odd prime factor of $3^n + 1$. Then

$$3^n + 1 \equiv 0 \pmod{p} \Rightarrow \left(3^{\frac{n+1}{2}}\right)^2 \equiv -3 \pmod{p} \Rightarrow \left(\frac{-3}{p}\right) = 1.$$

By Euler's Criterion, $\left(\dfrac{-1}{p}\right) = (-1)^{\frac{p-1}{2}}$. By the quadratic reciprocal law,

$$\left(\frac{3}{p}\right)\left(\frac{p}{3}\right) = (-1)^{\frac{3-1}{2} \cdot \frac{p-1}{2}} = (-1)^{\frac{p-1}{2}},$$

so

$$\left(\frac{p}{3}\right) = \left(\frac{p}{3}\right)\left(\frac{-3}{p}\right) = \left(\frac{p}{3}\right)\left(\frac{3}{p}\right)\left(\frac{-1}{p}\right) = \left[(-1)^{\frac{p-1}{2}}\right]^2 = 1.$$

Since $\left(\dfrac{1}{3}\right) = 1$ and $\left(\dfrac{2}{3}\right) = \left(\dfrac{-1}{3}\right) = -\left(\dfrac{1}{3}\right) = -1$, so p is not of the form $3k + 2$.

We now return to the original problem.

$$6m \mid [(2m+3)^n + 1] \Leftrightarrow 6m \mid [(2m)^n + 3^n + 1]$$
$$\Leftrightarrow \begin{cases} (2m)^n \equiv 2 \pmod 3 & (*) \\ 3^n + 1 \equiv 0 \pmod{2m}. & (**) \end{cases}$$

(The sufficiency of the last condition is obvious: the second congruence only holds when $(m, 3) = 1$. In this case, we can apply the Chinese Remainder Theorem.)

(1) If $6m \mid [(2m+3)^n + 1]$, then (∗) and (∗∗) hold at the same time. (∗) implies that $m \equiv 1 \pmod 3$ and n is odd. If m is even, since n is odd, then $3^n + 1 \equiv 0 \pmod 4$ and $3^n + 1 \equiv 4 \pmod 8$. Therefore $2^2 \| (3^n + 1)$. (∗∗) implies that $2^2 \| 2m$, so $2\|m$. Since $m \equiv 1 \pmod 3$, so $m = 6k_0 + 4$ for some $k_0 \in \mathbb{N}$, and $12k_0 + 8 = 2m \mid (3^n + 1)$.

The Lemma indicates that all odd prime factors of $3^n + 1$ are of the form $3k + 1$, hence all odd factors of $3^n + 1$ are of form $3k + 1$ also. However, $3k_0 + 2$ is an odd factor of $3^n + 1$, a contradiction. Thus, m is not even.

Since $2m \mid (3^n + 1)$ and $2^2 \mid (3^n + 1)$, $4m \mid (3^n + 1)$ is obtained at once.

(2) If $4m \mid (3^n + 1)$, then $4 \mid (3^n + 1) \Rightarrow n$ is odd. If m is even, then $8 \mid (3^n + 1)$. However $3^n + 1 \equiv 4 \pmod 8$, so m must be odd, thus m is of form $3k + 1$. From $2m \mid (3^n + 1)$ and $(2m)^n \equiv 2^n \equiv 2 \pmod 3$,

$$(2m + 3)^n + 1 \equiv (2m)^n + 3^n + 1 \equiv 0 \pmod{2m} \text{ and}$$
$$(2m + 3)^n + 1 \equiv (2m)^n + 3^n + 1 \equiv 0 \pmod 3,$$

therefore $6m \mid [(2m + 3)^n + 1]$.

5. The conclusion is clear for $n = 0$. For $n \geq 1$, the left hand side can be rewritten as follows:

$$\sum_{k=1}^{p-1} \left\{ \frac{k^{2^n}}{p} + \frac{1}{2} \right\} = 2 \sum_{k=1}^{\frac{p-1}{2}} \left\{ \frac{k^{2^n}}{p} + \frac{1}{2} \right\}$$

$$= 2 \left(\sum_{k=1}^{p-1} \left(\frac{k^{2^n}}{p} + \frac{1}{2} \right) - \sum_{k=1}^{p-1} \left\lfloor \frac{k^{2^n}}{p} + \frac{1}{2} \right\rfloor \right)$$

$$= \frac{p-1}{2} + 2 \left(\frac{1}{p} \sum_{k=1}^{\frac{p-1}{2}} k^{2^n} - \sum_{k=1}^{\frac{p-1}{2}} \left\lfloor \frac{k^{2^n}}{p} + \frac{1}{2} \right\rfloor \right).$$

Below we show that $\displaystyle\sum_{k=1}^{\frac{p-1}{2}} \left\lfloor \frac{k^{2^n}}{p} + \frac{1}{2} \right\rfloor = \frac{1}{p} \sum_{k=1}^{\frac{p-1}{2}} k^{2^n}$.

It is easy to see that $\displaystyle \left\lfloor \frac{k^{2^n}}{p} + \frac{1}{2} \right\rfloor = \left\lfloor \frac{2k^{2^n}}{p} \right\rfloor - \left\lfloor \frac{k^{2^n}}{p} \right\rfloor, 1 \leq k \leq \frac{p-1}{2}$.

For each $k \in \{1, 2, \ldots, (p-1)/2\}$, write $2k^{2^n} = p \left\lfloor \dfrac{2k^{2^n}}{p} \right\rfloor + r_k, r_k \in$

$\{1, 2, \ldots, p-1\}$. Since $p \equiv 7 \pmod 8$, 2 is a quadratic residue modulo p, so each r_k is also a quadratic residue modulo p.

Below we show that $r_k \neq r_l$ if $k \neq l$. Suppose that $r_k = r_l$, then

$$0 = r_k - r_l \equiv 2(k^{2^n} - l^{2^n}) \equiv 2(k - l) \prod_{j=0}^{n-1} (k^{2^j} + l^{2^j}) \pmod p.$$

Since $1 \leq k, l \leq \frac{p-1}{2}$ implies that $k + l \not\equiv 0 \pmod p$, and if $j \geq 1$ then $k^{2^j} + l^{2^j} \not\equiv 0 \pmod p$ since -1 is a quadratic non-residue modulo $p \equiv 7 \pmod 8$. Hence $k - l \equiv 0 \pmod p$. However, $1 \leq k, l \leq \frac{p-1}{2}$, so $k = l$, a contradiction.

Thus, we have proven that $\{r_k\}$ is the set of $\dfrac{p-1}{2}$ distinct quadratic residues modulo p.

Similarly, the remainders r'_k of k^{2^n} modulo p for $k = 1, 2, \ldots, \frac{p-1}{2}$ also form a system of quadratic residues modulo p, therefore $\displaystyle\sum_{k=1}^{\frac{p-1}{2}} r_k = \sum_{k=1}^{\frac{p-1}{2}} r'_k$.

Thus,

$$\sum_{k=1}^{\frac{p-1}{2}} \left\lfloor \frac{k^{2^n}}{p} + \frac{1}{2} \right\rfloor = \sum_{k=1}^{\frac{p-1}{2}} \left\lfloor \frac{2k^{2^n}}{p} \right\rfloor - \sum_{k=1}^{\frac{p-1}{2}} \left\lfloor \frac{k^{2^n}}{p} \right\rfloor$$

$$= \frac{1}{p} \sum_{k=1}^{\frac{p-1}{2}} (2k^{2^n} - r_k) - \frac{1}{p} \sum_{k=1}^{\frac{p-1}{2}} (k^{2^n} - r'_k) = \frac{1}{p} \sum_{k=1}^{\frac{p-1}{2}} k^{2^n}.$$

Solutions to Testing Questions   25

Testing Questions   (25-A)

1. Changing the inequality to the form $\dfrac{\sqrt{x} + \sqrt{y}}{\sqrt{2x + y}} \leq k$, the problem becomes to find the maximum value of $\dfrac{\sqrt{x} + \sqrt{y}}{\sqrt{2x + y}}$ when $x, y > 0$.

By the Cauchy-Schwartz inequality,

$$(\sqrt{x} + \sqrt{y})^2 = \left(\frac{1}{\sqrt{2}} \cdot \sqrt{2x} + 1 \cdot \sqrt{y} \right)^2 \leq \left(\frac{1}{2} + 1 \right)(2x + y),$$

namely $\dfrac{\sqrt{x}+\sqrt{y}}{\sqrt{2x+y}} \le \sqrt{\dfrac{3}{2}} = \dfrac{\sqrt{6}}{2}$, so $k \le \dfrac{\sqrt{6}}{2}$. If $k < \dfrac{\sqrt{6}}{2}$, $x = \dfrac{1}{4}$, $y = 1$ then gives

$$\sqrt{x}+\sqrt{y} = \frac{3}{2} = \frac{\sqrt{6}}{2} \cdot \sqrt{2x+y} > k\sqrt{2x+y},$$

so the range of k is $\left[\dfrac{\sqrt{6}}{2}, +\infty\right)$.

2. The Cauchy-Schwartz inequality yields

$$\frac{1}{x^2+y+1} \le \frac{1+y+z^2}{(x+y+z)^2}, \qquad \frac{1}{y^2+z+1} \le \frac{1+z+x^2}{(x+y+z)^2},$$
$$\frac{1}{z^2+x+1} \le \frac{1+x+y^2}{(x+y+z)^2}.$$

Adding these inequalities together, we obtain

$$\frac{1}{x^2+y+1} + \frac{1}{y^2+z+1} + \frac{1}{z^2+x+1} \le \frac{3+x+y+z+x^2+y^2+z^2}{(x+y+z)^2}.$$

Denote the right hand side of the last inequality by S, then it suffices to show that $S \le 1$. In fact,

$$S \le 1 \Leftrightarrow 3+x+y+z \le 2(xy+yz+zx) \Leftrightarrow 3 \le x+y+z,$$

and since $x+y+z = xy+yz+xz \le \dfrac{(x+y+z)^2}{3}$, the needed inequality is proven at once. Thus, the original inequality is proven.

3. Let $A = \dfrac{a^2}{ab^2(4-ab)} + \dfrac{b^2}{bc^2(4-bc)} + \dfrac{c^2}{ca^2(4-ca)}$, $B = \dfrac{b^2}{ab^2(4-ab)} +$
    $\dfrac{c^2}{bc^2(4-bc)} + \dfrac{a^2}{ca^2(4-ca)}$, then it suffices to show that $A \ge 1$ and $B \ge 1$.
    By the Cauchy-Schwartz inequality,

$$\left(\frac{4-ab}{a} + \frac{4-bc}{b} + \frac{4-ca}{c}\right) A \ge \left(\frac{1}{a}+\frac{1}{b}+\frac{1}{c}\right)^2.$$

Let $k = \dfrac{1}{a}+\dfrac{1}{b}+\dfrac{1}{c}$, then $A \ge \dfrac{k^2}{4k-3}$. The Cauchy-Schwartz inequality gives $(a+b+c)\left(\dfrac{1}{a}+\dfrac{1}{b}+\dfrac{1}{c}\right) \ge 9$, so $k = \dfrac{1}{a}+\dfrac{1}{b}+\dfrac{1}{c} \ge 3$. Hence

$$k^2 - (4k-3) = (k-3)(k-1) \ge 0 \Rightarrow k^2 \ge 4k-3 \Rightarrow A \ge 1.$$

Since $B = \dfrac{1}{a(4-ab)} + \dfrac{1}{b(4-bc)} + \dfrac{1}{c(4-ca)}$, so the Cauchy-Schwartz

inequality gives $\left(\dfrac{4-ab}{a} + \dfrac{4-bc}{b} + \dfrac{4-ca}{c}\right) B \geq \left(\dfrac{1}{a} + \dfrac{1}{b} + \dfrac{1}{c}\right)^2 =$

k^2. Hence $B \geq \dfrac{k^2}{4k-3} \geq 1$.

Thus, $A + 3B \geq 4$.

4. Note that $1 - \dfrac{a-bc}{a+bc} = \dfrac{2bc}{1-b-c+bc} = \dfrac{2bc}{(1-b)(1-c)}$, and similarly,

$1 - \dfrac{b-ca}{b+ca} = \dfrac{2ca}{(1-c)(1-a)}, 1 - \dfrac{c-ab}{c+ab} = \dfrac{2ab}{(1-a)(1-b)}$, so the orig-

inal inequality is equivalent to

$$\frac{2bc}{(1-b)(1-c)} + \frac{2ca}{(1-c)(1-a)} + \frac{2ab}{(1-a)(1-b)} \geq \frac{3}{2}.$$

By simplification, the above inequality is equivalent to

$$4(bc + ca + ab - 3abc) \geq 3(bc + ca + ab + 1 - a - b - c - abc),$$

namely $ab + bc + ca \geq 9abc$, or $\dfrac{1}{a} + \dfrac{1}{b} + \dfrac{1}{c} \geq 9$. By the Cauchy-Schwartz

inequality,

$$\frac{1}{a} + \frac{1}{b} + \frac{1}{c} = (a+b+c)\left(\frac{1}{a} + \frac{1}{b} + \frac{1}{c}\right) \geq 9.$$

5. Note that $\displaystyle\sum_{i=1}^{n} \frac{x_i(2x_i - x_{i+1} - x_{i+2})}{x_{i+1} + x_{i+2}} \geq 0 \Leftrightarrow \sum_{i=1}^{n}\left(\frac{2x_i^2}{x_{i+1}+x_{i+2}} - x_i\right) \geq 0$

$\Leftrightarrow \displaystyle\sum_{i=1}^{n} \frac{x_i^2}{x_{i+1}+x_{i+2}} \geq \frac{1}{2}\sum_{i=1}^{n} x_i$, where $x_{n+1} = x_1, x_{n+2} = x_2$. (∗)

By the Cauchy-Schwartz inequality,

$$\sum_{i=1}^{n} \frac{x_i^2}{x_{i+1}+x_{i+2}} \cdot \sum_{i=1}^{n}(x_{i+1} + x_{i+2}) \geq \left(\sum_{i=1}^{n} x_i\right)^2.$$

When both sides are divided by $2\displaystyle\sum_{i=1}^{n} x_i$, then (∗) is obtained.

6. By the Cauchy-Schwartz inequality, for any $x_1, x_2, \ldots, x_n > 0$, we have the inequality

$$\frac{1}{\sum_{i=1}^{n} x_i} \le \frac{1}{n^2} \sum_{i=1}^{n} \frac{1}{x_i}.$$

Since $a_1 + a_2 + \cdots + a_n = 1$ and $n > 2$,

$$\sum_{i=1}^{n} \frac{1}{a_i (a_i + n - 2)} = \sum_{i=1}^{n} \frac{1}{a_i \sum_{\substack{j=1, \\ j \ne i}}^{n} (1 - a_j)}$$

$$\le \sum_{i=1}^{n} \frac{1}{(n-1)^2} \sum_{\substack{j=1, \\ j \ne i}}^{n} \frac{1}{a_i (1 - a_j)} = \frac{1}{(n-1)^2} \sum_{j=1}^{n} \sum_{\substack{i=1, \\ i \ne j}}^{n} \frac{1}{a_i (1 - a_j)}.$$

Since $0 < a_i < 1$ for $i = 1, 2, \ldots, n$, so for any $i, j \in \{1, 2, \ldots, n\}, i \ne j, a_0 = a_n, a_{n+1} = a_1$,

$$a_i \ge \frac{\prod_{k=1}^{n} a_k}{a_{i-1} a_j} \quad \text{and} \quad a_{j+1} \ge \frac{\prod_{k=1}^{n} a_k}{a_{j-1} a_j}.$$

So $\displaystyle \sum_{\substack{i=1, \\ i \ne j}}^{n} a_i \ge \sum_{\substack{i=1, \\ i \ne j}}^{n} \frac{\prod_{k=1}^{n} a_k}{a_i a_j} \Rightarrow (1 - a_j) \cdot \frac{a_j}{\prod_{k=1}^{n} a_k} \ge \sum_{\substack{i=1, \\ i \ne j}}^{n} \frac{1}{a_i}$, namely

$$\sum_{\substack{i=1, \\ i \ne j}}^{n} \frac{1}{a_i (1 - a_j)} \le \frac{1}{\prod_{\substack{k=1, \\ k \ne j}}^{n} a_k}. \text{ Thus,}$$

$$\sum_{i=1}^{n} \frac{1}{a_i (a_i + n - 2)} \le \frac{1}{(n-1)^2} \sum_{j=1}^{n} \frac{1}{\prod_{\substack{k=1, \\ k \ne j}}^{n} a_k},$$

and hence

$$\sum_{i=1}^{n} \frac{\prod_{\substack{k=1, \\ k \ne i}}^{n} a_k}{a_i + n - 2} = \prod_{k=1}^{n} a_k \cdot \sum_{i=1}^{n} \frac{1}{a_i (a_i + n - 2)} \le \frac{\sum_{j=1}^{n} a_j}{(n-1)^2} = \frac{1}{(n-1)^2}.$$

7. By the AM-GM inequality, $B \ge \dfrac{1}{b} + \dfrac{1}{c} + \dfrac{1}{a}$, and

$$A - (a + b + c)$$
$$= \frac{a^4 + b^4 + c^4 - a^2 b^2 - b^2 c^2 - c^2 a^2}{(a+b)(b+c)(c+a)}$$
$$= \frac{1}{2(a+b)(b+c)(c+a)} \cdot [(a^2 - b^2)^2 + (b^2 - c^2)^2 + (c^2 - a^2)^2] \ge 0.$$

Thus, by the Cauchy-Schwartz inequality,

$$AB \geq (a+b+c) \cdot \left(\frac{1}{a} + \frac{1}{b} + \frac{1}{c}\right) \geq (1+1+1)^2 = 9.$$

8. By the Cauchy-Schwartz inequality,

$$(b^2+c+c^2+a+a^2+b) \cdot \left(\frac{a^4}{b^2+c} + \frac{b^4}{c^2+a} + \frac{c^4}{a^2+b}\right) \geq (a^2+b^2+c^2)^2.$$

Thus,

$$\frac{a^4}{b^2+c} + \frac{b^4}{c^2+a} + \frac{c^4}{a^2+b} \geq \frac{(a^2+b^2+c^2)^2}{a^2+b^2+c^2+3}.$$

Let $a^2 + b^2 + c^2 = x$, then $x \geq \frac{1}{3}(a+b+c)^2 = 3$, hence

$$\frac{x^2}{3+x} \geq \frac{3}{2} \Leftrightarrow 2x^2 \geq 9 + 3x \Leftrightarrow 2x^2 - 3x - 9 \geq 0$$
$$\Leftrightarrow (2x+3)(x-3) \geq 0,$$

which obviously holds since $x \geq 3$. Thus, the original inequality is proven.

9. We may assume that $x > y > z$. denote $x = z + a + b$, $y = z + b$, where $a, b > 0$. Then

$$(xy + yz + zx)\left[\frac{1}{(x-y)^2} + \frac{1}{(y-z)^2} + \frac{1}{(z-x)^2}\right]$$

$$= [(z+a+b)(z+b) + z(2z+2b+a)] \cdot \left[\frac{1}{a^2} + \frac{1}{b^2} + \frac{1}{(a+b)^2}\right]$$

$$\geq (a+b)b\left[\frac{1}{a^2} + \frac{1}{b^2} + \frac{1}{(a+b)^2}\right], \qquad (30.36)$$

where the equality holds if and only if $z = 0$. It suffices to show that

$$(a+b)b\left[\frac{1}{a^2} + \frac{1}{b^2} + \frac{1}{(a+b)^2}\right] \geq 4. \qquad (30.37)$$

Let $a = \lambda b$ where $\lambda > 0$. Then

$$(30.37) \Leftrightarrow (\lambda + 1)\left[\frac{1}{\lambda^2} + 1 + \frac{1}{(\lambda+1)^2}\right] \geq 4. \qquad (30.38)$$

After simplification, we find that

$$(30.38) \Leftrightarrow \lambda^4 - 2\lambda^3 - \lambda^2 + 2\lambda + 1 \geq 0 \Leftrightarrow (\lambda^2 - \lambda - 1)^2 \geq 0,$$

which is obvious. Further, the equality holds if and only if $\lambda = \dfrac{1 + \sqrt{5}}{2}$.

In summary, we have proven the original inequality, and the equality holds if and only if $\{x, y, z\} = \{0, t, \frac{1+\sqrt{5}}{2}t\}$, where $t > 0$.

10.   The AM-GM inequality implies $x^2 + y^2 + z^2 \geq xy + yz + zx$, so

$$(x + y + z)^2 \geq 3(xy + yz + zx) \Rightarrow \frac{1}{9} + \frac{2}{27}(xy + yz + zx) \leq \frac{1}{3}.$$

Hence it suffices to show $\dfrac{x^3}{y^3 + 8} + \dfrac{y^3}{z^3 + 8} + \dfrac{z^3}{x^3 + 8} \geq \dfrac{1}{3}$.

Since $(x^3 + y^3 + z^3)^2 \geq 3(x^3 y^3 + y^3 z^3 + z^3 x^3)$, by the Cauchy-Schwartz inequality and the AM-GM inequality,

$$
\begin{aligned}
&\frac{x^3}{y^3 + 8} + \frac{y^3}{z^3 + 8} + \frac{z^3}{x^3 + 8} \\
&\geq \frac{(x^3 + y^3 + z^3)^2}{x^3 y^3 + y^3 z^3 + z^3 x^3 + 8(x^3 + y^3 + z^3)} \\
&\geq \frac{(x^3 + y^3 + z^3)^2}{\frac{1}{3}(x^3 + y^3 + z^3)^2 + 8(x^3 + y^3 + z^3)} \\
&= \frac{x^3 + y^3 + z^3}{\frac{1}{3}(x^3 + y^3 + z^3) + 8}.
\end{aligned}
$$

Since $\dfrac{x^3 + y^3 + z^3}{\frac{1}{3}(x^3 + y^3 + z^3) + 8} \geq \dfrac{1}{3} \Leftrightarrow x^3 + y^3 + z^3 \geq 3$, by the Cauchy-Schwartz inequality, $(x + y + z)(x^3 + y^3 + z^3) \geq (x^2 + y^2 + z^2)^2$ and $(1 + 1 + 1)(x^2 + y^2 + z^2) \geq (x + y + z)^2$, so

$$(x + y + z)(x^3 + y^3 + z^3) \geq (x^2 + y^2 + z^2)^2 \geq \left[\frac{(x + y + z)^2}{3}\right]^2 = 9,$$

$$\therefore x^3 + y^3 + z^3 \geq 3,$$

as desired.

Testing Questions (25-B)

1. From Schur's inequality with $r = 2$, $\sum x^2(x - y)(x - z) \geq 0$, it follows that

$$\sum x^4 \geq \sum x^3(y + z) - xyz \sum x,$$

$$\left(\sum x^2\right)^2 \geq \sum \left[x(y^3 + z^3) + x^2(y^2 + z^2 - yz)\right],$$

$$\left(\sum x^2\right)^2 \geq \sum \left[x(y^2 + z^2 - yz)\right] \cdot \sum x. \qquad (*)$$

Let $x = \dfrac{1}{a}, y = \dfrac{1}{b}, z = \dfrac{1}{c}$ and substitute them into $(*)$, we obtain

$$\left(\sum \frac{1}{a^2}\right)^2 \geq \sum \left[\frac{1}{a}\left(\frac{1}{b^2} + \frac{1}{c^2} - \frac{1}{bc}\right)\right] \cdot \sum \frac{1}{a}.$$

Applying the Cauchy-Schwartz inequality yields

$$\sum \frac{b^2c^2}{a^3(b^2 - bc + c^2)} \cdot \sum \frac{b^2 - bc + c^2}{ab^2c^2} \geq \left(\sum \frac{1}{a^2}\right)^2,$$

$$\therefore \sum \frac{b^2c^2}{a^3(b^2 - bc + c^2)} \geq \sum \frac{1}{a}.$$

It suffices to show that $\left(\sum \dfrac{1}{a}\right) \cdot \left(\sum ab\right) \geq 3 \sum a = 3$. For this we have

$$\left(\sum \frac{1}{a}\right) \cdot \left(\sum ab\right) - 3\left(\sum a\right) = \frac{1}{abc}\left[\left(\sum ab\right)^2 - abc \sum a\right]$$

$$= \frac{1}{2abc}[(ab - bc)^2 + (bc - ca)^2 + (ca - ab)^2] \geq 0.$$

2. By the Cauchy-Schwartz inequality and the AM-GM inequality,

$$y^2(2x + z)^2 \leq y^2(x^2 + z^2)(2^2 + 1^2) = 5y^2(1 - y^2) \leq \frac{5}{4},$$

i.e., $|2xy + yz| \leq \dfrac{\sqrt{5}}{2}$, so $\lambda_{max} \geq 2$.

On the other hand, letting $x = \dfrac{\sqrt{10}}{5}, y = \dfrac{\sqrt{2}}{2}, z = \dfrac{\sqrt{10}}{10}$, then

$$\left|\frac{\lambda}{\sqrt{5}} + \frac{1}{2\sqrt{5}}\right| \leq \frac{\sqrt{5}}{2} \Rightarrow |2\lambda + 1| \leq 5 \Rightarrow \lambda \leq 2.$$

Hence $\lambda_{max} = 2$.

3. There are a few ways to prove this inequality. Here, we introduce a method similar to that used to prove Schur's inequality.

We first note that

$$
\frac{x^2 + yz}{\sqrt{2x^2(y+z)}} = \frac{x^2 - x(y+z) + yz}{\sqrt{2x^2(y+z)}} + \frac{x(y+z)}{\sqrt{2x^2(y+z)}}
$$

$$
= \frac{(x-y)(x-z)}{\sqrt{2x^2(y+z)}} + \sqrt{\frac{y+z}{2}}
$$

$$
\geq \frac{(x-y)(x-z)}{\sqrt{2x^2(y+z)}} + \frac{\sqrt{y} + \sqrt{z}}{2}. \tag{30.39}
$$

Similarly, we have

$$
\frac{y^2 + zx}{\sqrt{2y^2(z+x)}} \geq \frac{(y-z)(y-x)}{\sqrt{2y^2(z+x)}} + \frac{\sqrt{z} + \sqrt{x}}{2}, \tag{30.40}
$$

$$
\frac{z^2 + xy}{\sqrt{2z^2(x+y)}} \geq \frac{(z-x)(z-y)}{\sqrt{2z^2(x+y)}} + \frac{\sqrt{x} + \sqrt{y}}{2}. \tag{30.41}
$$

We now add (30.39), (30.40) and (30.41) to get

$$
\frac{x^2 + yz}{\sqrt{2x^2(y+z)}} + \frac{y^2 + zx}{\sqrt{2y^2(z+x)}} + \frac{z^2 + xy}{\sqrt{2z^2(x+y)}}
$$

$$
\geq \frac{(x-y)(x-z)}{\sqrt{2x^2(y+z)}} + \frac{(y-z)(y-x)}{\sqrt{2y^2(z+x)}} + \frac{(z-x)(z-y)}{\sqrt{2z^2(x+y)}} + \sqrt{x} + \sqrt{y} + \sqrt{z}
$$

$$
= \frac{(x-y)(x-z)}{\sqrt{2x^2(y+z)}} + \frac{(y-z)(y-x)}{\sqrt{2y^2(z+x)}} + \frac{(z-x)(z-y)}{\sqrt{2z^2(x+y)}} + 1.
$$

Thus, it suffices to show that

$$
\frac{(x-y)(x-z)}{\sqrt{2x^2(y+z)}} + \frac{(y-z)(y-x)}{\sqrt{2y^2(z+x)}} + \frac{(z-x)(z-y)}{\sqrt{2z^2(x+y)}} \geq 0. \tag{30.42}
$$

Now, assume without loss of generality, that $x \geq y \geq z$. We then have

$$
\frac{(x-y)(x-z)}{\sqrt{2x^2(y+z)}} \geq 0
$$

and

$$
\frac{(z-x)(z-y)}{\sqrt{2z^2(x+y)}} + \frac{(y-z)(y-x)}{\sqrt{2y^2(z+x)}} = \frac{(y-z)(x-z)}{\sqrt{2z^2(x+y)}} - \frac{(y-z)(x-y)}{\sqrt{2y^2(z+x)}}
$$

$$\geq \frac{(y-z)(x-y)}{\sqrt{2z^2(x+y)}} - \frac{(y-z)(x-y)}{\sqrt{2y^2(z+x)}}$$

$$= (y-z)(x-y)\left(\frac{1}{\sqrt{2z^2(x+y)}} - \frac{1}{\sqrt{2y^2(z+x)}}\right).$$

The last quantity is non-negative due to the fact that

$$y^2(z+x) = y^2z + y^2x \geq yz^2 + z^2x = z^2(x+y).$$

This completes the proof.

4. From $x^2 + y^2 + z^2 = 3$ and the Cauchy-Schwartz inequality,

$$(x^2 + y^2 + z^2)^2 = 3(x^2 + y^2 + z^2) \geq (x+y+z)^2,$$
$$x^2 + y^2 + z^2 \geq x + y + z. \qquad (*)$$

By the Cauchy-Schwartz inequality, $(x^2+y+z)(1+y+z) \geq (x+y+z)^2$, hence it suffices to show that

$$\frac{x\sqrt{1+y+z} + y\sqrt{1+z+x} + z\sqrt{1+x+y}}{x+y+z} \leq \sqrt{3}.$$

Using another application of the Cauchy-Schwartz inequality and $(*)$,

$$(x\sqrt{1+y+z} + y\sqrt{1+z+x} + z\sqrt{1+x+y})^2$$
$$= (\sqrt{x}\cdot\sqrt{x+xy+zx} + \sqrt{y}\cdot\sqrt{y+yz+xy} + \sqrt{z}\cdot\sqrt{z+zx+yz})^2$$
$$\leq (x+y+z)[(x+xy+xz) + (y+yz+xy) + (z+zx+yz)]$$
$$= (x+y+z)[(x+y+z) + 2(xy+yz+zx)]$$
$$\leq (x+y+z)[(x^2+y^2+z^2) + 2(xy+yz+zx)] = (x+y+z)^3,$$

hence

$$\frac{x\sqrt{1+y+z} + y\sqrt{1+z+x} + z\sqrt{1+x+y}}{x+y+z} \leq \sqrt{x+y+z}.$$

Using $(*)$ again, the original inequality is obtained at once.

5. Let p, q, r be the three positive roots of $\varphi(x)$. By Viete's Theorem,

$$-\frac{b}{a} = p+q+r, \quad \frac{c}{a} = pq+qr+rp, \quad -\frac{d}{a} = pqr.$$

Since $\varphi(0) < 0 \Leftrightarrow d < 0$, so $a > 0$. Label

$$2b^3 + 9a^2d - 7abc \leq 0, \qquad (30.43)$$

then (30.43) $\Leftrightarrow 2(p+q+r)^3 - 7(p+q+r)(pq+qr+rp) + 9pqr \geq 0$.
Schur's inequality yields

$$(p+q+r)^3 - 4(p+q+r)(pq+qr+rp) + 9pqr \geq 0. \quad (30.44)$$

The AM-GM inequality yields $(p+q+r)^2 \geq 3(pq+qr+rp)$, namely

$$(p+q+r)^3 - 3(p+q+r)(pq+qr+rp) \geq 0. \quad (30.45)$$

Thus, (30.44) + (30.45) yields

$$2(p+q+r)^3 - 7(p+q+r)(pq+qr+rp) + 9pqr \geq 0,$$

as desired.

Solutions to Testing Questions 26

Testing Questions (26-A)

1. When $a \geq b$, then $a^p \geq b^p$ and $a^q \geq b^q$, Hence by the rearrangement inequality, $a^p \cdot a^q + b^p \cdot b^q \geq a^p b^q + a^q b^p$, and the equality holds if and only if $a^p = b^p$ or $a^q = b^q$, i.e., $a = b$.

    When $a \leq b$, then $a^p \leq b^p$ and $a^q \leq b^q$, by the rearrangement inequality, $a^p \cdot a^q + b^p \cdot b^q \geq a^p b^q + a^q b^p$, and the equality holds if and only if $a^p = b^p$ or $a^q = b^q$, i.e., $a = b$.

2. Without loss of generality it can be assumed that $a \geq b \geq c$. Then
    $$b(c+a-b) - a(b+c-a) = a^2 - b^2 + bc - ac = (a-b)(a+b-c) \geq 0.$$
    $$c(a+b-c) - b(c+a-b) = b^2 - c^2 + ca - ab = (b-c)(b+c-a) \geq 0.$$

    Therefore $a(b+c-a) \leq b(c+a-b) \leq c(a+b-c)$. By the rearrangement inequality,

    $$\begin{aligned} &a^2(b+c-a) + b^2(c+a-b) + c^2(a+b-c) \\ &\quad \leq ba(b+c-a) + cb(c+a-b) + ac(a+b-c), \\ &a^2(b+c-a) + b^2(c+a-b) + c^2(a+b-c) \\ &\quad \leq ca(b+c-a) + ab(c+a-b) + bc(a+b-c). \end{aligned}$$

    By adding these two inequalities, the desired inequality is obtained.

3. Assume that $a \geq b \geq c$. Then $ac + bc + c \leq bc + ab + b \leq ab + ac + a$ and

$$\frac{c}{ac + bc + c} \leq \frac{b}{ab + bc + b} \leq \frac{a}{ab + ac + a}.$$

Note that $\displaystyle\sum_{a,b,c} \frac{1}{a + b + 1} = 1 \Leftrightarrow \sum_{a,b,c} \frac{c}{ac + bc + c} = 1$, and so by Chebyshev's inequality,

$$\sum_{a,b,c} \frac{c}{ac + bc + c} \cdot \sum_{a,b,c} (ac + bc + c) \leq 3 \sum_{a,b,c} c$$

$$\Leftrightarrow 3 \sum_{a,b,c} c \geq \sum_{a,b,c} (ac + bc + c) \Leftrightarrow \sum_{a,b,c} c \geq \sum_{a,b,c} ab.$$

4. Assume that $x \geq y \geq z$. Then $x^k \geq y^k \geq z^k$. By Chebyshev's inequality,

$$3(x^{k+1} + y^{k+1} + z^{k+1}) \geq (x + y + z)(x^k + y^k + z^k). \quad (30.46)$$

Since $0 < z \leq y \leq x < 1$ and $x^{k-1} \geq y^{k-1} \geq z^{k-1}$,

$$x(1-x) - y(1-y) = x(y+z) - y(z+x) = z(x-y) \geq 0$$
$$\Rightarrow x(1-x) \geq y(1-y) \Rightarrow x^k(1-x) \geq y^k(1-y)$$
$$\Rightarrow x^{k+1} + y^k + z^k \leq y^{k+1} + z^k + x^k,$$

and similarly, $y^{k+1} + z^k + x^k \leq z^{k+1} + x^k + y^k$, therefore

$$\frac{x^{k+1}}{x^{k+1} + y^k + z^k} \geq \frac{y^{k+1}}{y^{k+1} + z^k + z^k} \geq \frac{z^{k+1}}{z^{k+1} + x^k + y^k}.$$

By Chebyshev's inequality again,

$$\frac{x^{k+2}}{x^{k+1} + y^k + z^k} + \frac{y^{k+2}}{y^{k+1} + z^k + z^k} + \frac{z^{k+2}}{z^{k+1} + x^k + y^k}$$

$$\geq \frac{1}{3}(x+y+z) \left(\frac{x^{k+1}}{x^{k+1} + y^k + z^k} + \frac{y^{k+1}}{y^{k+1} + z^k + z^k} + \frac{z^{k+1}}{z^{k+1} + x^k + y^k} \right)$$

$$= \frac{1}{3} \sum_{x,y,z} \frac{x^{k+1}}{x^{k+1} + y^k + z^k} \cdot \sum_{x,y,z} (x^{k+1} + y^k + z^k) \cdot \frac{1}{\displaystyle\sum_{x,y,z} x^{k+1} + 2 \sum_{x,y,z} x^k}$$

$$\geq (x^{k+1} + y^{k+1} + z^{k+1}) \cdot \frac{1}{\displaystyle\sum_{x,y,z} x^{k+1} + 2 \sum_{x,y,z} x^k}$$

$$= \frac{x^{k+1} + y^{k+1} + z^{k+1}}{x^{k+1} + y^{k+1} + z^{k+1} + 2(x+y+z)(x^k + y^k + z^k)}$$

$$\geq \frac{x^{k+1} + y^{k+1} + z^{k+1}}{x^{k+1} + y^{k+1} + z^{k+1} + 2\cdot 3(x^{k+1} + y^{k+1} + z^{k+1})} = \frac{1}{7}.$$

The equality holds if and only if $x = y = z = \dfrac{1}{3}$.

5. Write $BC = a, CA = b, AB = c, QR = p, RP = q, PQ = r$. Let $AR = x, BP = y, CQ = z$ and $s = \dfrac{1}{2}(a + b + c) = \dfrac{1}{2}L$. Then

$$x = s - a, \qquad y = s - b, \qquad z = s - c.$$

By applying the cosine rule to triangles ABC and ARQ respectively,

$$a^2 = b^2 + c^2 - 2bc \cos A = (b - c)^2 + 2bc(1 - \cos A),$$
$$p^2 = 2x^2(1 - \cos A) = 2(s - a)^2(1 - \cos A).$$

Eliminating $1 - \cos A$ from them gives

$$p^2 = (s-a)^2 \frac{a^2 - (b-c)^2}{bc} = \frac{4(s-a)(s-b)(s-c)}{abc} a(s-a). \quad (30.47)$$

Note that $4(s-a)(s-b) = (b+c-a)(a-b+c) = c^2 - (a-b)^2 \leq c^2$ and similarly

$$4(s-b)(s-c) \leq a^2, \qquad 4(s-c)(s-a) \leq b^2.$$

Multiplying the 3 inequalities together and taking square roots, we obtain $8(s-a)(s-b)(s-c) \leq abc$. From (30.47), this yields

$$p^2 \leq \frac{1}{2}a(s-a) \quad \text{or} \quad \left(\frac{a}{p}\right)^3 \geq 2\sqrt{2}\left(\frac{a}{s-a}\right)^{\frac{3}{2}},$$

and similarly

$$\left(\frac{b}{q}\right)^3 \geq 2\sqrt{2}\left(\frac{b}{s-b}\right)^{\frac{3}{2}}, \qquad \left(\frac{c}{r}\right)^3 \geq 2\sqrt{2}\left(\frac{c}{s-c}\right)^{\frac{3}{2}}.$$

Use M to denote the left hand side of the original inequality, then by the power mean inequality,

$$M \geq 2\sqrt{2}\left[\left(\frac{a}{s-a}\right)^{\frac{3}{2}} + \left(\frac{b}{s-b}\right)^{\frac{3}{2}} + \left(\frac{c}{s-c}\right)^{\frac{3}{2}}\right]$$

$$\geq \frac{2\sqrt{2}}{\sqrt{3}}\left(\frac{a}{s-a} + \frac{b}{s-b} + \frac{c}{s-c}\right)^{\frac{3}{2}}. \qquad (30.48)$$

We can assume that $a \geq b \geq c$, so $\dfrac{1}{s-a} \geq \dfrac{1}{s-b} \geq \dfrac{1}{s-c}$. By Chebyshev's inequality and the AM-GM inequality,

$$\frac{a}{s-a} + \frac{b}{s-b} + \frac{c}{s-c} \geq \frac{1}{3}(a+b+c)\left(\frac{1}{s-a} + \frac{1}{s-b} + \frac{1}{s-c}\right)$$

$$\geq \frac{a+b+c}{[(s-a)(s-b)(s-c)]^{\frac{1}{3}}} = \frac{(a+b+c)s^{\frac{1}{3}}}{[s(s-a)(s-b)(s-c)]^{\frac{1}{3}}}$$

$$= 2^{-\frac{1}{3}}\left(\frac{L^2}{T}\right)^{\frac{2}{3}}. \tag{30.49}$$

By combining (30.48) and (3.49), $M \geq \dfrac{2\sqrt{2}}{\sqrt{3}} \cdot \dfrac{1}{\sqrt{2}} \cdot \dfrac{L^2}{T} = \dfrac{2}{\sqrt{3}} \cdot \dfrac{L^2}{T}$.

6. *Lemma* (**Young's Inequality**): For positive real numbers a and b and $p, q > 1$ with $\dfrac{1}{p} + \dfrac{1}{q} = 1$, the inequality

$$ab \leq \frac{1}{p}a^p + \frac{1}{q}b^q$$

holds.

Proof of Lemma: Since $f(x) = \ln x, x > 0$ is a continuous concave function, by the weighted Jensen's inequality,

$$\ln(ab) = \frac{1}{p}\ln a^p + \frac{1}{q}\ln b^q \leq \ln\left(\frac{1}{p}a^p + \frac{1}{q}b^q\right),$$

therefore $ab \leq \dfrac{1}{p}a^p + \dfrac{1}{q}b^q$.

We now return to the proof of Hölder's inequality. Let $A = \displaystyle\sum_{i=1}^{n} x_i^p$, $B = \displaystyle\sum_{j=1}^{n} y_j^q$, then Young's inequality gives

$$\frac{x_i}{A^{\frac{1}{p}}} \cdot \frac{y_i}{B^{\frac{1}{q}}} \leq \frac{1}{p} \cdot \frac{x_i^p}{A} + \frac{1}{q} \cdot \frac{y_i^q}{B}, \quad i = 1, 2, \cdots, n.$$

Therefore

$$\sum_{i=1}^{n} \frac{x_i}{A^{\frac{1}{p}}} \cdot \frac{y_i}{B^{\frac{1}{q}}} \leq \frac{1}{p}\sum_{i=1}^{n} \frac{x_i^p}{A} + \frac{1}{q}\sum_{i=1}^{n} \frac{y_i^q}{B} = \frac{1}{p} + \frac{1}{q} = 1.$$

Thus

$$\sum_{i=1}^{n} x_i y_i \le A^{\frac{1}{p}} \cdot B^{\frac{1}{q}} = \left(\sum_{i=1}^{n} x_i^p\right)^{\frac{1}{p}} \cdot \left(\sum_{i=1}^{n} y_i^q\right)^{\frac{1}{q}}.$$

7. (i) By using Jensen's inequality on the convex function $f(x) = x^2$,

$$\left[\frac{3 + 2(x + y + z)}{3}\right]^2 = \left[\frac{(1 + x + y) + (1 + y + z) + (1 + x + z)}{3}\right]^2$$
$$\le \frac{(1 + x + y)^2 + (1 + y + z)^2 + (1 + x + z)^2}{3},$$

where the equality holds if and only if $1 + x + y = 1 + y + z = 1 + x + z$, namely $x = y = z = 1$. By the AM-GM inequality,

$$3 + 2(x + y + z) \ge 3 + 6\sqrt[3]{xyz} \ge 9,$$

so $27 \le (1 + x + y)^2 + (1 + y + z)^2 + (1 + x + z)^2$.

(ii) $(1 + x + y)^2 + (1 + y + z)^2 + (1 + x + z)^2 \le 3(x + y + z)^2$
$\Leftrightarrow 3 + 2(x^2 + y^2 + z^2) + 2(xy + yz + zx) + 4(x + y + z) \le 3(x^2 + y^2 + z^2) + 6(xy + yz + zx)$
$\Leftrightarrow 3 + 4(x + y + z) \le x^2 + y^2 + z^2 + 4(xy + yz + zx) = (x + y + z)^2 + 2(xy + yz + zx)$
$\Leftrightarrow 7 \le (u - 2)^2 + 2v$, where $u = x + y + z, v = xy + yz + zx$.

By AM-GM inequality, $u \ge 3$ and $v \ge 3$, so $(u - 2)^2 + 2v \ge 1 + 6 = 7$, and the equality holds if and only if $x = y = z = 1$.

Testing Questions (26-B)

1. (1) When $a_1 a_2 + a_2 a_3 + \cdots + a_n a_1 \ge \dfrac{1}{n}$, since $a_1, a_2, \ldots, a_n$ and $\dfrac{1}{a_1^2 + a_1}$, $\dfrac{1}{a_2^2 + a_2}, \ldots, \dfrac{1}{a_n^2 + a_n}$ have reverse orders, so by the rearrangement inequality and Cauchy-Schwartz inequality,

$$\sum_{i=1}^{n} \frac{a_i}{a_{i+1}^2 + a_{i+1}} \ge \sum_{i=1}^{n} \frac{a_i}{a_i^2 + a_i} = \sum_{i=1}^{n} \frac{1}{1 + a_i}$$
$$\ge \frac{(1 + 1 + \cdots + 1)^2}{(1 + a_1) + (1 + a_2) + \cdots + (1 + a_n)} = \frac{n^2}{n + 1}.$$

Thus,

$$\sum_{i=1}^{n}(a_i a_{i+1}) \cdot \sum_{i=1}^{n} \frac{a_i}{a_{i+1}^2 + a_{i+1}} \geq \frac{n}{n+1}.$$

(2) When $\sum_{i=1}^{n} a_i a_{i+1} < \frac{1}{n}$, by the rearrangement inequality and Cauchy-Schwartz inequality,

$$\sum_{i=1}^{n} a_i a_{i+1} \cdot \sum_{j=1}^{n} \frac{a_j}{a_{j+1}^2 + a_{j+1}} = \sum_{1 \leq i,j \leq n} a_i a_{i+1} \cdot \frac{a_j}{a_{j+1}^2 + a_{j+1}}$$

$$= \frac{1}{2} \sum_{1 \leq i,j \leq n} \left(a_i a_{i+1} \cdot \frac{a_j}{a_{j+1}^2 + a_{j+1}} + a_j a_{j+1} \cdot \frac{a_i}{a_{i+1}^2 + a_{i+1}} \right)$$

$$= \frac{1}{2} \sum_{1 \leq i,j \leq n} a_i a_j \left(\frac{a_{i+1}}{a_{j+1}^2 + a_{j+1}} + \frac{a_{j+1}}{a_{i+1}^2 + a_{i+1}} \right)$$

$$\geq \frac{1}{2} \sum_{1 \leq i,j \leq n} a_i a_j \left(\frac{a_{j+1}}{a_{j+1}^2 + a_{j+1}} + \frac{a_{i+1}}{a_{i+1}^2 + a_{i+1}} \right)$$

$$= \frac{1}{2} \sum_{1 \leq i,j \leq n} a_i a_j \left(\frac{1}{a_{j+1} + 1} + \frac{1}{a_{i+1} + 1} \right) = \sum_{1 \leq i,j \leq n} a_i a_j \cdot \frac{1}{a_{i+1} + 1}$$

$$= \sum_{i=1}^{n} \frac{a_i}{a_{i+1} + 1} \geq \frac{(\sum_{i=1}^{n} a_i)^2}{\sum_{i=1}^{n} a_i (a_{i+1} + 1)} = \frac{1}{\sum_{i=1}^{n} a_i a_{i+1} + 1}$$

$$> \frac{1}{\frac{1}{n} + 1} = \frac{n}{n+1}.$$

2. Without loss of generality, we can assume that $a \geq b \geq c$. Then

$$\frac{(c-a)(c-b)}{3(c+ab)} \geq 0$$

and

$$\frac{(a-b)(a-c)}{3(a+bc)} + \frac{(b-a)(b-c)}{3(b+ac)} = \frac{c(a-b)^2}{3} \left[\frac{1+a+b-c}{(a+bc)(b+ac)} \right] \geq 0.$$

Therefore $\sum_{a,b,c} \frac{(a-b)(a-c)}{3(a+bc)} \geq 0$. Since

$$\sum_{a,b,c} \frac{1}{a+bc} \leq \frac{9}{2(ab+bc+ca)}$$

$$\Leftrightarrow \sum_{a,b,c} \frac{1}{a(a+b+c)+3bc} \leq \frac{3}{2(ab+bc+ca)}$$

$$\Leftrightarrow \sum_{a,b,c} \left[\frac{3}{2(ab+bc+ca)} - \frac{1}{a(a+b+c)+3bc} \right] \geq 0$$

$$\Leftrightarrow \sum_{a,b,c} \frac{(a-b)(a-c)}{a(a+b+c)+3bc} = \sum_{a,b,c} \frac{(a-b)(a-c)}{3(a+bc)} \geq 0,$$

so $\displaystyle \sum_{a,b,c} \frac{1}{a+bc} \leq \frac{9}{2(ab+bc+ca)}.$

On the other hand, from the AM-GM inequality,

$$\frac{1}{a\sqrt{2(a^2+bc)}} = \frac{\sqrt{b+c}}{\sqrt{2a}\cdot\sqrt{(ab+ac)(a^2+bc)}} \geq \frac{\sqrt{2(b+c)}}{\sqrt{a}(a+c)(a+b)}.$$

Hence it suffices to show $\displaystyle \sum_{a,b,c} \sqrt{\frac{b+c}{2a}}\cdot\frac{1}{(a+c)(a+b)} \geq \frac{9}{4(ab+bc+ca)}.$

Since

$$\sqrt{\frac{b+c}{2a}} \leq \sqrt{\frac{a+c}{2b}} \leq \sqrt{\frac{a+b}{2c}}$$

and

$$\frac{1}{(a+c)(a+b)} \leq \frac{1}{(b+c)(a+b)} \leq \frac{1}{(a+c)(c+b)},$$

by Chebyshev's inequality,

$$\sum_{a,b,c} \sqrt{\frac{b+c}{2a}}\cdot\frac{1}{(a+c)(a+b)} \geq \frac{1}{3}\sum_{a,b,c}\sqrt{\frac{b+c}{2a}}\cdot\sum_{a,b,c}\frac{1}{(a+c)(a+b)}$$

$$= \frac{2}{(a+b)(b+c)(c+a)}\sum_{a,b,c}\sqrt{\frac{b+c}{2a}}.$$

Therefore it suffices to show that $\displaystyle \sum_{a,b,c}\sqrt{\frac{b+c}{2a}} \geq \frac{9(a+b)(b+c)(c+a)}{8(ab+bc+ca)}.$

Let $t = \sqrt[6]{\dfrac{(a+b)(b+c)(c+a)}{8abc}} \geq 1$ (by the AM-GM inequality), then

$$\frac{9(a+b)(b+c)(c+a)}{8(ab+bc+ca)} = \frac{27t^6}{8t^6+1}.$$

By the AM-GM inequality, $\displaystyle \sum_{a,b,c}\sqrt{\frac{b+c}{2a}} \geq 3t$ so

$$3t \geq \frac{27t^6}{8t^6 + 1} \Leftrightarrow 8t^6 - 9t^5 + 1 \geq 0 \Leftrightarrow (t-1)(8t^5 - t^4 - t^3 - t^2 - t - 1) \geq 0.$$

The last inequality holds clearly when $t \geq 1$. Thus, the original inequality is proven.

3.   It is obvious that $\sqrt[4]{\dfrac{(a^2 + b^2)(a^2 - ab + b^2)}{2}} \leq \dfrac{a^2 + b^2}{a + b}$

$\Leftrightarrow (a+b)^4(a^2 - ab + b^2) \leq 2(a^2 + b^2)^3 \Leftrightarrow (a-b)^4(a^2 + ab + b^2) \geq 0.$
Thus, it suffices to show that

$$\sum_{cyc} \frac{a^2 + b^2}{a + b} \leq \frac{2}{3}\left(\sum_{cyc} a^2\right)\sum_{cyc}\frac{1}{a + b}.$$

Without loss of generality we may assume that $a \geq b \geq c$. Then

$$a^2 + b^2 \geq a^2 + c^2 \geq b^2 + c^2, \quad \frac{1}{a+b} \leq \frac{1}{a+c} \leq \frac{1}{b+c}.$$

By Chebyshev's inequality,

$$\sum_{cyc} \frac{a^2 + b^2}{a + b} \leq \frac{1}{3}\cdot\sum_{cyc}(a^2 + b^2)\cdot\sum_{cyc}\frac{1}{a+b} = \frac{2}{3}\sum_{cyc}a^2\cdot\sum_{cyc}\frac{1}{a+b},$$

as desired.

4. $\sqrt{a^3 + a} + \sqrt{b^3 + b} + \sqrt{c^3 + c} \geq 2\sqrt{a + b + c}$
$\Leftrightarrow \sqrt{a^3 + a(ab + bc + ca)} + \sqrt{b^3 + b(ab + bc + ca)}$
$\quad + \sqrt{c^3 + c(ab + bc + ca)} \geq 2\sqrt{(a + b + c)(ab + bc + ca)}$
$\Leftrightarrow \sqrt{a(a + b)(c + a)} + \sqrt{b(a + b)(b + c)} + \sqrt{c(c + a)(b + c)}$
$\quad \geq 2\sqrt{(a + b + c)(ab + bc + ca)}$
$\Leftrightarrow \displaystyle\sum_{cyc} \frac{a}{\sqrt{a(b + c)}} \geq 2\sqrt{\frac{(a + b + c)(ab + bc + ca)}{(a + b)(b + c)(c + a)}}.$

By applying the weighted Jensen's inequality to the function $f(x) = \dfrac{1}{\sqrt{x}}$, $x > 0$, it is obtained that

$$\frac{a}{\sqrt{a(b + c)}} + \frac{b}{\sqrt{b(c + a)}} + \frac{c}{\sqrt{c(a + b)}} \geq \frac{a + b + c}{\sqrt{\frac{\sum_{cyc} a^2(b+c)}{a+b+c}}}.$$

Thus, it suffices to show that

$$(a+b+c)^2[a^2(b+c)+b^2(c+a)+c^2(a+b)+2abc]$$
$$\geq 4(ab+bc+ca)[a^2(b+c)+b^2(c+a)+c^2(a+b)]$$

$$\Leftrightarrow (a+b+c)^2[(a+b+c)(ab+bc+ca)-abc]$$
$$\geq 4(ab+bc+ca)[(a+b+c)(ab+bc+ca)-3abc]$$

$$\Leftrightarrow (a+b+c)^3(ab+bc+ca)-abc(a+b+c)^2$$
$$\geq 4(a+b+c)(ab+bc+ca)^2-12abc(ab+bc+ca)$$

$$\Leftrightarrow (\textstyle\sum ab)[(\sum a)^3-4(\sum a)(\sum ab)+9abc]$$
$$\geq abc[(a+b+c)^2-3(ab+bc+ca)]$$

$$\Leftrightarrow (\textstyle\sum ab)[a(a-b)(a-c)+b(b-c)(b-a)+c(c-a)(c-b)]$$
$$\geq abc[(a-b)(a-c)+(b-a)(b-c)+(c-a)(c-b)]$$

$$\Leftrightarrow a^2(b+c)(a-b)(a-c)+b^2(c+a)(b-c)(b-a)$$
$$+c^2(a+b)(c-a)(c-b) \geq 0.$$

Without loss of generality we may assume that $c = \min\{a,b,c\}$, then the last inequality is equivalent to

$$(a-b)^2(a^2b+ab^2+a^2c+b^2c-ac^2-bc^2)+c^2(a+b)(c-a)(c-b) \geq 0,$$

and it is obvious.

5. Let $a = y+z, b = z+x, c = x+y$, then x, y, z are all positive, and

$$B = \sum_{cyc}\frac{1}{2\sqrt{zx}} = \sum_{cyc}\frac{1}{2\sqrt{yz}}, \quad A = \sum_{cyc}\frac{x^2+y^2+z^2+xy+zx+3yz}{2x+y+z},$$

$$AB = \left(\sum_{cyc}\frac{1}{2\sqrt{yz}}\right) \cdot \left(\sum_{cyc}\frac{x^2+y^2+z^2+xy+zx+3yz}{2x+y+z}\right)$$

$$\geq \left(\sum_{cyc}\sqrt{\frac{1}{2\sqrt{yz}}\cdot\frac{x^2+y^2+z^2+xy+zx+3yz}{2x+y+z}}\right)^2.$$

It suffices to show that each number under a square root sign is greater than or equal to 1. Below, as an example, we show that the expression under the first square root sign is greater than or equal to 1.

$$\frac{1}{2\sqrt{yz}}\cdot\frac{x^2+y^2+z^2+xy+zx+3yz}{2x+y+z} \geq 1$$
$$\Leftrightarrow (x^2+y^2+z^2+xy+3yz+zx)^2 \geq 4yz(2x+y+z)^2$$
$$\Leftrightarrow x^4+y^4+z^4+3x^2y^2+3x^2z^2+3y^2z^2+2x^3y+2xy^3+2x^3z$$
$$+2xz^3+2y^3z+2yz^3 \geq 8xy^2z+8x^2yz+8xyz^2. \qquad (*)$$

By the power mean inequality,

$$x^4 + y^4 + z^4 \geq \frac{1}{27}(x + y + z)^4$$
$$\geq xyz(x + y + z) = xy^2z + x^2yz + xyz^2.$$

By the AM-GM inequality,

$$3x^2y^2 + 3x^2z^2 + 3y^2z^2$$
$$\geq (xy + yz + xz)^2 \geq 3(xy^2z + x^2yz + xyz^2).$$

Besides,

$$x^3y + xy^3 + x^3z + xz^3 + y^3z + yz^3 - 2xyz(x + y + z)$$
$$= (x^3y + yz^3 - xyz^2 - x^2yz) + (y^3z + x^3z - x^2yz - xy^2z)$$
$$+(z^3x + xy^3 - xy^2z - xyz^2)$$
$$= y(x + z)(x - z)^2 + z(x + y)(x - y)^2 + x(y + z)(y - z)^2$$
$$\geq 0.$$

By adding up the first two inequalities with 2 times the third inequality, $(*)$ is obtained.

Solutions to Testing Questions   27

Testing Questions   (27-A)

1. First of all, we show that

$$\frac{1}{1 + a^2} + \frac{1}{1 + b^2} \leq \frac{2}{1 + ab}. \qquad (30.50)$$

The inequality above is true because

(30.50)
$$\Leftrightarrow 1 + ab + b^2 + ab^3 + 1 + ab + a^2 + a^3b \leq 2 + 2a^2 + 2b^2 + 2a^2b^2$$
$$\Leftrightarrow ab(a^2 - 2ab + b^2) \leq a^2 - 2ab + b^2 \Leftrightarrow (ab - 1)(a - b)^2 \leq 0.$$

Since $ab \leq 1$, so (30.50) is proven.

Next, by the Cauchy-Schwartz inequality and (30.50),

$$\sum \frac{1}{\sqrt{a^2 + 1}} \leq \sqrt{2\left(\frac{1}{1 + a^2} + \frac{1}{1 + b^2}\right)} \leq \sqrt{\frac{4}{1 + ab}} = \frac{2}{\sqrt{1 + ab}}.$$

Thus, the desired inequality is proven.

2. $a + b + c = 1$ and $a^2 + b^2 \geq 2ab$ implies that

$$
\begin{aligned}
(\textstyle\sum ab)^2 &= a^2b^2 + b^2c^2 + c^2a^2 + 2a^2bc + 2b^2ca + 2c^2ab \\
&= (a^2 + b^2)c^2 + a^2b^2 + 2abc(a + b + c) \\
&\geq 2abc^2 + a^2b^2 + 2abc
\end{aligned}
$$

therefore $\dfrac{1}{ab + 2c^2 + 2c} \geq \dfrac{ab}{(\sum ab)^2}$. Similarly,

$$
\frac{1}{bc + 2a^2 + 2a} \geq \frac{bc}{(\sum ab)^2}, \quad \frac{1}{ca + 2b^2 + 2b} \geq \frac{ca}{(\sum ab)^2}.
$$

By adding them up, the conclusion is proven.

3.   The left inequality can be obtained directly from the Power Mean inequality. Here, we derive it by algebraic manipulations as follows:

$$
\frac{a + b + c}{3} \leq \sqrt{\frac{a^2 + b^2 + c^2}{3}} \Leftrightarrow (a + b + c)^2 \leq 3(a^2 + b^2 + c^2)
$$
$$
\Leftrightarrow a^2 + b^2 + c^2 + 2(ab + bc + ca) \leq 3(a^2 + b^2 + c^2)
$$
$$
\Leftrightarrow 2(ab + bc + ca) \leq 2(a^2 + b^2 + c^2)
$$
$$
\Leftrightarrow 0 \leq (a - b)^2 + (b - c)^2 + (c - a)^2.
$$

The equality holds if and only if $a = b = c$.

To Prove the right inequality, we have

$$
\sqrt{\frac{a^2 + b^2 + c^2}{3}} \leq \frac{\frac{ab}{c} + \frac{bc}{a} + \frac{ca}{b}}{3} \Leftrightarrow 3(\textstyle\sum a^2) \leq \left(\sum \frac{ab}{c}\right)^2
$$

$$
\Leftrightarrow a^2 + b^2 + c^2 \leq \frac{a^2b^2}{c^2} + \frac{b^2c^2}{a^2} + \frac{c^2a^2}{b^2}
$$

$$
\Leftrightarrow \left(\frac{bc}{a}\right)\left(\frac{ca}{b}\right) + \left(\frac{bc}{a}\right)\left(\frac{ab}{c}\right) + \left(\frac{ca}{b}\right)\left(\frac{ab}{c}\right) \leq \left(\frac{ab}{c}\right)^2 + \left(\frac{bc}{a}\right)^2 +
$$
$$
\left(\frac{ac}{b}\right)^2 \Leftrightarrow \left(\frac{ab}{c} - \frac{bc}{a}\right)^2 + \left(\frac{bc}{a} - \frac{ac}{b}\right)^2 + \left(\frac{ab}{c} - \frac{ac}{b}\right)^2 \geq 0.
$$

The last inequality is clear, and the equality holds if and only if $a^2 = b^2 = c^2$, i.e. $a = b = c$.

4.   For the denominator, let $a - b = x, b - c = y, c - a = -(x + y)$, then
$(a - b)^3 + (b - c)^3 + (c - a)^3 = x^3 + y^3 + z^3 = -3xy(x + y) = 3(a - b)(b - c)(c - a)$.

Similarly, for the numerator we obtain

$$(a^2 - b^2)^3 + (b^2 - c^2)^3 + (c^2 - a^2)^3 = 3(a^2 - b^2)(b^2 - c^2)(c^2 - a^2),$$

therefore, by the AM-GM inequality,

$$\frac{(a^2 - b^2)^3(b^2 - c^2)^3 + (c^2 - a^2)^3}{(a - b)^3 + (b - c)^3 + (c - a)^3} = (a + b)(b + c)(c + a)$$
$$\geq 2\sqrt{ab} \cdot 2\sqrt{bc} \cdot 2\sqrt{ca} = 8abc.$$

5. Let $S = \dfrac{(x + 1)(y + 1)^2}{3\sqrt[3]{z^2 x^2} + 1} + \dfrac{(y + 1)(z + 1)^2}{3\sqrt[3]{x^2 y^2} + 1} + \dfrac{(z + 1)(x + 1)^2}{3\sqrt[3]{y^2 z^2} + 1}$. Since

$zx + z + x + 1 \geq 3\sqrt[3]{z^2 x^2}$, $xy + x + y \geq 3\sqrt[3]{x^2 y^2}$, $yz + y + z \geq 3\sqrt[3]{y^2 z^2}$,

it follows that

$$S \geq \frac{(x + 1)(y + 1)^2}{(z + 1)(x + 1)} + \frac{(y + 1)(z + 1)^2}{(x + 1)(y + 1)} + \frac{(z + 1)(x + 1)^2}{(y + 1)(z + 1)}.$$

Let $a = x + 1, b = y + 1, c = z + 1$, then $S \geq \dfrac{b^2}{c} + \dfrac{c^2}{a} + \dfrac{a^2}{b}$.

By the Cauchy-Schwartz inequality,

$$(a + b + c)S \geq (a + b + c)\left(\frac{b^2}{c} + \frac{c^2}{a} + \frac{a^2}{b}\right) \geq (a + b + c)^2$$
$$\Rightarrow S \geq \frac{(a + b + c)^2}{a + b + c} = a + b + c = x + y + z + 3.$$

6. Let $a + c - b = x, a + b - c = y, b + c - a = z$. Then $a = \dfrac{x + y}{2}, b = \dfrac{y + z}{2}, c = \dfrac{z + x}{2}$, and $a + b + c = x + y + z$. If we let K denote the left hand side of the original inequality, then

$$K = \frac{2x^4}{y(x + y)} + \frac{2y^4}{z(y + z)} + \frac{2z^4}{x(z + x)}.$$

By the Cauchy-Schwartz inequality,

$$\frac{K}{2}[y(x + y) + z(y + z) + x(z + x)] \geq (x^2 + y^2 + z^2)^2$$
$$\Rightarrow \frac{K}{2} \geq \frac{(x^2 + y^2 + z^2)^2}{x^2 + y^2 + z^2 + xy + yz + zx}$$
$$\Rightarrow \frac{K}{2} \geq \frac{(x^2 + y^2 + z^2)^2}{2(x^2 + y^2 + z^2)} = \frac{1}{2}(x^2 + y^2 + z^2)$$
$$\Rightarrow K \geq x^2 + y^2 + z^2 \geq \frac{(x + y + z)^2}{3} = \frac{(a + b + c)^2}{3} \geq \sum ab.$$

7. Since $\dfrac{(1+a^2)(1+b^2)(1+c^2)}{(1+a)(1+b)(1+c)} = \displaystyle\prod_{cyc} \dfrac{1+a^2}{1+a}$, it is natural to localize the inequality to dealing with $\dfrac{1+a^2}{1+a}$.

$$\left(\frac{1+a^2}{1+a}\right)^3 = \frac{(1+a^2)^3}{(1+a)^3} = \frac{1}{2}\left[\frac{(1+a^3)(1+a)^3 + (1-a^3)(1-a)^3}{(1+a)^3}\right]$$
$$\geq \frac{1}{2}(1+a^3),$$

where the equality holds if and only if $a = 1$. Similarly,

$$\left(\frac{1+b^2}{1+b}\right)^3 \geq \frac{1}{2}(1+b^3), \quad \left(\frac{1+c^2}{1+c}\right)^3 \geq \frac{1}{2}(1+c^3),$$

where the equalities holds if and only if $b = c = 1$, by times up them and using the AM-GM inequality,

$$\prod_{cyc}\frac{1+a^2}{1+a} \geq \frac{1}{2}\left[\prod_{cyc}(1+a^3)\right]^{\frac{1}{3}} = \frac{1}{2}\left[1 + \sum_{cyc}a^3 + \sum_{cyc}a^3b^3 + a^3b^3c^3\right]^{\frac{1}{3}}$$
$$\geq \frac{1}{2}(1 + 3abc + 3a^2b^2c^2 + a^3b^3c^3)^{\frac{1}{3}} = \frac{1}{2}(1+abc),$$

where the equality holds if and only if $a = b = c = 1$.

8. Note that

$$\frac{x^2 - xy + y^2}{x^2 + xy + y^2} \geq \frac{1}{3} \Leftrightarrow 3(x^2 - xy + y^2) \geq x^2 + xy + y^2$$
$$= 2(x-y)^2 \geq 0,$$

therefore

$$\frac{x^3 + y^3}{x^2 + xy + y^2} = \frac{(x+y)(x^2 - xy + y^2)}{x^2 + xy + y^2} \geq \frac{x+y}{3}.$$

Similar inequalities are obtained for the other two terms also. Thus,

$$\frac{x^3 + y^3}{x^2 + xy + y^2} + \frac{y^3 + z^3}{y^2 + yz + z^2} + \frac{z^3 + x^3}{z^2 + zx + x^2}$$
$$\geq \frac{x+y}{3} + \frac{y+z}{3} + \frac{z+x}{3} = \frac{2(x+y+z)}{3} \geq 2\sqrt[3]{xyz} = 2.$$

9. We prove the conclusion by induction on n as follows:

For $n = 6$, $\left(\dfrac{6}{2}\right)^6 > 6! > \left(\dfrac{6}{3}\right)^6 \Leftrightarrow 729 > 720 > 64$, the conclusion is true.

Assume that the conclusion is true for $n = k$ $(k \geq 6)$, i.e., $\left(\dfrac{k}{2}\right)^k > k! > \left(\dfrac{k}{3}\right)^k$, then for $n = k + 1$, to show $(k + 1)! < \left(\dfrac{k+1}{2}\right)^{k+1}$, it suffices to show that $(k + 1)\left(\dfrac{k}{2}\right)^k \leq \left(\dfrac{k+1}{2}\right)^{k+1}$, i.e. $2 \leq \left(1 + \dfrac{1}{k}\right)^k$.

The Binomial Expansion gives $\left(1 + \dfrac{1}{k}\right)^k = 2 + \displaystyle\sum_{i=2}^{k} \binom{k}{i}\dfrac{1}{k^i} > 2$, so $(k + 1)! < \left(\dfrac{k+1}{2}\right)^{k+1}$ is proven.

Similarly, to prove $(k+1)! > \left(\dfrac{k+1}{3}\right)^{k+1}$ it suffices to show $\left(1 + \dfrac{1}{k}\right)^k \leq 3$. Since $\binom{k}{i} \cdot \dfrac{1}{k^i} < \dfrac{1}{i!} \leq \dfrac{1}{i(i-1)} = \dfrac{1}{i-1} - \dfrac{1}{i}$ for any $2 \leq i \leq n$, it follows that

$$\left(1 + \dfrac{1}{k}\right)^k = 2 + \sum_{i=2}^{k} \binom{k}{i}\dfrac{1}{k^i} < 2 + \sum_{i=2}^{k}\left(\dfrac{1}{i-1} - \dfrac{1}{i}\right) < 3.$$

Thus, the inequality $(k + 1)! > \left(\dfrac{k+1}{3}\right)^{k+1}$ is proven also.

Testing Questions (27-B)

1. $\dfrac{a^2b(b-c)}{a+b} + \dfrac{b^2c(c-a)}{b+c} + \dfrac{c^2a(a-b)}{c+a} \geq 0$

$\Leftrightarrow \dfrac{a^2b^2}{a+b} + \dfrac{b^2c^2}{b+c} + \dfrac{c^2a^2}{c+a} \geq abc\left(\dfrac{a}{a+b} + \dfrac{b}{b+c} + \dfrac{c}{c+a}\right)$

$\Leftrightarrow \dfrac{ab}{c(a+b)} + \dfrac{bc}{a(b+c)} + \dfrac{ca}{b(c+a)} \geq \dfrac{a}{a+b} + \dfrac{b}{b+c} + \dfrac{c}{c+a}$

$$\Leftrightarrow (ab + bc + ca)\left(\frac{1}{ac + bc} + \frac{1}{ab + ac} + \frac{1}{bc + ab}\right) \geq \frac{ac}{ac + bc} +$$
$$\frac{ab}{ab + ac} + \frac{bc}{bc + ab} + 3.$$

Now use substitutions as follows: let

$$\begin{cases} x = ab + ac, \\ y = bc + ba, \\ z = ca + cb \end{cases} \Rightarrow \begin{cases} ac = \dfrac{x + z - y}{2}, \\ ab = \dfrac{x + y - z}{2}, \\ bc = \dfrac{y + z - x}{2} \end{cases} \Rightarrow ab + bc + ca = \dfrac{x + y + z}{2}.$$

Then, the original inequality is equivalent to

$$\frac{1}{2}(x + y + z)\left(\frac{1}{x} + \frac{1}{y} + \frac{1}{z}\right) \geq \frac{x + z - y}{2z} + \frac{x + y - z}{2x} + \frac{y + z - x}{2y} + 3$$

$$\Leftrightarrow (x + y + z)\left(\frac{1}{x} + \frac{1}{y} + \frac{1}{z}\right) \geq \frac{x - y}{z} + \frac{y - z}{x} + \frac{z - x}{y} + 9$$

$$\Leftrightarrow 3 + \frac{y}{x} + \frac{z}{x} + \frac{x}{y} + \frac{z}{y} + \frac{x}{z} + \frac{y}{z} \geq \frac{x - y}{z} + \frac{y - z}{x} + \frac{z - x}{y} + 9$$

$$\Leftrightarrow \frac{2y}{z} + \frac{2z}{x} + \frac{2x}{y} \geq 6 \Leftrightarrow \frac{y}{z} + \frac{z}{x} + \frac{x}{y} \geq 3.$$

The last inequality is obtained from the AM-GM inequality at once.

2. Let $\dfrac{a}{a - b} = x$, $\dfrac{b}{b - c} = y$, $\dfrac{c}{c - a} = z$. Then

$$\sum_{cyc}\left(\frac{2a - b}{a - b}\right)^2 \geq 5 \Leftrightarrow \sum_{cyc}(1 + x)^2 \geq 5 \Leftrightarrow 2\sum_{cyc}x + \sum_{cyc}x^2 \geq 2.$$

Since $\dfrac{1}{x} - 1 = -\dfrac{b}{a}$, $\dfrac{1}{y} - 1 = -\dfrac{c}{b}$, $\dfrac{1}{z} - 1 = -\dfrac{a}{c}$, so

$$\left(\frac{1}{x} - 1\right)\left(\frac{1}{y} - 1\right)\left(\frac{1}{z} - 1\right) = -1 \Rightarrow \frac{1}{xyz} - \sum_{cyc}\frac{1}{xy} + \sum_{cyc}\frac{1}{x} = 0$$

$$\Rightarrow 1 - \sum_{cyc}x + \sum_{cyc}xy = 0 \Rightarrow \sum_{cyc}xy = \sum_{cyc}x - 1,$$

therefore

$$2\sum_{cyc} x + \sum_{cyc} x^2 \geq 2 \Leftrightarrow 2\sum_{cyc} x + \left(\sum_{cyc} x\right)^2 - 2\sum_{cyc} xy \geq 2$$

$$\Leftrightarrow 2\sum_{cyc} x + \left(\sum_{cyc} x\right)^2 - 2\left(\sum_{cyc} x - 1\right) \geq 2 \Leftrightarrow \left(\sum_{cyc} x\right)^2 \geq 0.$$

The last inequality is clear, so the original inequality holds.

3. Let $a = 1, b = \dfrac{1}{n}, c = \dfrac{1}{n^2}$, then $\lim\limits_{n\to+\infty} \left(\dfrac{a+b}{a+2b} + \dfrac{b+c}{b+2c} + \dfrac{c+a}{c+2a}\right)$

$$= \lim_{n\to+\infty} \left(\frac{1+(1/n)}{1+(2/n)} + \frac{(1/n)+(1/n^2)}{(1/n)+((2/n^2)} + \frac{(1/n^2)+1}{(1/n^2)+2}\right)$$

$$= \lim_{n\to+\infty} \left(1 + 1 + \frac{1}{2}\right) = \frac{5}{2}, \text{ therefore } D \geq \frac{5}{2}.$$

Below we show that $\dfrac{a+b}{a+2b} + \dfrac{b+c}{b+2c} + \dfrac{c+a}{c+2a} < \dfrac{5}{2}$ for any $a, b, c > 0$.

$$\frac{a+b}{a+2b} + \frac{b+c}{b+2c} + \frac{c+a}{c+2a} < \frac{5}{2} \Leftrightarrow \frac{b}{a+2b} + \frac{c}{b+2c} + \frac{a}{c+2a} > \frac{1}{2}$$

$$\Leftrightarrow \frac{1}{2+(a/b)} + \frac{1}{2+(b/c)} + \frac{1}{2+(c/a)} > \frac{1}{2}.$$

Let $\dfrac{a}{b} = x, \dfrac{b}{c} = y, \dfrac{c}{a} = z$. It suffices to show that if $xyz = 1$ then

$$\frac{1}{2+x} + \frac{1}{2+y} + \frac{1}{2+z} > \frac{1}{2}.$$

Since $\dfrac{1}{2+x} + \dfrac{1}{2+y} + \dfrac{1}{2+z} > \dfrac{1}{2} \Leftrightarrow \dfrac{1}{2+x} + \dfrac{1}{2+y} > \dfrac{z}{4+2z}$

$$\Leftrightarrow \frac{1}{2+x} + \frac{1}{2+y} > \frac{1}{2+4xy} \Leftrightarrow (4+x+y)(2+4xy) > (2+x)(2+y)$$

$$\Leftrightarrow 4 + 15xy + 4xy(x+y) > 0, \text{ and the last inequality is obvious.}$$

Thus, the minimum value of D is $\dfrac{5}{2}$.

4. Notice that

$$\frac{a^2+2}{2} = \frac{(a^2-a+1)+(a+1)}{2} \geq \sqrt{(a+1)(a^2-a+1)} = \sqrt{a^3+1},$$

hence it suffices to show that $\sum \dfrac{a^2}{(a^2+2)(b^2+2)} \geq \dfrac{1}{3}$ i.e.

$$3\sum a^2(c^2+2) \geq (a^2+2)(b^2+2)(c^2+2). \qquad (30.51)$$

By expanding and using the fact that $abc = 8$, (30.51) is equivalent to

$$6(a^2+b^2+c^2)+3(a^2b^2+b^2c^2+c^2a^2) \geq$$
$$a^2b^2c^2+2(a^2b^2+b^2c^2+c^2a^2)+4(a^2+b^2+c^2)+8,$$

$$2(a^2+b^2+c^2)+a^2b^2+b^2c^2+c^2a^2 \geq 72. \qquad (30.52)$$

By the AM-GM inequality, we have

$$2(a^2+b^2+c^2) \geq 6\sqrt[3]{(abc)^2} = 24, \qquad (30.53)$$
$$a^2b^2+b^2c^2+c^2a^2 \geq 3\sqrt[3]{(abc)^4} = 48, \qquad (30.54)$$

therefore (30.52) is proven.

5. Label the original inequality:

$$\sum_{i=1}^{n} a_i^{2k+1} \geq \left(\sum_{i=1}^{n} a_i^k\right)^2. \qquad (30.55)$$

First, we prove by induction on n that

$$2\sum_{i=1}^{n} a_i^k \leq (a_n+1)^k a_n. \qquad (30.56)$$

For $n = 1$, (30.56) is clear. Assume that (30.56) is true for $n = m$, i.e.,

$$2\sum_{i=1}^{m} a_i^k \leq (a_m+1)^k a_m,$$

then for $n = m+1$,

$$2\sum_{i=1}^{m+1} a_i^k = 2\sum_{i=1}^{m} a_i^k + 2a_{m+1}^k \leq (a_m+1)^k a_m + 2a_{m+1}^k$$
$$\leq a_{m+1}^k(a_{m+1}-1)+2a_{m+1}^k = a_{m+1}^k(a_{m+1}+1)$$
$$\leq (a_{m+1}+1)^k a_{m+1}.$$

so (30.56) is true for $n = m+1$.

Below we prove (30.55) by induction on n. The inequality is clearly true for $n = 1$. Assume that (30.55) is true for $n = m$, i.e.,

$$\left(\sum_{i=1}^{m} a_i^k \right)^2 \leq \sum_{i=1}^{m} a_i^{2k+1}.$$

Then for $n = m + 1$,

$$
\begin{aligned}
\left(\sum_{i=1}^{m+1} a_i^k \right)^2 &= \left(\sum_{i=1}^{m} a_i^k \right)^2 + 2 \left(\sum_{i=1}^{m} a_i^k \right) a_{m+1}^k + a_{m+1}^{2k} \\
&\leq \sum_{i=1}^{m} a_i^{2k+1} + 2 \left(\sum_{i=1}^{m} a_i^k \right) a_{m+1}^k + a_{m+1}^{2k} \\
&\leq \sum_{i=1}^{m} a_i^{2k+1} + (a_m + 1)^k a_m a_{m+1}^k + a_{m+1}^{2k} \\
&\leq \sum_{i=1}^{m} a_i^{2k+1} + a_{m+1}^k [a_{m+1}^k (a_{m+1} - 1) + a_{m+1}^k] \\
&= \sum_{i=1}^{m+1} a_i^{2k+1}.
\end{aligned}
$$

Thus, (30.55) is true for $n = m + 1$ also.

Solutions to Testing Questions   28

Testing Questions   (28-A)

1. By using the multiplication principle, we choose the ten s_i according to the order $s_9, s_8, \cdots, s_3, s_2, s_1, s_0$, then the number of choices are

$$^3C_1, \ ^3C_1, \ \cdots, \ ^3C_1, \ 3, \ 2, \ 1,$$

respectively. Therefore the number of desired permutations is $(^3C_1)^7 \cdot 3 \cdot 2 \cdot 1 = 2 \cdot 3^8 = 13122$.

2. Let A be the set of all positive integers having required property.

In A there are 2 two digit numbers each containing one digit 3, so 3 appears a total of 2 times in them.

In A there are 3 three digit numbers each containing two digits 3, so 3 appears a total of 6 times in them.

In A there are 3 three digit numbers each containing one digit 3, so 3 appears a total of 3 times in them.

In A there are 12 four digit numbers each containing one digit 3, so 3 appears a total of 12 times in them.

In A there are 5 five digit numbers each containing one digit 3, so 3 appears a total of 5 times.

Thus, the digit 3 appears a total of 28 times in the numbers in A.

3. Let $A_i, i = 0, 1, 2$ be the subset of S consisting by all the numbers in S which have remainder i when divided by 3. Then

$$|A_0| = 16, \qquad |A_1| = 17, \qquad |A_2| = 17.$$

$\{a, b, c\}$ is good if and only if a, b, c all come from the same A_i or come from distinct A_i. Thus, the number of good subsets is

$$\binom{16}{3} + 2\binom{17}{3} + 16 \cdot 17 \cdot 17 = 6544.$$

4. There are $6^3 = 216$ possible distinct products, where $5^3 = 125$ products do not have 5 as a factor, $3^3 = 27$ products do not have 2 as a factor, and $2^3 = 8$ products do not have 2 and 5 as factors, so the number of products which are divisible by 10 is

$$216 - 125 - 27 + 8 = 72.$$

5. The number of triangles with the required property is

$$\frac{1}{3}\left(\binom{23}{2} - 22 - 21 - 20\right) \cdot 30 = 10(253 - 63) = 1900.$$

6. We may assume that $b_M = \max\{b_1, b_2, \ldots, b_n\}$ and the lines $A_1 A_M, A_2 A_M,$ $\ldots, A_{b_M} A_M$ are all blue. Then the lines $A_i A_j$ where $1 \le i < j \le b_M$ are all not blue, and $b_k \le b_M$ for $k > b_M$, so the total number of blue lines is less than or equal to $(n - b_M)b_M$. Then the AM-GM inequality yields

$$(n - b_M)b_M \le \left(\frac{n - b_M + b_M}{2}\right)^2 = \frac{n^2}{4},$$

so

$$b_1 + b_2 + \cdots + b_n \le 2 \times \frac{n^2}{4} = \frac{n^2}{2}.$$

7. For the first step we arrange the three music books. We have $3! = 6$ ways for this.

    Next, arrange the mathematical book and the English books.

    There are 2 ways to arrange the math book when it is arranged at the left or right side of the music books. In this case the English books must be arranged in the two gaps between the music books, so we have 2 ways for this. Therefore there are $2 \times 2 = 4$ ways to arrange the mathematical book and the English books for this case.

    If the math book is arranged in a gap between music books, then there are two ways for this. Then there are $2 \times \binom{4}{1} = 8$ ways to arrange the English books, so we have a total of $2 \times 8 = 16$ ways in this case.

    Thus, the number of ways satisfying the requirements is $6 \times (4 + 16) = 120$.

8. By adding $6, 5, 4, 3, 2, 1$ marks to the marks of Issac's answers to the first, second, third, fourth, fifth, and sixth questions respectively, then the sequence of Isaac's marks becomes a strictly increasing sequence taking values from $\{1, 2, ..., 16\}$. This correspondence is one-to-one.

    Thus, the number of possible results is $\binom{16}{6} = 8008$.

9. Let the permutation of the n objects be $a_1, a_2, \cdots, a_n$. If the subset

    $$\{a_{i_1}, a_{i_2}, \cdots, a_{i_k}\}$$

    is a distant subset of capacity k, where $1 \leq i_1 < i_2 < \cdots < i_k \leq n$, then we have $i_{j+1} - i_j \geq 2$ for $j = 1, 2, \cdots, k - 1$, therefore

    $$1 \leq i_1 < i_2 - 1 < i_3 - 2 < \cdots < i_k - (k - 1) \leq n - (k - 1),$$

    i.e. the sequence $i_1, i_2 - 1, i_3 - 2, \cdots, i_k - (k - 1)$ is an strictly increasing subsequence of $1, 2, \cdots, n - k + 1$. Conversely, for any strictly increasing subsequence $j_1 < j_2 < \cdots < j_k$ of the sequence $1, 2, \cdots, n - k + 1$, the sequence

    $$a_{j_1}, a_{j_2+1}, \cdots, a_{j_k+k-1}$$

    form a distant subset of capacity k. Thus, the number of distant subsets with capacity k is $\binom{n-k+1}{k}$.

Testing Questions   (28-B)

1. First, arrange the three b's and three c's in a row. By symmetry, we only consider the cases that b where b is the leftmost character. Arrangements have the following possible cases:

   (i) "b c b c b c". In this case, the number of ways to insert four a's is $\binom{7}{4} = 35$;

   (ii) "b c c b c b" or "b c b c c b". In this case, the number of ways to insert four a's is $2 \times \binom{6}{3} = 40$;

   (iii) "b b c c b c" or "b c c b b c" or "b b c b c c" or "b c b b c c". In this case, the number of ways to insert four a's is $4 \times \binom{5}{2} = 40$;

   (iv) "b c c c b b" or "b b c c c b". In this case, the number of ways to insert four a's is $2 \times \binom{4}{1} = 8$;

   (v) "b b b c c c". In this case, the number of ways to insert four a's is $\binom{4}{4} = 1$;

   Thus, the total number of desired permutations is $2(35+40+40+8+1) = 248$.

2. In a plane use 100 points on a circle to denote the 100 people: $A_1, B_1, A_2, B_2, \ldots, A_{50}, B_{50}$, where the points A_i and B_i denote the two people from the ith country. Use a real line segment to connect A_i and B_i.

   These 100 points can be partitioned as 50 *neighbor pairs*. Use a dotted line to connect the two points in a neighbor pair. Below we show that it is possible to color these 100 points with red and blue such that the two ends of each real segment are of distinct colors and the two ends of each dotted line is so also. Then the points of the same color can be taken as one group.

   The following operations can realize our purpose. First color A_1 red and B_1 blue. take A_2 be the point connected with B_1 by a dotted line, and color A_2 red and B_2 blue, and continue this process until we color the point which is connected with A_1 by a dotted line. Note that it cannot be red. Otherwise, it is an end point of a dotted line, so it cannot be the endpoint of a dotted line emitted from A_1 also. Hence we can color it blue.

   If all the 100 point have colored, the purpose is reached; if there are point not colored yet, then start the process again from any remaining point, until all the points are colored. Since two endpoints of each real segment and each dotted segment have different colors, it satisfies the requirement.

3. Bob can win as follows.

    Claim After each of his moves, Bob can ensure that in that maximum number in each row is a square in $A \cup B$, where

    $$A = \{(1; 1); (2; 1); (3; 1); (1; 2); (2; 2); (3; 2); (1; 3); (2; 3)\}$$

    and

    $$B = \{(5; 3); (4; 4); (5; 4); (6; 4); (4; 5); (5; 5); (6; 5); (4; 6); (5; 6); (6; 6)\}.$$

    Proof. Bob pairs each square of $A \cup B$ with a square in the same row that is not in $A \cup B$, so that each square of the grid is in exactly one pair. Whenever Alice plays in one square of a pair, Bob will play in the other square of the pair on his next turn. If Alice moves with x in $A \cup B$, Bob writes y with $y < x$ in the paired square. If Alice moves with x not in $A \cup B$, Bob writes z with $z > x$ in the paired square in $A \cup B$. So after Bob's turn, the maximum of each pair is in $A \cup B$, and thus the maximum of each row is in $A \cup B$.

    Thus, when all the numbers are written, the maximum square in row 6 is in B and the maximum square in row 1 is in A. Since there is no path from B to A that stays in $A \cup B$, Bob wins.

4. We consider the cases $n = 1, 2, 3, \ldots$. It's easy to see that $f(1) = 0$, $f(2) = 1$, $f(3) = 5$. Below we find the relation between $f(n)$ and $f(n-1)$.

    $f(n)$ is clearly equal to $f(n-1)$ plus the number of those equilateral triangles with 1 vertex or 2 vertices on the nth row. Let C_1 be those with two vertices on the nth row, then it is obvious that $|C_1| = \binom{n}{2}$.

    Let C_2 consist of those equilateral triangles with only 1 vertex on the nth row. Let XYZ be the big array with X above the horizontal segment YZ (where Y is left to Z). If $\triangle ABC$ is such a triangle with A on the nth row and A, B, C are arranged in clockwise direction. Then the line which is parallel to XY and passes through B and the line which is parallel to XZ and passes through C must intersect the nth row at two points P and Q, so the $\triangle ABC$ determines an unique ordered triple (P, A, Q) on the nth row.

    Conversely, given any ordered triple (P, A, Q) (where A is between P and Q) on the nth row, then taking P, Q as two vertices one can form an equilateral triangle inside the array first, and next by taking A as a vertex to form an inscribed equilateral triangle of the resulting triangle. By symmetry, the inscribed triangle exists uniquely.

Thus, $|C_2| = \binom{n}{3}$, hence $f(n) = f(n-1) + \binom{n}{2} + \binom{n}{3}$. Using the fact that $f(1) = 0$,

$$
\begin{aligned}
f(n) &= \sum_{i=2}^{n} \left(\binom{i}{2} + \binom{i}{3} \right) = \frac{1}{6} \sum_{i=2}^{n} (i^3 - i) \\
&= \frac{1}{6} \left(\sum_{i=1}^{n} i^3 - \sum_{i=1}^{n} i \right) = \frac{1}{24} n^2 (n+1)^2 - \frac{1}{12} n(n+1) \\
&= \frac{(n-1)n(n+1)(n+2)}{24}.
\end{aligned}
$$

5. Let n be the number of participants at the conference. We proceed by induction on n.

    If $n = 1$, then we have one participant who can eat in either room; that gives us total of $2 = 2^1$ options.

    Let $n \geq 2$. The case in which some participant, P, has no friends is trivial. In this case, P can eat in either of the two rooms, so the total number of ways to split n participants is twice as many as the number of ways to split $(n-1)$ participants besides the participant P. By induction, the latter number is a power of two, 2^k, hence the number of ways to split n participants is $2 \times 2^k = 2^{k+1}$, also a power of two. So we assume from here on that every participant has at least one friend.

    We consider two different cases separately: the case when some participant has an odd number of friends, and the case when each participant has an even number of friends:

    Case 1: *Some participant, Z, has an odd number of friends.*

    Remove Z from consideration and for each pair (X, Y) of Z's friends, reverse the relationship between X and Y (from friends to strangers or vice versa).

    Claim. *The number of possible seatings is unchanged after removing Z and reversing the relationship between X and Y in each pair (X, Y) of Z's friends.*

    Proof of the claim. Suppose we have an arrangement prior to Z's departure. By assumption, Z has an even number of friends in the room with him. If this number is 0, the room composition is clearly still valid after Z leaves the room. If this number is positive, let X be one of Z's friends in the room with him. By assumption, person X also has an <u>even</u> number of friends in the same room. Remove Z from the room; then X will have an odd number of friends left in the room, and there will be an odd number of Z's friends in this room besides X. Reversing the relationship between X and each of Z's friends in this room will therefore restore the parity to even.

The same reasoning applies to any of Z's friends in the other dining room. Indeed, there will be an odd number of them in that room, hence each of them will reverse relationships with an even number of individuals in that room, preserving the parity of the number of friends present.

Moreover, a legitimate seating without Z arises from <u>exactly one</u> arrangement including Z, because in the case under consideration, only one room contains an even number of Z's friends.

Thus, we have to double the number of seatings for $(n - 1)$ participants which is, by the induction hypothesis, a power of 2. Consequently, for n participants we will get again a power of 2 for the number of different arrangements.

Case 2: *Each participant has an <u>even</u> number of friends.*

In this case, each valid split of participants in two rooms gives us an even number of friends in either room.

Let (A, B) be any pair of friends. Remove this pair from consideration and for each pair (C, D), where C is a friend of A and D is a friend of B, change the relationship between C and D to the opposite; do the same if C is a friend of B and D is a friend of A. Note that if C and D are friends of both A and B, their relationship will be reversed twice, leaving it unchanged.

Consider now an arbitrary participant X different from A and B and choose one of the two dining rooms. [Note that in the case under consideration, the total number of participants is at least 3, so such a triplet $(A, B; X)$ can be chosen.] Let A have m friends in this room and let B have n friends in this room; both m and n are even. When the pair (A, B) is removed, X's relationship will be reversed with either n, or m, or $m + n - 2k$ (for k the number of mutual friends of A and B in the chosen room), or 0 people within the chosen room (depending on whether he/she is a friend of only A, only B, both, or neither). Since m and n are both even, the parity of the number of X's friends in that room will be therefore unchanged in any case.

Again, a legitimate seating without A and B will arise from <u>exactly one</u> arrangement that includes the pair (A, B): just add each of A and B to the room with an odd number of the other's friends, and then reverse all of the relationships between a friend of A and a friend of B. In this way we create a one-to-one correspondence between all possible seatings before and after the (A, B) removal.

Since the number of arrangements for n participants is twice as many as that for $(n - 2)$ participants, and that number for $(n - 2)$ participants is, by the induction hypothesis, a power of 2, we get in turn a power of 2 for the number of arrangements for n participants. The problem is completely solved.

Solutions to Testing Questions   29

Testing Questions   (29-A)

1. Since every $n - 2$ persons take totally 3^k times of calls, and every two persons are in the $\binom{n-2}{n-4}$ distinct groups of $n - 2$ persons, therefore, if m is the total number of calls taken by the n persons, then, from the assumptions,

$$m = \frac{\binom{n}{n-2}3^k}{\binom{n-2}{n-4}} = \frac{\binom{n}{2}3^k}{\binom{n-2}{2}} = \frac{n(n-1)3^k}{(n-2)(n-3)}.$$

(i)   When $(3, n) = 1$, then $(n - 3, n) = 1$ and $(n - 3, 3^k) = 1$. Since $((n - 1), (n - 2)) = 1$, we have $(n - 3) \mid (n - 1)$, i.e. $\dfrac{n-1}{n-3} = 1 + \dfrac{2}{n-3}$ is a positive integer. Therefore $n - 3 \mid 2$ i.e. $n = 4$ or 5. Since $\binom{n-2}{2} \geq 3$, therefore $n \geq 5$, hence $n = 5$.

(ii)   When $3 \mid n$, then $3 \mid n - 3$ and hence $(3, n - 2) = 1$. Since $(n - 2, n - 1) = 1$, we have $n - 2 \mid n$, hence $n - 2 \mid 2$. Therefore $n = 3$ or 4 which contradicts $n \geq 5$.

Thus $n = 5$ is the unique solution.

2.   Consider the set A formed by all nonnegative integers whose digits are not greater than the corresponding digits of n on each digit place. Therefore there are only two choices (0 and 1) for the places where the digit of n is 1, and three choices (0, 1 and 2) for the places where the digit of n is 2, $\cdots$, and there are a total of 10 choices for the places where the digit of n is 9. Thus, $|A| = 2^{\alpha_1} 3^{\alpha_2} \cdots 9^{\alpha_8} 10^{\alpha_9}$. It is obvious that every number in A is not greater than n, therefore $|A| \leq n + 1$, i.e.

$$2^{\alpha_1} 3^{\alpha_2} \cdots 9^{\alpha_8} 10^{\alpha_9} \leq n + 1.$$

3.   If some two judges give the same judgment to some one contestant, we say that there is a "same judgment". Every two fixed judges can make at most k "same judgments", therefore

The total number of the "same judgments" is not greater than $k\binom{b}{2}$.   (∗)

On the other hand, for a contestant i ($i = 1, 2, \cdots, a$), if there are n_i and $b - n_i$ judges who give "pass" and "fail" respectively, then the total number of "same judgments" mentioning the contestant is given by

$$
\binom{n_i}{2} + \binom{b - n_i}{2}
$$
$$
= \frac{n_i(n_i - 1) + (b - n_i)(b - n_i - 1)}{2} = \frac{n_i^2 + (n_i - b)^2 - b}{2}
$$
$$
= \frac{2n_i^2 - 2n_i b + b^2 - b}{2} = \frac{4n_i^2 - 4n_i b + b^2 + b^2 - 2b}{4}
$$
$$
= \frac{(2n_i - b)^2 + (b - 1)^2 - 1}{4} \geq \left(\frac{b-1}{2}\right)^2
$$

since $(2n_i - b)^2 \geq 1$ for odd b, therefore,

$$
k\binom{b}{2} \geq \sum_{i=1}^{a}\left[\binom{n_i}{2} + \binom{b - n_i}{2}\right] \geq a\left(\frac{b-1}{2}\right)^2,
$$
$$
\frac{k}{a} \geq \frac{b-1}{2b}.
$$

4.   Let the number of ways be $a_n(m)$, or a_n, for short.

   (i)   When $n = 2$, then $a_2(m) = {}^m P_2 = m(m-1)$.

   (ii)   Find a recurrence relation. Since there are m ways to colour S_1, $(m - 1)$ ways to colour S_2, $\cdots$, $(m - 1)$ ways to colour S_{n-1}, and if we still use $m - 1$ ways to color S_n, then we have $m(m - 1)^{n-1}$ ways to colour $S_1, S_2, \cdots, S_n$, here S_1 and S_n may be coloured with same colour.

We have $a_n(m)$ ways of colourings such that S_1 and S_n are of different colours, and when S_1 and S_n have the same colour, then the number of ways of colouring is one for colouring $n - 1$ sectors, so there are $a_{n-1}(m)$ ways for this. Hence for $n \geq 2$ we have

$$
a_n(m) + a_{n-1}(m) = m(m - 1)^{n-1}. \tag{30.57}
$$

(iii)   To solve for a_n define $b_n = \dfrac{a_n}{(m-1)^n}$. By (30.57),

$$b_n + \frac{1}{m-1}b_{n-1} = \frac{m}{m-1} \quad \text{or} \quad b_n(m) - 1 = -\frac{1}{m-1}(b_{n-1} - 1),$$

therefore

$$b_n(m) - 1$$

$$= (b_2(m) - 1)\left(-\frac{1}{m-1}\right)^{n-2} = \left[\frac{m(m-1)}{(m-1)^2} - 1\right]\frac{(-1)^n}{(m-1)^{n-2}}$$

$$= \frac{1}{m-1} \cdot \frac{(-1)^n}{(m-1)^{n-2}} = \frac{(-1)^n}{(m-1)^{n-1}},$$

$$\Rightarrow a_n = (m-1)^n b_n(m) = (-1)^n(m-1) + (m-1)^n.$$

5.   Label the 24 partition points by $1, 2, \cdots, 24$ in that order, and arrange the 24 numbers in the table as follows:

1	4	7	10	13	16	19	22
9	12	15	18	21	24	3	6
17	20	23	2	5	8	11	14

Then the length of arc between any two adjacent numbers (including the first and the last) in each row is 3, and the length of arc between any two adjacent numbers (including the first and the last) in each column is 8. Therefore from each column we need to select exactly one number, and any two numbers in two adjacent columns (including the first and the columns) cannot be in same row.

When we consider each column as a sector and each row as one colour, then we have 8 sectors and three colours, the question becomes to find the number of ways for colouring the 8 sectors with three colours such that any two adjacent sectors are coloured by different colours.

From the result of Q4 above, we have

$$a_8(3) = (-1)^8(3-1) + (3-1)^8 = 2 + 2^8 = 258.$$

6.   For $n \geq 2$ use $g(n)$ to denote the number of wave numbers $\overline{a_1 a_2 \cdots a_n}$ satisfying the condition $a_n > a_{n-1}$. By symmetry, we have $2g(n) = f(n)$.

Use $m(i)$ to denote the number of wave numbers $\overline{a_1 a_2 \cdots a_{n-1}}$ satisfying with $a_{n-1} = i, a_{n-2} > a_{n-1}$, then

$$m(4) = 0 \quad \text{and} \quad m(1) + m(2) + m(3) = g(n-1).$$

If $a_{n-1} = 1$ then $a_{n-2} = 2, 3, 4$, so $m(1) = g(n-2)$. If $a_{n-1} = 3$, then $a_{n-2} = 4, a_{n-3} = 1, 2, 3$, so $m(3) = g(n-3)$. Thus,

$$g(n) = 3m(1) + 2m(2) + m(3) = 2g(n-1) + g(n-2) - g(n-3).$$

It is easy to find that $g(2) = 6, g(3) = 14, g(4) = 31$, and so from above recursive formula, we have the results as shown in the following table.

n	5	6	7	8	9	10
$g(n)$	70	157	353	793	1782	4004

Therefore $f(10) = 2g(10) = 8008$.

Note that the sequence of remainders of $g(n)$ mod 13 has the following pattern:

$$6, 1, 5, 5, 1, 2, 0, 1, 0, 1, 1, 3, 6, 1, 5, 5, \cdots,$$

so it is a periodic sequence with the minimum period 12. Therefore

$$g(2008) \equiv g(4) \equiv 5 \pmod{13} \Rightarrow f(2008) \equiv 10 \pmod{13}.$$

7. Let x be the number of distances that appeared exactly once, y the number of distances that appeared exactly twice. We want to find a lower bound for $x + y$.

Let the points be $P_1, P_2, \ldots, P_n$ from left to right, then P_1 is the left endpoint of $n - 1$ distinct distances. For the point P_2, it is the left endpoint of $n - 2$ distinct distances, where some may have appeared in the previous $n - 1$ distances, but this repeat can appear at most once:

If $P_1 P_i = P_2 P_j$ and $P_1 P_k = P_2 P_l$, then $P_1 P_2 = P_i P_j = P_k P_l$, so it contradicts the fact that a same distance can appear at most twice. Thus, among the $(n-2)$ distinct distances starting from P_2, at least $n-3$ are new.

Similarly, among the $n - 3$ distinct distances starting from P_3, at least $n - 3 - 2 = n - 5$ are new. Thus, a lower bound of $x + y$ is given by

$$(n-1) + (n-3) + (n-5) + \cdots = \begin{cases} \dfrac{n^2 - 1}{4} & \text{for odd } n \\ \dfrac{n^2}{4} & \text{for even } n. \end{cases}$$

Thus, $x + y \geq \left\lfloor \dfrac{n^2}{4} \right\rfloor$ for $n \in \mathbb{N}$, and $2x + 2y \geq \left\lfloor \dfrac{n^2}{2} \right\rfloor$. Since $x + 2y = \dfrac{n(n-1)}{2}$, so

$$x \geq \left\lfloor \frac{n^2}{2} \right\rfloor - \left(\frac{n^2}{2} - \frac{n}{2} \right) = \left\lfloor \frac{n}{2} \right\rfloor.$$

8. Let a_i be the number of i-digit integers formed by some or all of the four digits, namely, $0, 1, 2$, and 3, such that these numbers contain none of the two blocks 12 and 21, and b_{1i} and b_{2i} be the number of i-digit integers formed by some or all of the four digits, such that the last digit is 1 and 2 respectively and these integers contain none of the two blocks 12 and 21. Then, by symmetry we have $b_{1i} = b_{2i}$, which we denote by b_i.

    By considering the last digit of an n-digit number, we have the reduction formulas: for $n \geq 3$,

    $$a_n = 2a_{n-1} + 2b_n \qquad \text{and} \qquad b_n = 2a_{n-2} + b_{n-1}.$$

    Therefore for $n \geq 3$

    $$a_n - 2a_{n-1} = 2b_n = 4a_{n-2} + 2b_{n-1} = 4a_{n-2} + (a_{n-1} - 2a_{n-2}),$$
    $$a_n = 3a_{n-1} + 2a_{n-2}.$$

    It is obvious that $a_1 = 3$ (since the first digit is not zero), $a_2 = 3 \times 4 - 2 = 10$, so

n	3	4	5	6	7	8	9
a_n	36	128	456	1624	5784	20600	73368

9. For each convex polygon P whose vertices are in S, let $c(P)$ be the number of points of S which are inside P, so that $a(P) + b(P) + c(P) = n$, the total number of points in S. Denoting $1 - x$ by y,

    $$\sum_P x^{a(P)} y^{b(P)} = \sum_P x^{a(P)} y^{b(P)} (x + y)^{c(P)}$$
    $$= \sum_P \sum_{i=0}^{c(P)} \binom{c(P)}{i} x^{a(P)+i} y^{b(P)+c(P)-i}.$$

    View this expression as a homogeneous polynomial of degree n in two independent variables x, y. In the expanded form, it is the sum of terms $x^r y^{n-r}$ ($0 \leq r \leq n$) multiplied by some nonnegative integer coefficients.

    For a fixed r, the coefficient of $x^r y^{n-r}$ represents the number of ways of choosing a convex polygon P and then choosing some of the points of S inside P so that the number of vertices of P and the number of chosen points inside P jointly add up to r.

    This corresponds to just choosing an r-element subset of S. The correspondence is bijective because every set T of points from S splits in exactly one way into the union of two disjoint subsets, of which the first is the set of vertices of a convex polygon — namely, the convex hull of T — and the second consists of some points inside that polygon.

Thus, the coefficient of $x^r y^{n-r}$ equals $\binom{n}{r}$. The desired result follows:

$$\sum_P x^{a(P)} y^{b(P)} = \sum_{r=0}^{n} \binom{n}{r} x^r y^{n-r} = (x + y)^n = 1.$$

Testing Questions   (29-B)

1. Use n to replace 10001, and use two ways to count the ordered triples (a, R, S), where a, R, S denote a student, a club and a society such that $a \in R$ and $R \in S$. Such a triple is called "admissible".

    Fix a student a and a society S. Then, by condition (ii), there exists a unique club R such that the triple (a, R, S) is admissible. Since there are nk ways to choose the ordered pair (a, S), so there are nk admissible triples in total.

    Fix a club R, and use $|R|$ to denote the number of students in R. Then, by condition (iii), R belongs to $\dfrac{|R| - 1}{2}$ societies. Therefore there are a total of $\dfrac{|R|(|R| - 1)}{2}$ admissible triples containing the R. Letting M be the set of all clubs, then the total number of all admissible triples is $\sum_{R \in M} |R|(|R| - 1)/2$. Therefore

    $$nk = \sum_{R \in M} \frac{|R|(|R| - 1)}{2}.$$

    As $\displaystyle\sum_{R \in M} \frac{|R|(|R| - 1)}{2} = \sum_{R \in M} \binom{|R|}{2}$, by condition (1) we have $\displaystyle\sum_{R \in M} \binom{|R|}{2} = \binom{n}{2}$. Thus, $nk = n(n - 1)/2$ i.e. $k = \dfrac{n - 1}{2}$. For $n = 10001, k = 5000$.

2. Let N_c be the set of good numbers of which each is the maximum value of its column and is the median of its row. Since they are in distinct rows, so $|N_c| \leq m$.

    Let a be the maximum element in N_c, and let $a_1, a_2, \cdots, a_{\frac{n-1}{2}}$ be the numbers greater than a in the same row as a. Hence there are no elements of N_c in the columns that contain some a_i. Hence, $|N_c| \leq \dfrac{n + 1}{2}$. Thus,

    $$|N_c| \leq \min\left\{m, \frac{n + 1}{2}\right\}.$$

Similarly consider the good numbers of which each is the maximum of its row and the median of its column. It then follows that the total number of good numbers is not greater than

$$\min\left\{m, \frac{n+1}{2}\right\} + \min\left\{n, \frac{m+1}{2}\right\}.$$

Below we prove that for any odd integers $m, n > 1$ there must exist a method to fill in the numbers such that the number of good numbers is equal to $\min\left\{m, \frac{n+1}{2}\right\} + \min\left\{n, \frac{m+1}{2}\right\}$.

(i) When $m \neq n$, without loss of generality we always assume that $1 < m < n$.

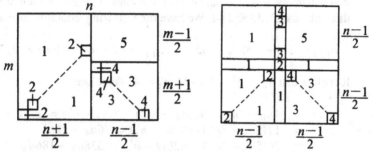

As shown in the left figure above, partition the squares in the rectangle of size $m \times n$ as regions 1 to 5, and then sequentially in the regions 1 to 5 write down the numbers $1, 2, 3, \ldots, mn$. Then the numbers in regions 2 and 4 are good numbers, and the total number of such numbers is

$$\min\left\{m, \frac{n+1}{2}\right\} + \frac{m+1}{2} = \min\left\{m, \frac{n+1}{2}\right\} + \min\left\{n, \frac{m+1}{2}\right\}.$$

(ii) When $m = n > 3$, as shown in the right figure above, write down the numbers $1, 2, \ldots, mn$ sequentially in the regions 1 to 5, then the numbers in regions 2 and 4 are good numbers, and the total number of such numbers is

$$\frac{n-1}{2} + \frac{n-1}{2} + 2 = n + 1.$$

(iii) When $m = n = 3$, as shown in the right figure, the filled numbers $5, 6, 7, 8$ are good numbers, so the number of good numbers is

$$4 = 2\min\left\{n, \frac{m+1}{2}\right\}.$$

1	7	9
2	6	4
5	3	8

In summary, $\min\left\{m, \dfrac{n+1}{2}\right\} + \min\left\{n, \dfrac{m+1}{2}\right\}$ is the maximum value we seek.

3. *Solution 1*   Let n_i be the number of permutations of i-digits formed by some or all of the five digits, namely, $0, 1, 2, 3,$ and 4, such that these permutations contain none of the three blocks $22, 33$ and 44, and $n'_{2i}, n'_{3i}, n'_{4i}$ be the number of i-digit integers formed by some or all of the five digits, such that the first digit is $2, 3, 4$ respectively and these integers contain none of the three blocks $22, 33$ and 44. Then, by symmetry we have $n'_{2i} = n'_{3i} = n'_{4i}$, denoted by n'_i.

Let x be the number of 7-digit integers formed by some or all of the five digits, namely, $0, 1, 2, 3,$ and 4, such that these integers contain none of the three blocks $22, 33$ and 44. We have the following reduction formulas

$$n_{i+1} = 5n_i - 3n'_i \quad \text{and} \quad n'_{i+1} = n_i - n'_i \quad \text{for } i = 2, 3, 4, 5.$$

It is clear that $n_2 = 5^2 - 3 = 22, n'_2 = 4$. Therefore

$$
\begin{aligned}
x &= 4n_6 - 3n'_6 = 4(5n_5 - 3n'_5) - 3(n_5 - n'_5) = 17n_5 - 9n'_5 \\
&= 17(5n_4 - 3n'_4) - 9(n_4 - n'_4) = 76n_4 - 42n'_4 \\
&= 76(5n_3 - 3n'_3) - 42(n_3 - n'_3) = 338n_3 - 186n'_3 \\
&= 338(5n_2 - 3n'_2) - 186(n_2 - n'_2) = 1504n_2 - 828n'_2 \\
&= 1504 \cdot 22 - 828 \cdot 4 = 29776.
\end{aligned}
$$

Solution 2   Let a_i be the number of i-digit integers formed by some or all of the five digits, namely, $0, 1, 2, 3,$ and 4, such that these numbers contain none of the three blocks $22, 33$ and 44, and b_{2i}, b_{3i}, b_{4i} be the number of i-digit integers formed by some or all of the five digits, such that the last digit is $2, 3, 4$ respectively and these integers contain none of the three blocks $22, 33$ and 44, then, by the symmetry, we have $b_{2i} = b_{3i} = b_{4i}$, denoted by b_i.

By considering the last digit of an n-digit number, we have the reduction formulas: for $n \geq 3$,

$$a_n = 2a_{n-1} + 3b_n \quad \text{and} \quad b_n = 2a_{n-2} + 2b_{n-1}.$$

Therefore for $n \geq 3$

$$
\begin{aligned}
a_n - 2a_{n-1} &= 3b_n = 6a_{n-2} + 6b_{n-1} = 6a_{n-2} + 2(a_{n-1} - 2a_{n-2}), \\
a_n &= 4a_{n-1} + 2a_{n-2}.
\end{aligned}
$$

It is obvious that $a_1 = 4, a_2 = 4 \times 5 - 3 = 17$, so

$$a_3 = 4 \times 17 + 2 \times 4 = 76,$$
$$a_4 = 4 \times 76 + 2 \times 17 = 338,$$
$$a_5 = 4 \times 338 + 2 \times 76 = 1504,$$
$$a_6 = 4 \times 1504 + 2 \times 338 = 6692,$$
$$a_7 = 4 \times 6692 + 2 \times 1504 = 29776.$$

4. For any natural number n, let a_n be the number of distinct codes, b_n be the number of codes such that A_1 and A_2 have same digit and same colour, and c_n be the number of distinct codes such that A_1 and A_2 have the same digit only or same color only, and d_n be the number of codes such that only one pair of adjacent two vertices, say A_1 and A_2 have different digits and different colors. Then

$$a_n + d_n = 4 \times 3^{n-1}, \quad a_n = b_n + c_n, \quad b_n = a_{n-1}.$$

According to the status of the pair (A_1, A_3), the following relations hold:

(i)   When A_1 and A_3 have the same digit and the same color, since A_2 has three choices in this case, the number of such codes is $3b_{n-1}$.

(ii)   When A_1 and A_3 have the same digit only or the same color only, since A_2 has two choices in this case, the number of such codes is $2c_{n-1}$.

(iii)   When A_1 and A_3 have different digits and different colors, since A_2 has two choices in this case, the number of such codes is $2d_{n-1}$.

As a result,

$$\begin{aligned}
a_n &= 3b_{n-1} + 2c_{n-1} + 2d_{n-1} \\
&= b_{n-1} + 2(b_{n-1} + c_{n-1}) + 2(4 \times 3^{n-2} - a_{n-1}) \\
&= a_{n-2} + 8 \times 3^{n-2}.
\end{aligned}$$

Since $a_1 = 4, a_2 = 4 \times 3 = 12, a_3 = 4 + 8 \times 3 = 28$ and $a_4 = 12 + 8 \times 9 = 84$, by induction, we have

$$a_n = \begin{cases} 3^n + 1, & \text{for odd } n, \\ 3^n + 3, & \text{for even } n. \end{cases}$$

5. Let a_n be the number of $2 \times n$ grids with stranded black squares and b_n be the number of that with no stranded black squares. Then

$$a_n + b_n = 2^{2n} = 4^n \quad \text{and} \quad a_1 = 0.$$

Below we use the recurrence method to find the expression of a_n in terms of n.

According to the definition, the stranded black squares can be only on the second row.

(1)   When there is a stranded black square in the first $n-1$ columns, then there must be stranded blacks when we color the two squares in the nth column, so there are $2^2 a_{n-1} = 4a_{n-1}$ possible cases.

(2)   When the first $n-1$ squares in the second row are not all black and the first $n-1$ columns do not contain a stranded square, then the black square on the nth column and second row is the unique stranded black square. If the first $n-1$ squares in the second row are all black, then the first $n-1$ squares in the first row can be arbitrarily coloured without producing a stranded square, so they cover 2^{n-1} possible cases. Hence, the number of cases where the first $n-1$ squares in the second row are not all black there is no stranded black square is

$$b_{n-1} - 2^{n-1} = 2^{2(n-1)} - a_{n-1} - 2^{n-1}.$$

Thus,

$$a_n = 4a_{n-1} + 4^{n-1} - a_{n-1} - 2^{n-1} = 3a_{n-1} + 4^{n-1} - 2^{n-1}.$$

Since $a_1 = 0$, by solving this recursive equation,

$$a_n = 4^n + 2^n - 6 \times 3^{n-1} = 4^n + 2^n - 2 \times 3^n \Rightarrow b_n = 2 \times 3^n - 2^n.$$

Solutions to Test Questions   30

Testing Questions   (30-A)

1.   There are more than one function that satisfy the given equation. For example, the function f given by

$$f(n) = \begin{cases} n, & n \geq 0; \\ -3n, & n < 0 \end{cases}$$

is one such functions:

(i)   When $n > 0$, then $f(f(n) - 2n) = f(-n) = 3n = 2f(n) + n$.

(ii)   When $n < 0$, then $f(f(n) - 2n) = f(-5n) = -5n = 2f(n) + n$.

(iii)   When $n = 0$, then $f(f(n) - 2n) = f(0) = 0 = 2f(n) + n$.

2. Let $y = -f(x + 1)$ in (30.18), then $f(x + f(-f(x + 1))) = 0$, so there exists a real number a such that $f(a) = 0$.

(i) When $a \neq 1$, then letting $y = x + 1$ in the given equation yields

$$f(x + f(x + 1)) = x + 1 + f(x + 1).$$

Let $g(x) = x + f(x + 1)$, then g is continuous and $f(g(x)) = g(x) + 1$. Let $y = a$ in (30.18), then $f(x) = a + f(x + 1)$ or

$$g(x - 1) - g(x) = a - 1.$$

Thus, $g(x)$ is unbounded, so the range of g is $\mathbb{R}$. Then $f(x) = x+1, x \in \mathbb{R}$. By checking, $f(x) = x + 1, x \in \mathbb{R}$ satisfies the conditions in the question.

(ii) When $a = 1$, i.e., $f(1) = 0$, letting $x = 0$ in (30.18) yields $f(f(y)) = y$. Let $y = f(1 - x)$ in (30.18), then

$$f(1-x) + f(x+1) = f(x + f(f(1-x))) = f(x+1-x) = f(1) = 0.$$

Let $y = 1 - x$ in (30.18), then

$$f(x+f(1-x)) = 1-x+f(x+1) = 1-x-f(1-x) = 1-(x+f(1-x)).$$

Let $h(x) = x + f(1 - x)$, then $f(h(x)) = 1 - h(x)$. Let $y = 1$ and use $-x$ to replace x in the original equation, then

$$f(-x) = 1 + f(1 - x),$$

so $h(x + 1) - h(x) = 2$, i.e. h is an unbounded continuous function, so its range is all real numbers. Thus, $f(h(x)) = 1 - h(x)$ implies that $f(x) = 1 - x, x \in \mathbb{R}$. By checking $f(x) = 1 - x$ satisfies the original equation. Thus,

$$f(x) = 1 + x \quad \text{or} \quad f(x) = 1 - x.$$

3. First we will show that f is injective. If $a \neq b$ but $f(a) = f(b)$, then for each n we have $f(a) + f(n) \mid (a + n)^k$ and $f(a) + f(n) = f(b) + f(n) \mid (b+n)^k$. Thus, $f(a) + f(n)$ is a common divisor of $(a+n)^k$ and $(b+n)^k$. If n satisfies the condition that $\gcd(a+n, b+n) = 1$ then this can not happen. But $\gcd(a + n, b + n) = (a + n, b - a)$ and if $b - a \neq 0$ there is number $a + n$ that is relatively prime to it. (For example, we can choose n such that $a + n$ is a very big prime)

Now let b be a natural number. For every n we have $f(n) + f(b) \mid (n+b)^k$ and $f(n) + f(b + 1) \mid (n + b + 1)^k$. But $(n + b)$ and $(n + b + 1)$ are relatively prime to each other, so $1 = (f(n) + f(b), f(n) + f(b + 1)) = (f(n) + f(b), f(b + 1) - f(b))$.

We want to show that $f(b + 1) - f(b) = \pm 1$. If it is not equal to ± 1, then there is a prime p which divides it. Let a be such that $p^a > b$. Now put $n = p^a - b$. We have $f(n) + f(b) \mid (n + b)^k = p^{ak}$, so $p \mid f(n) + f(b)$. We had $p \mid f(b + 1) - f(b)$, so $p \mid \gcd(f(n) + f(b), f(b + 1) - f(b))$ which contradicts the previous argument.

Hence, $f(b+1) - f(b) = \pm 1$ for every number b. But since f is injective, it is either always equal to 1 or always equal to -1. (This is because for two consecutive b's it cannot change sign.)

However it cannot always be equal to -1, because f takes only positive integer values. So $f(b + 1) - f(b) = 1$ for every number b. Hence there is a number c such that $f(n) = n + c$. c is non-negative because $f(1) = 1 + c$ is positive. If c is positive, then take a prime p greater than $2c$. Now $f(1) + f(p - 1) \mid p^k$ which shows that $p \mid f(1) + f(p - 1) = p + 2c$. But this is a contradiction because $2c < p$.

In conclusion, $c = 0$ and the function must be $f(n) = n$, which obviously satisfies the conditions of the problem statement.

4. (a) Let $L_1 := \{2k : k > 0\}$, $E_1 := \{0\} \cup \{4k + 1 : k \geq 0\}$, $G_1 := \{4k + 3 : k \geq 0\}$.

We will show that $L_1 = L$, $E_1 = E$, and $G_1 = G$. It suffices to verify that $L_1 \subseteq L$, $E_1 \subseteq E$, and $G_1 \subseteq G$ because L_1, E_1, and G_1 are mutually disjoint and $L_1 \cup E_1 \cup G_1 = \mathbb{N}_0$.

Firstly, if $k > 0$, then $f(2k) - f(2k + 1) = -k < 0$ and therefore $L_1 \subseteq L$.

Secondly, $f(0) = 0$ and

$$
\begin{aligned}
f(4k + 1) &= 2k + 2f(2k) = 2k + 4f(k), \\
f(4k + 2) &= 2f(2k + 1) = 2(k + 2f(k)) = 2k + 4f(k)
\end{aligned}
$$

for all $k \geq 0$. Thus, $E_1 \subseteq E$.

Lastly, in order to prove $G_1 \subseteq G$, we claim that $f(n + 1) - f(n) \leq n$ for all n. (In fact, one can prove a stronger inequality : $f(n + 1) - f(n) \leq n/2$.) This is clearly true for even n from the definition of f since for $n = 2t$,

$$f(2t + 1) - f(2t) = t \leq n.$$

If $n = 2t + 1$ is odd, then (assuming inductively that the result holds for all nonnegative $m < n$), we have

$$
\begin{aligned}
f(n + 1) - f(n) &= f(2t + 2) - f(2t + 1) = 2f(t + 1) - t - 2f(t) \\
&= 2(f(t + 1) - f(t)) - t \leq 2t - t = t < n.
\end{aligned}
$$

For all $k \geq 0$,

$$
\begin{aligned}
f(4k+4) - f(4k+3) &= f(2(2k+2)) - f(2(2k+1)+1) \\
&= 4f(k+1) - (2k+1+2f(2k+1)) \\
&= 4f(k+1) - (2k+1+2k+4f(k)) \\
&= 4(f(k+1) - f(k)) - (4k+1) \\
&\leq 4k - (4k+1) < 0.
\end{aligned}
$$

This proves $G_1 \subseteq G$.

(b) Note that $a_0 = a_1 = f(1) = 0$. Let $k \geq 2$ and let $N_k = \{0, 1, 2, \ldots, 2^k\}$. First we claim that the maximum a_k occurs at the largest number in $G \cap N_k$, that is, $a_k = f(2^k - 1)$. We use mathematical induction on k to prove the claim. Note that $a_2 = f(3) = f(2^2 - 1)$.

Now let $k \geq 3$. For every even number $2t$ with $2^{k-1} + 1 < 2t \leq 2^k$,

$$ f(2t) = 2f(t) \leq 2a_{k-1} = 2f(2^{k-1}-1) \tag{†} $$

by the induction hypothesis. For every odd number $2t+1$ with $2^{k-1} + 1 \leq 2t + 1 < 2^k$,

$$
\begin{aligned}
f(2t+1) = t + 2f(t) &\leq 2^{k-1} - 1 + 2f(t) \\
&\leq 2^{k-1} - 1 + 2a_{k-1} = 2^{k-1} - 1 + 2f(2^{k-1} - 1)
\end{aligned} \tag{‡}
$$

by the induction hypothesis. Combining (†), (‡) and

$$ f(2^k - 1) = f(2(2^{k-1} - 1) + 1) = 2^{k-1} - 1 + 2f(2^{k-1} - 1), $$

we may conclude that $a_k = f(2^k - 1)$ as desired.
Furthermore, we obtain

$$ a_k = 2a_{k-1} + 2^{k-1} - 1 $$

for all $k \geq 3$. Note that this recursive formula for a_k also holds for $k = 1$ and 2. Unwinding this recursive formula, we get

$$
\begin{aligned}
a_k &= 2a_{k-1} + 2^{k-1} - 1 = 2(2a_{k-2} + 2^{k-2} - 1) + 2^{k-1} - 1 \\
&= 2^2 a_{k-2} + 2 \cdot 2^{k-1} - 2 - 1 \\
&= 2^2(2a_{k-3} + 3^{k-3} - 1) + 2 \cdot 2^{k-1} - 2 - 1 \\
&= 2^3 a_{k-3} + 3 \cdot 2^{k-1} - 2^2 - 2 - 1 \\
&\ \ \vdots \\
&= 2^k a_0 + k2^{k-1} - 2^{k-1} - 2^{k-2} - \cdots - 2 - 1 \\
&= k2^{k-1} - 2^k + 1
\end{aligned}
$$

for all $k \geq 0$.

5. Let $v = u$ be any real value, then

$$f(2u) = f(2u)f(0) + f(0)f(-2u). \qquad (30.58)$$

(i) If $f(0) = 0$, then $f(2u) = 0$ for any $u \in \mathbb{R}$, so $f(u) \equiv 0$ identically. By checking, the function satisfies the requirement of problem.

(ii) If $f(0) = c \neq 0$, then (30.58) yields

$$f(-2u) = \frac{1-c}{c}f(2u) \qquad \text{for all } u \in \mathbb{R}. \qquad (30.59)$$

Letting $u = 0$ in (30.59), then $c = 1 - c$, i.e., $c = \dfrac{1}{2}$, and $f(u) = f(-u)$, $u \in \mathbb{R}$. Then the given equation becomes

$$f(2u) = f(u+v)f(u-v) + f(u-v)f(u+v) = 2f(u-v)f(u+v),$$

where $u, v \in \mathbb{R}$. Letting $u = 0$ in it then gives $\dfrac{1}{2} = 2f(v)^2$, so $f(v) = \frac{1}{2}$, $v \in \mathbb{R}$ since $f(v) \geq 0$ for all real v.

Thus, f is 0 identically or $\dfrac{1}{2}$ identically.

6. In this solution the label (30.19) means the equation (30.19) of lecture 30. Let $y = x$ in (30.19), we obtain

$$f(x^3) = xf(x^2). \qquad (30.60)$$

Substituting (30.60) into (30.19), we obtain

$$(x + y)f(xy) = xf(y^2) + yf(x^2). \qquad (30.61)$$

Let $y = -x$ in (30.19), then $f(x^3) = -f(-x^3)$, so

$$f(x) = -f(-x). \qquad (30.62)$$

Use $-y$ to replace y in (30.61), and by (30.62), we have

$$-(x - y)f(xy) = xf(y^2) - yf(x^2). \qquad (30.63)$$

(30.61) + (30.63) gives that

$$yf(xy) = xf(y^2). \qquad (30.64)$$

Letting $y = 1$ in (30.64), we obtain $f(x) = f(1)x$. By checking $f(x) = ax$ where $a \in \mathbb{R}$ satisfies the equation (30.19), so it is a solutions.

7. Let $x = 2$ in (30.21) of lecture 30, then $3f(1) = 2f(2)$, therefore $f(1)$ is even. It is easy to see by induction that

$$f(n) = \frac{1}{2}(n + 1)f(1), \quad n \in \mathbb{N}.$$

For any $p, q \in \mathbb{Z}^+$ with $(p, q) = 1$, since $f\left(\frac{p}{q}\right) = f\left(\frac{q}{p}\right)$, by using (ii) repeatedly, it's easy to prove by induction on q that

$$f\left(\frac{p}{q}\right) = \frac{1}{2}(p + q)f(1).$$

When $p > q$, then

$$\frac{p}{q}f\left(\frac{p}{q}\right) = \frac{p+q}{q}f\left(\frac{p-q}{q}\right) = \frac{p+q}{2q}(p - q + q)f(1).$$

When $p < q$, then $f\left(\frac{p}{q}\right) = f\left(\frac{q}{p}\right)$. Thus, The desired f is given by

$$f\left(\frac{p}{q}\right) = (p + q)m, \quad \text{for } p, q \in \mathbb{Z}^+,$$

where $m \in \mathbb{N}$.

8. Let $y = 0$ in (30.22) of lecture 30, then $f(x) = f(x)g(0) + f(0)$, namely $f(x)(1 - g(0)) = f(0)$. If $g(0) \neq 1$, then $f(x) = \frac{f(0)}{1 - g(0)}$ which is a constant, contradicting the fact that f is strictly increasing. Therefore $g(0) = 1$ and $f(0) = 0$.

Since $f(x)g(y) + f(y) = f(x+y) = f(y)g(x) + f(x) \Rightarrow f(x)(g(y) - 1) = f(y)(g(x) - 1)$. f is strictly increasing and $f(0) = 0$ implies that $f(x) \neq 0$ if $x \neq 0$. Therefore, there is some constant C such that

$$\frac{g(y) - 1}{f(y)} = \frac{g(x) - 1}{f(x)} = C, x, y \neq 0,$$

i.e., $g(x) - 1 = Cf(x)$ for $x \neq 0$. Since $g(0) = 1$, $f(0) = 0$, so above equality holds also for $x = 0$. Then for any $x, y \in \mathbb{R}$,

$$g(x + y) = 1 + Cf(x + y) = 1 + Cf(x)g(y) + Cf(y)$$
$$= g(y) + Cf(x)g(y) = g(y)(1 + Cf(x)) = g(x)g(y).$$

so $g(nx) = g^n(x)$ for all $n \in \mathbb{Z}$. Since f is strictly monotone, so is $g(x) = 1 + Cf(x)$. Letting $n = 2$ and replacing x with $\frac{x}{2}$ in the equation above, we have $g(x) = g^2\left(\dfrac{x}{2}\right) > 0, x \in \mathbb{R}$. Let $g(1) = a\ (a > 0)$, then $g(n) = a^n$.

$$g(x) = (g(nx))^{\frac{1}{n}} \Rightarrow g\left(\frac{m}{n}\right) = (g(m))^{\frac{1}{n}} = a^{\frac{m}{n}},$$

so $g(x) = a^x$ for $x \in \mathbb{Q}$. Since g is strictly monotone, so $g(x) = a^x, x \in \mathbb{R}$.

It is easy to verify that the function $g(x) = a^x, x \in \mathbb{R}$ with $a > 0$ satisfies the condition. In fact, let $f(x) = 1 - a^x$ if $0 < a < 1$; let $f(x) = x$ if $a = 1$; and let $f(x) = a^x - 1$ is $a > 1$.

9. In this solution the label (30.23) means the equation (30.23) of lecture 30.

Let $f(0) = a$. Let $x = y = 0$ in (30.23), then $f\left(\dfrac{a}{2}\right) = f(a) - a$. In (30.23) let $x = \frac{1}{2}a, y = 0$, then $f\left(\dfrac{f(a) - \frac{a}{2}}{2}\right) = 2a$. In (30.23) let $x = a, y = 0$, then $f\left(\dfrac{f(a) + a}{2}\right) = 2a$. Since f is non-decreasing,

$$f(t) = 2a, \qquad \text{for all } t \in \left[\frac{f(a)-\frac{a}{2}}{2}, \frac{f(a)+a}{2}\right],$$

therefore $f\left(\dfrac{f(a) + \frac{a}{2}}{2}\right) = 2a$. On the other hand, In (30.23) let $x = y = \dfrac{a}{2}$, then

$$f\left(\frac{f(a) + \frac{a}{2}}{2}\right) = 2a - f(a) + f(f(a) - a) \Rightarrow f(a) = f(f(a) - a).$$

In (30.23)* let $x = 0, y = \dfrac{a}{2}$, then $f(a) = -a + f(f(a) - a)$, hence $-a = 0$, i.e., $f(0) = 0$.

Now in (30.23) let $x = 0$, then $f(y) = f(f(y))$. Again in (30.23) let $x = y$, then

$$f\left(\frac{3}{2}x + \frac{1}{2}f(x)\right) = 2x - f(x) + f(f(x)) = 2x,$$

so f is a surjection. Thus, for any $x \in \mathbb{R}^+ \cup \{0\}$,

$$2x = f\left(\frac{3}{2}x + \frac{1}{2}f(x)\right) = f\left(f\left(\frac{3}{2}x + \frac{1}{2}f(x)\right)\right) = f(2x),$$

therefore $f(x) = x$ for any $x \in \mathbb{R}^+ \cup \{0\}$.

Test Questions (30-B)

1. Label the given equation:

$$f(x^2 + yf(z)) = xf(x) + zf(y). \qquad (30.65)$$

Substituting $x = y = 0$ into (30.65) yields $f(0) = zf(0), z \in \mathbb{R}$, so $f(0) = 0$. Substituting $x = 0$ into (30.65) yields

$$f(yf(z)) = zf(y). \qquad (30.66)$$

Substituting $y = 1$ into (30.66) yields

$$f(f(z)) = zf(1), \qquad \text{for all } z \in \mathbb{R}. \qquad (30.67)$$

Substituting $y = 0$ into (30.65) yields

$$f(x^2) = xf(x). \qquad (30.68)$$

By using (30.68) and then (30.67),

$$f(xf(x)) = f(f(x^2)) = x^2 f(1). \qquad (30.69)$$

Letting $y = z = x$ in (30.66) and using (30.69), it follows that

$$xf(x) = f(xf(x)) = x^2 f(1), x \in \mathbb{R} \Rightarrow f(x) = f(1)x, x \in \mathbb{R}.$$

Let $f(1) = C$. Substituting it into (30.65) gives $C(x^2 + Cyz) = Cx^2 + Cyz$, so $C^2 = C$, i.e., $C = 0$ or $C = 1$. Taking $C = 1$ to get the maximum value of $f(12345)$, we have

$$\max_{f \in S} f(12345) = 12345.$$

2. First of all we show that f is a bijection. In fact, for any fixed $n \in S$, if $f(m_1) = f(m_2)$ for some $m_1, m_2 \in S$, then

$$m_1^2 + 2n^2 = f(f^2(m_1) + 2f^2(n)) = f(f^2(m_2) + 2f^2(n)) = m_2^2 + 2n^2$$
$$\Rightarrow m_1^2 = m_2^2 \Rightarrow m_1 = m_2.$$

f is clearly a surjection, so f is a bijection.

Consider the identity $(x + 3)^2 + 2x^2 = (x - 1)^2 + 2(x + 2)^2, x \in S$. Therefore

$$f(f^2(x + 3) + 2f^2(x)) = (x + 3)^2 + 2x^2 = (x - 1)^2 + 2(x + 2)^2$$
$$= f(f^2(x - 1) + 2f^2(x + 2)),$$

hence

$$f^2(x+3) + 2f^2(x) = f^2(x-1) + 2f^2(x+2). \tag{30.70}$$

Denote $f^2(n), n \in S$ by a_n, then $a_{n+3} + 2a_n = a_{n-1} + 2a_{n+2}$ for $n \geq 2$. By solving its characteristic equation, we obtain

$$f^2(n) = a_n = an^2 + bn + c + (-1)^n d, \quad \text{where } a, b, c, d \in \mathbb{Q}.$$

Since

$$f(f^2(m) + 2f^2(n)) = m^2 + 2n^2$$
$$\Rightarrow f^2(f^2(m) + 2f^2(n)) = (m^2 + 2n^2)^2$$
$$\Rightarrow f^2(am^2 + bm + 2an^2 + 2bn + 3c + (-1)^m d + 2(-1)^n d)$$
$$= (m^2 + 2n^2)^2,$$

therefore

$$a[am^2 + bm + 2an^2 + 2bn + 3c + (-1)^m d + 2(-1)^n d]^2$$
$$+b[am^2 + bm + 2an^2 + 2bn + 3c + (-1)^m d + 2(-1)^n d]$$
$$+c + (-1)^{am^2 + bm + c + (-1)^m d} d = m^4 + 4n^4 + 4m^2 n^2.$$

The equality above holds for any $m, n \in S$, so it is an identity. The comparison of coefficients of m^4 gives $a = 1$. Substituting it into the identity,

$$[m^2 + bm + 2n^2 + 2bn + 3c + (-1)^m d + 2(-1)^n d]^2$$
$$+b[m^2 + bm + 2n^2 + 2bn + 3c + (-1)^m d + 2(-1)^n d]$$
$$+c + (-1)^{m^2 + bm + c + (-1)^m d} d = m^4 + 4n^4 + 4m^2 n^2.$$

The comparison of the coefficients of m^3 gives $b = 0$. Then

$$[m^2 + 2n^2 + 3c + (-1)^m d + 2(-1)^n d]^2$$
$$+c + (-1)^{m^2 + c + (-1)^m d} d = m^4 + 4n^4 + 4m^2 n^2.$$

The comparison of the coefficients of m^2 and the comparison of the constant terms then gives

$$\begin{cases} 3c + (-1)^m d + 2(-1)^n d = 0, \\ c + (-1)^{m^2 + c + (-1)^m d} d = 0, \end{cases}$$

for all $m, n \in S$. Solving them gives $c = d = 0$. Thus, $f^2(n) = n^2$ or $f(n) = n$. It is clear that such f satisfies all the conditions in question.

3. First we show that f is a bijection. Since f is strictly monotone on $(0, +\infty)$, if $x_1, x_2 > 0$ are such that $f(x_1) = f(x_2)$, then

$$f(f(x_1)) = f(f(x_2)). \tag{30.71}$$

In the given equation use x_1, x_2 to replace x, y respectively, then

$$f\left(\frac{f(x_1)}{f(x_2)}\right) = \frac{f(f(x_1))}{x_2} \Leftrightarrow f(1) = \frac{f(f(x_1))}{x_2} \Leftrightarrow x_2 = \frac{f(f(x_1))}{f(1)}.$$
(30.72)

Similarly, if we use x_2, x_1 to replace x, y respectively, the given equation gives

$$x_1 = \frac{f(f(x_2))}{f(1)}.$$
(30.73)

Combining (30.71), (30.72) and (30.73), we obtain $x_1 = x_2$, so f is an injection. It is obvious that f is surjective, so f is a bijection.

In the given equation let $x = y = 1$, then $f(1) = \dfrac{f(f(1))}{1}$ or $f(1) = f(f(1))$, so $f(1) = 1$. Thus, $f(f(x)) = x, x > 0$.

Let $\mu > 0$, then there exists $\lambda > 0$ such that $\lambda = f(\mu)$. Hence $f(\lambda) = f(f(\mu)) = \mu$. Since $f\left(\dfrac{f(x)}{f(y)}\right) = \dfrac{f(f(x))}{y}$ and $f(f(x)) = x$,

$$f\left(\frac{f(x)}{f(y)}\right) = \frac{x}{y} \Rightarrow f\left(f\left(\frac{f(x)}{f(y)}\right)\right) = f\left(\frac{x}{y}\right) \Rightarrow f\left(\frac{x}{y}\right) = \frac{f(x)}{f(y)}.$$

Let $x = 1$, then $f\left(\dfrac{1}{y}\right) = \dfrac{1}{f(y)}, y \in \mathbb{R}^+$.

Since $f\left(x \cdot \dfrac{1}{y}\right) = f(x) \cdot f\left(\dfrac{1}{y}\right)$ for all $x, y > 0$, so

$$f(xy) = f(x)f(y), \qquad x, y > 0.$$

Let $g(u) = \ln f(e^u), u \in \mathbb{R}$, then

$$g(u + v) = \ln f(e^u \cdot e^v) = \ln[f(e^u) \cdot f(e^v)] = g(u) + g(v).$$

The monotonicity of f implies that of g on $\mathbb{R}^+$, so $g(u) = au, u \in \mathbb{R}$. Thus,

$$au = \ln f(e^u) \Leftrightarrow f(e^u) = e^{au} = (e^u)^a \Leftrightarrow f(x) = x^a, x > 0.$$

Hence for all $x > 0$,

$$f(f(x)) = x \Rightarrow f(x^a) = x \Rightarrow x^{a^2} = x \Rightarrow x^{a^2-1} = 1 \Rightarrow a^2 = 1$$
$$\Leftrightarrow a = \pm 1 \Rightarrow f(x) = x, x > 0 \text{ or } f(x) = f(x) = \frac{1}{x}, x > 0.$$

It is easy to verify that these two solutions satisfy the conditions in question.

4. Throughout the solution, we will use the notation $g_k(x) = \overbrace{g(g(\cdots g(x)\cdots))}^{k}$, including $g_0(x) = x$ as well.

   Suppose that there exists a Spanish Couple (f, g) on the set $\mathbb{N}$. From property (i) we have $f(x) \geq x$ and $g(x) \geq x$ for all $x \in \mathbb{N}$.

   We claim that $g_k(x) \leq f(x)$ for all $k \geq 0$ and all positive integers x. The proof is done by induction on k. We already have the base case $k = 0$ since $x \leq f(x)$. For induction step from k to $k + 1$, apply the induction hypothesis on $g_2(x)$ instead of x, then apply (ii):

   $$g(g_{k+1}(x)) = g_k(g_2(x)) \leq f(g_2(x)) < g(f(x)).$$

   Since g is increasing, it follows that $g_{k+1}(x) < f(x)$. The claim is proven.

   If $g(x) = x$ for all $x \in \mathbb{N}$ then $f(g(g(x))) = f(x) = g(f(x))$, and we have a contradiction with (ii). Therefore one can choose an $x_0 \in \mathbb{N}$ for which $x_0 < g(x_0)$. Now consider the sequence $x_0, x_1, \ldots$ where $x_k = g_k(x_0)$. The sequence is strictly increasing. Indeed, we have $x_0 < g(x_0) = x_1$, and $x_k < x_{k+1}$ implies $x_{k+1} = g(x_k) < g(x_{k+1}) = x_{k+2}$.

   Hence, we obtain a strictly increasing sequence $x_0 < x_1 < \cdots$ of positive integers which has an upper bound, namely $f(x_0)$. This cannot happen in the set $\mathbb{N}$ of positive integers, thus no Spanish Couple exists on $\mathbb{N}$.

5. It is not hard to see that the two functions $f(x) = x$ and $f(x) = -x$ for all real x respectively solve the functional equation. In the remainder of the solution, we prove that there are no further solutions.

   Let f be a function satisfying the given equation. It is clear that f cannot be a constant. Let us first show that $f(0) = 0$. Suppose that $f(0) \neq 0$. For any real t, substituting $(x, y) = (0, \frac{t}{f(0)})$ into the given functional equation, we obtain

   $$f(0) = f(t), \tag{30.74}$$

   contradicting the fact that f is not a constant function. Therefore, $f(0) = 0$. Next, for any t, substituting $(x, y) = (t, 0)$ and $(x, y) = (t, -t)$ into the given equation, we get

   $$f(tf(t)) = f(0) + t^2 = t^2,$$

   and

   $$f(tf(0)) = f(-tf(t)) + t^2,$$

   respectively. Therefore, we conclude that

   $$f(tf(t)) = t^2, \quad f(-tf(t)) = -t^2, \quad \text{for every real } t. \tag{30.75}$$

Consequently, for every real v, there exists a real u such that $f(u) = v$. We also see that if $f(t) = 0$, then $0 = f(tf(t)) = t^2$ so that $t = 0$, and thus 0 is the only real number satisfying $f(t) = 0$.

We next show that for any real number s,

$$f(-s) = -f(s). \tag{30.76}$$

This is clear if $f(s) = 0$. Suppose now $f(s) < 0$, then we can find a number t for which $f(s) = -t^2$. As $t \neq 0$ implies $f(t) \neq 0$, we can also find a number a such that $af(t) = s$. Substituting $(x, y) = (t, a)$ into the given equation, we get

$$f(tf(t + a)) = f(af(t)) + t^2 = f(s) + t^2 = 0,$$

and therefore, $tf(t + a) = 0$, which implies $t + a = 0$, and hence $s = -tf(t)$. Consequently, $f(-s) = f(tf(t)) = t^2 = -(-t^2) = -f(s)$ holds in this case.

Finally, suppose $f(s) > 0$ holds. Then there exists a real number $t \neq 0$ for which $f(s) = t^2$. Choose a number a such that $tf(a) = s$. Substituting $(x, y) = (t, a - t)$ into the given equation, we get $f(s) = f(tf(a)) = f((a - t)f(t)) + t^2 = f((a - t)f(t)) + f(s)$. Thus, we have $f((a - t)f(t)) = 0$, from which we conclude that $(a - t)f(t) = 0$. Since $f(t) \neq 0$, we get $a = t$ so that $s = tf(t)$ and thus we see $f(-s) = f(-tf(t)) = -t^2 = -f(s)$ holds in this case also. This observation finishes the proof of (30.76).

By substituting $(x, y) = (s, t)$, $(x, y) = (t, -s - t)$ and $(x, y) = (-s - t, s)$ into the given equation, we obtain

$$f(sf(s + t))) = f(tf(s)) + s^2,$$
$$f(tf(-s)) = f((-s - t)f(t)) + t^2,$$

and

$$f((-s - t)f(-t)) = f(sf(-s - t)) + (s + t)^2,$$

respectively. Using the fact that $f(-x) = -f(x)$ holds for all x to rewrite the second and the third equation, and rearranging the terms, we obtain

$$f(tf(s)) - f(sf(s + t)) = -s^2,$$
$$f(tf(s)) - f((s + t)f(t)) = -t^2,$$
$$f((s + t)f(t)) + f(sf(s + t)) = (s + t)^2.$$

Adding up these three equations now yields $2f(tf(s)) = 2ts$, and therefore, we conclude that $f(tf(s)) = ts$ holds for every pair of real numbers

s, t. By fixing s so that $f(s) = 1$, we obtain $f(x) = sx$. In view of the given equation, we see that $s = \pm 1$. It is easy to check that both functions $f(x) = x$ and $f(x) = -x$ satisfy the given functional equation, so these are the desired solutions.

Appendices

Appendix A

Theorem on Second Order Recursive Sequences

Theorem I. *For a sequence $\{a_n\}$ given by*

$$a_{n+1} = pa_n + qa_{n-1}, \qquad n = 2, 3, 4, \cdots,$$

where a_1, a_2 are given as the initial values, if its characteristic equation $t^2 = pt + q$ has two real roots α and β, then for $n \geq 2$,

(i) $\qquad a_n = A\alpha^n + B\beta^n$ *if* $\alpha \neq \beta$;

(ii) $\qquad a_n = (An + B)\alpha^n$ *if* $\alpha = \beta$,

where A, B are constants determined by the initial values a_1 and a_2.

Proof. We write the given recursive formula into the following new form:

$$a_{n+1} - \alpha a_n = \beta(a_n - \alpha a_{n-1}),$$

then it is easy to obtain $a_{n+1} - \alpha a_n = \beta^{n-1}(a_2 - \alpha a_1)$.
(i) When $\alpha \neq \beta$, then $\alpha\beta = -q \neq 0$ implies $\alpha, \beta \neq 0$. Since

$$
\begin{aligned}
a_n - \alpha a_{n-1} &= \beta^{n-2}(a_2 - \alpha a_1), \\
\alpha(a_{n-1} - \alpha a_{n-2}) &= \alpha\beta^{n-3}(a_2 - \alpha a_1), \\
\alpha^2(a_{n-2} - \alpha a_{n-3}) &= \alpha^2\beta^{n-4}(a_2 - \alpha a_1), \\
&\cdots \qquad\qquad \cdots \\
\alpha^{n-2}(a_2 - \alpha a_1) &= \alpha^{n-2}(a_2 - \alpha a_1),
\end{aligned}
$$

by adding up these equalities, we obtain

$$a_n - \alpha^{n-1}a_1 = (a_2 - \alpha a_1)(\beta^{n-2} + \alpha\beta^{n-3} + \cdots + \alpha^{n-2}),$$

261

therefore

$$
\begin{aligned}
a_n &= a_1 \cdot \alpha^{n-1} + (a_2 - \alpha a_1)\frac{\beta^{n-1} - \alpha^{n-1}}{\beta - \alpha} \\
&= \left[a_1 - \frac{a_2 - \alpha a_1}{\beta - \alpha} \right]\alpha^{n-1} + \frac{a_2 - \alpha a_1}{\beta - \alpha}\beta^{n-1} \\
&= \frac{-a_2 + \beta a_1}{\alpha(\beta - \alpha)}\alpha^n + \frac{a_2 - \alpha a_1}{\beta(\beta - \alpha)}\beta^n.
\end{aligned}
$$

(ii)   When $\alpha = \beta$, then

$$
\begin{aligned}
a_n - \alpha^{n-1}a_1 &= (a_2 - \alpha a_1)(\beta^{n-2} + \alpha\beta^{n-3} + \cdots + \alpha^{n-2}) \\
&= (a_2 - \alpha a_1) \cdot (n-1)\alpha^{n-2},
\end{aligned}
$$

therefore

$$
a_n = \left[\left(\frac{a_1}{\alpha} - \frac{a_2 - \alpha a_1}{\alpha^2} \right) + n\left(\frac{a_2 - \alpha a_1}{\alpha^2} \right) \right]\alpha^n.
$$

Thus, the theorem is proven.   □

Appendix B

Proofs of Theorems On Pell's Equation

Theorem I. *The equation*

$$x^2 - dy^2 = 1,$$ (B.1)

where d is a positive non-square integer, has at least one positive integer solution.

Lemma 1. Let β be an irrational number. Then for any positive integer $p > 1$ there exist $x, y \in \mathbb{N}$ with $1 \leq y \leq p$ such that

$$|x - y\beta| < \frac{1}{p}.$$ (B.2)

Proof of Lemma 1. Since the $p + 1$ distinct numbers $0, \{\beta\}, \{2\beta\}, \ldots, \{p\beta\}$ are all in the interval $[0, 1)$, so there are i, j with $0 \leq i < j \leq p$ such that $0 < |\{j\beta\} - \{i\beta\}| < \frac{1}{p}$, namely

$$|(\lfloor j\beta \rfloor - \lfloor i\beta \rfloor) - (j - i)\beta| < \frac{1}{p}.$$

hence it is enough to take $x = \lfloor j\beta \rfloor - \lfloor i\beta \rfloor, y = j - i$.

Consequence 1: There are infinitely many ordered pairs (x, y) of two positive integers such that

$$|x - y\beta| < \frac{1}{y}.$$ (B.3)

From Lemma 1, there is an ordered pair (x_1, y_1) of two positive integers such that (B.3) holds. Take positive integer $p_1 > 1$ satisfying $\frac{1}{p_1} < |x_1 - y_1\beta| < \frac{1}{y_1}$ (p_1 must exist since $|x_1 - y_1\beta|$ is irrational), then Lemma 1 implies that there exists ordered pair (x_2, y_2) of two positive integers such that $|x_2 - y_2\beta| < \frac{1}{p_1} \leq$

$\dfrac{1}{y_2}$. By continuing this process infinitely many times, we can get infinitely many
desired pairs $\{(x_i, y_i)\}, i = 1, 2, \ldots$.

Lemma 2.   When d is a non-square positive integer, there must exist infinitely
many ordered pairs (x, y) of two positive integers, such that

$$|x^2 - dy^2| < 1 + 2\sqrt{d}.$$

Proof of Lemma 2.   By Consequence 1, there exist infinitely many ordered
pairs (x, y) of two positive integers such that $|x - \sqrt{d}\,y| < \dfrac{1}{y}$, hence

$$
\begin{aligned}
|x^2 - dy^2| &= |x - \sqrt{d}\,y| \cdot |x + \sqrt{d}\,y| < \frac{1}{y} \cdot |x + \sqrt{d}\,y| \\
&\le \frac{1}{y}(|x - \sqrt{d}\,y| + 2\sqrt{d}\,y) < \frac{1}{y^2} + 2\sqrt{d} \le 1 + 2\sqrt{d}.
\end{aligned}
$$

Consequence 2.   When d is a non-square positive integer, there must exist an
integer k with $0 < |k| < 1 + 2\sqrt{d}$, such that the Pell-type equation

$$x^2 - dy^2 = k \tag{B.4}$$

has infinitely many positive integer solutions (x, y).

Note that there are only finitely many integers k with absolute value less than
$1 + 2\sqrt{d}$, so there must be at least one such k such that the number of solutions
(x, y) of (B.4) is infinite.

Proof.   Consequence 2 implies that there are two solutions of (B.4), $(x_1, y_1) \ne$
(x_2, y_2), such that

$$x_1 \equiv x_2 \pmod{|k|}, \qquad y_1 \equiv y_2 \pmod{|k|},$$

then

$$(x_1^2 - dy_1^2)(x_2^2 - dy_2^2) = (x_1x_2 - dy_1y_2)^2 - d(x_1y_2 - x_2y_1)^2 = k^2,$$

and $x_1x_2 - dy_1y_2 \equiv x_1^2 - dy_1^2 = k \equiv 0, x_1y_2 - x_2y_1 \equiv x_1y_1 - x_1y_1 \equiv 0$
(mod $|k|$), so there are nonnegative integers x, y such that

$$|x_1x_2 - dy_1y_2| = x|k| \quad \text{and} \quad |x_1y_2 - x_2y_1| = y|k|,$$

hence $x^2 - dy^2 = 1$.

Below we show that $y > 0$ (and hence $x > 0$). If $y = 0$, then $x_1y_2 = x_2y_1$.
Let $\dfrac{x_1}{x_2} = \dfrac{y_1}{y_2} = q > 0$, then (B.4) yields

$$k = x_1^2 - dy_1^2 = q^2(x_2^2 - dy_2^2) = q^2 k \Rightarrow q = 1 \Rightarrow (x_1, y_1) = (x_2, y_2),$$

which contradicts $(x_1, y_1) \ne (x_2, y_2)$. Thus, we have proven that (B.1) has at
least one positive integer solution. $\square$

Theorem II. *Let (a, b) be the minimum solution of (B.1), then (x, y) is a positive integer solution of (B.1) if and only if there is $n \in \mathbb{N}$ such that $x + \sqrt{d}\,y = (a + \sqrt{d}\,b)^n$.*

Proof. Suppose that $x + \sqrt{d}\,y = (a + \sqrt{d}\,b)^n$ for some $n \in \mathbb{N}$, then $x - \sqrt{d}\,y = (a - \sqrt{d}\,b)^n$, so

$$x^2 - dy^2 = (a + \sqrt{d}\,b)^n \cdot (a - \sqrt{d}\,b)^n = (a^2 - db^2)^n = 1,$$

i.e., (x, y) is a positive integer solution of (B.1).

Conversely, if (x, y) is a positive integer solution of (B.1), and there is no $n \in \mathbb{N}$ such that $x + \sqrt{d}\,y = (a + \sqrt{d}\,b)^n$, then $x + \sqrt{d}\,y > a + \sqrt{d}\,b$ implies that there is $n \in \mathbb{N}$ such that $(a + \sqrt{d}\,b)^n < x + \sqrt{d}\,y < (a + \sqrt{d}\,b)^{n+1}$, therefore

$$1 < (x + \sqrt{d}\,y)(a - \sqrt{d}\,b)^n < a + \sqrt{d}\,b. \tag{B.5}$$

Let $(x + \sqrt{d}\,y)(a - \sqrt{d}\,b)^n = u + \sqrt{d}\,v$. It is easy to see that (u, v) is an integer solution of (B.1).

Since $u + \sqrt{d}\,v > 1 \Rightarrow 0 < u - \sqrt{d}\,v = \dfrac{1}{u + \sqrt{d}\,v} < 1 \Rightarrow 2u > 1 \Rightarrow u > 0$.

Also $2\sqrt{d}\,v > 1 - 1 = 0 \Rightarrow v > 0$, so (u, v) is a positive integer solution of (B.1) and $u + \sqrt{d}\,v < a + \sqrt{d}\,b$, which contradicts the fact that $a + \sqrt{d}\,b$ is the minimum solution of (B.1). Thus, the conclusion of the Theorem II is proven. $\square$

Consequence 3. The Pell's equation (B.1) must have infinitely many positive integer solutions (x, y). If (a, b) is the minimum solution of (B.1), then all the positive integer solutions (x_n, y_n) are given by

$$\begin{cases} x_n &=& \dfrac{1}{2}[(a + \sqrt{d}\,b)^n + (a - \sqrt{d}\,b)^n], \\[2mm] y_n &=& \dfrac{1}{2\sqrt{d}}[(a + \sqrt{d}\,b)^n - (a - \sqrt{d}\,b)^n]. \end{cases} \tag{B.6}$$

Below we introduce without proof another theorem on the Pell-type equation $x^2 - dy^2 = -1$.

Theorem III. *When d is a non-square positive integer, and if the equation*

$$x^2 - dy^2 = -1 \tag{B.7}$$

has a positive integer solution, then equation (B.7) has infinitely many positive integer solutions, and if (a, b) is the positive integer solution with minimum value of $x + \sqrt{d}\,y$ among all positive integer solutions (x, y), then all the positive integer solutions (x, y) of (B.7) can be expressed as

$$x + \sqrt{d}\,y = (a + \sqrt{d}\,b)^{2n+1},$$

and if (x_0, y_0) is the minimum solution of (B.1), then

$$x_0 + \sqrt{d}\, y_0 = (a + \sqrt{d}\, b)^2.$$

Appendix C

Theorems On Quadratic Residues

Theorem I. *For any odd prime p, the numbers of non-zero quadratic residues modulo p and quadratic non-residues modulo p are both $\dfrac{p-1}{2}$.*

Proof. For an integer a with $(a, p) = 1$, when $x^2 \equiv a \pmod{p}$ has a solution for x, then a must be one of the following $p - 1$ numbers modulo p:

$$\pm 1, \ \pm 2, \cdots, \ \pm \frac{p-1}{2},$$

hence its square is one of $\{1^2, 2^2, \cdots, (p-1)/2)^2\}$. It's obvious that each of these $(p-1)/2$ numbers is a quadratic residue modulo p. Further, any two of these $(p-1)/2$ numbers are not congruent modulo p: Otherwise, there are integers $1 \leq l < k \leq (p-1)/2$ such that $l^2 \equiv k^2 \pmod{p}$, then

$$k^2 - l^2 = (k + l)(k - l) \equiv 0 \pmod{p}.$$

However, $1 \leq k - l < k + l < p - 1$ implies p cannot divide both $k + l$ and $k - l$, a contradiction. Note that, the solution of $x^2 \equiv l^2 \pmod{p}$ and that of $x^2 \equiv k^2 \pmod{p}$ for $1 \leq l < k \leq (p-1)/2$ must be incongruent modulo p.

On the other hand, for each of the quadratic residues l^2 in $\{1^2, 2^2, \cdots, (p-1)/2)^2\}$, there are two solutions x_0 and $-x_0$, and $x_0 \not\equiv -x_0 \pmod{p}$ holds: Otherwise, $2x_0 \equiv 0 \pmod{p}$ implies $x_0 \equiv 0 \pmod{p} \Rightarrow l^2 \equiv 0 \pmod{p}$, a contradiction. Hence corresponding to each of the $(p-1)/2$ quadratic congruence equations of which each has two incongruent solutions, and these $p - 1$ solutions form a complete residue system modulo p. This means that the total number of quadratic residues modulo p is exactly $(p - 1)/2$.

Since there are $p - 1$ distinct residues modulo p, from these $p - 1$ numbers remove the $(p - 1)/2$ quadratic residues, the remaining $(p - 1)/2$ numbers are just the quadratic non-residues modulo p. $\qquad \square$

Theorem II. (Euler's Criterion) *Let p be an odd prime, a an integer with $(a, p) = 1$. Then a is a quadratic residue of p if and only if*

$$a^{\frac{p-1}{2}} \equiv 1 \quad (\text{mod } p).$$

Consequence I: $\left(\dfrac{a}{p}\right) \equiv a^{\frac{p-1}{2}} (\text{mod } p).$ *Consequence II:* $\left(\dfrac{a}{p}\right)\left(\dfrac{b}{p}\right) = \left(\dfrac{ab}{p}\right).$

Proof.   If the equation $x^2 \equiv a \ (\text{mod } p)$ has a solution x_0, then we can let $x_0 > 0$. Since $(x_0^2, p) = (a, p) = 1$, $a^{\frac{p-1}{2}} = x_0^{p-1} \equiv 1 \ (\text{mod } p)$ by Fermat's Little theorem.

On the other hand, if a is a quadratic non-residue of p, namely the equation $x^2 \equiv a \ (\text{mod } p)$ has no solutions, then, for each i with $1 \le i \le p - 1$, the equation $iy \equiv a \ (\text{mod } p)$ for y has a unique solution j with $1 \le j \le p - 1$ and $i \ne j$, such that $ij \equiv a \ (\text{mod } p)$. It is clear that different i correspond to different j, so that $1, 2, \cdots, p - 1$ can be matched into $(p - 1)/2$ such pairs. By taking their product, it is obtained that

$$a^{\frac{p-1}{2}} \equiv 1 \cdot 2 \cdots (p - 1) \equiv (p - 1)! \equiv -1 \quad (\text{mod } p),$$

by Wilson's Theorem.

Consequence (I) is just Euler's Criterion written in terms of the Legendre symbol, and Consequence (II) is obtained at once from

$$\left(\frac{ab}{p}\right) = (ab)^{\frac{p-1}{2}} = a^{\frac{p-1}{2}} \cdot b^{\frac{p-1}{2}} = \left(\frac{a}{p}\right)\left(\frac{b}{p}\right).$$

$\square$

Theorem III. (Gauss' Lemma)   *For an odd prime number p and an integer a with $(a, p) = 1$, define the set S by*

$$S = \left\{ a, 2a, 3a, \cdots, \left(\frac{p-1}{2}\right)a \right\}.$$

Among the remainders of the numbers in S mod p if n numbers are greater than $p/2$, then $\left(\dfrac{a}{p}\right) = (-1)^n$.

Proof.   It is clear that any two of the $(p - 1)/2$ numbers in S are not congruent modulo p. Let $r_1, r_2, \cdots, r_m$ be the remainders less than $p/2$ and $s_1, s_2, \cdots, s_n$ be those remainders that are greater than $p/2$. Then $0 < r_i < p/2$ and $p/2 < s_i < p$, and $n + m = (p - 1)/2$. Therefore

$$r_1, r_2, \cdots, r_m, p - s_1, p - s_2, \cdots, p - s_n$$

are all positive and less than $p/2$.

We now prove by contradiction that these $(p-1)/2$ values are distinct. Suppose that $p - s_i = r_j$ for some choice of i and j, then there exist integers u, v with $1 \le u, v \le (p-1)/2$, such that

$$s_i \equiv ua \pmod{p} \quad \text{and} \quad r_j \equiv va \pmod{p}.$$

Hence $(u+v)a \equiv s_i + r_j = p \equiv 0 \pmod{p}$ which implies $p \mid (u+v)$, a contradiction, since $2 \le u+v < p$. Thus, the $(p-1)/2$ numbers are actually the numbers $1, 2, \cdots, (p-1)/2$. As a result,

$$\left(\frac{p-1}{2}\right)!$$
$$= r_1 r_2 \cdots r_m (p - s_1) \cdots (p - s_n) \equiv r_1 \cdots r_m(-s_1) \cdots (-s_n) \pmod{p}$$
$$\equiv (-1)^n r_1 \cdots r_m s_1 \cdots s_n \equiv (-1)^n a \cdot 2a \cdot 3a \cdots (p - 1/2)a \pmod{p}$$
$$\equiv (-1)^n a^{(p-1)/2} \cdot \left(\frac{p-1}{2}\right)! \pmod{p}$$

Therefore $a^{(p-1)/2} \equiv (-1)^n \pmod{p}$, i.e., $\left(\dfrac{a}{p}\right) = (-1)^n$. $\qquad\square$

Theorem IV. *(Quadratic Reciprocity Law) For distinct odd primes p and q,*

$$\left(\frac{p}{q}\right)\left(\frac{q}{p}\right) = (-1)^{\frac{p-1}{2} \cdot \frac{q-1}{2}}.$$

Proof. First of all we need a lemma as follows.

Lemma. Using the notations of Gauss' Lemma, if a is odd and p is an odd prime with $(a, p) = 1$, then

$$\left(\frac{a}{p}\right) = (-1)^n = (-1)^{\sum_{k=1}^{(p-1)/2}[ka/p]}.$$

Proof of Lemma. Using the notations of Gauss' Lemma, for each number ka $(1 \le k \le (p-1)/2)$ in the set S,

$$ka = \left[\frac{ka}{p}\right]p + t_k, \qquad 1 \le t_k \le p - 1. \qquad (C.1)$$

The t_k is one of $r_1, r_2, \cdots, r_m$ if $t_k < p/2$, and is one of $s_1, s_2, \cdots, s_n$ if $t_k > p/2$, therefore

$$\sum_{k=1}^{(p-1)/2} ka = \sum_{k=1}^{(p-1)/2} \left[\frac{ka}{p}\right]p + \sum_{i=1}^{m} r_i + \sum_{j=1}^{n} s_j. \qquad (C.2)$$

On the other hand, since $r_1, \cdots, r_m, p-s_1, p-s_2, \cdots, p-s_n$ is a permutation of the first $(p-1)/2$ natural numbers,

$$\sum_{k=1}^{(p-1)/2} k = \sum_{i=1}^{m} r_i + \sum_{j=1}^{n} (p - s_j) = pn + \sum_{i=1}^{m} r_i - \sum_{j=1}^{n} s_j. \qquad (C.3)$$

It suffices to show $n \equiv \sum_{k=1}^{(p-1)/2} [ka/p] \pmod 2$. By (C.2) − (C.3), then

$$(a-1) \sum_{k=1}^{(p-1)/2} k = p \left(\sum_{k=1}^{(p-1)/2} \left[\frac{ka}{p} \right] - n \right) + 2 \sum_{j=1}^{n} s_j. \qquad (C.4)$$

By taking modulo 2 to both sides of (C.4), we obtain $\sum_{k=1}^{(p-1)/2} \left[\frac{ka}{p} \right] - n \equiv 0$ (mod 2), i.e.

$$n \equiv \sum_{k=1}^{(p-1)/2} \left[\frac{ka}{p} \right] \pmod 2.$$

Now we return to the proof of the theorem.

Consider the rectangle in the xy-coordinate plane whose vertices are $(0,0)$,

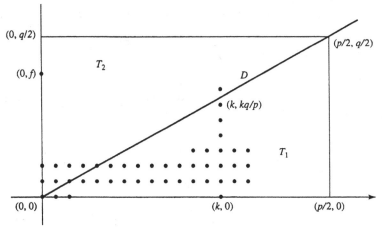

$(p/2, 0)$, $(0, q/2)$, and $(p/2, q/2)$. Let R denote the region inside this rectangle, excluding its boundary lines. We count the number of lattice points inside R in two different ways below. Since p and q are both odd, the lattice points in R consist of all points (n, m), where $1 \le n \le (p-1)/2$ and $1 \le m \le (q-1)/2$; clearly, the number of such points is

$$\frac{p-1}{2} \cdot \frac{q-1}{2}.$$

Now the diagonal line ℓ from $(0,0)$ to $(p/2, q/2)$ has the equation $y = (q/p)x$, or equivalently, $py = qx$. Since $\gcd(p, q) = 1$, none of the lattice points inside R will lie on ℓ, since x must be a multiple of p and y must be a multiple of q if the lattice point (x, y) is on ℓ, but there are no such points in R. Suppose that T_1 denotes the portion of R which is below the diagonal ℓ, and T_2 the portion above ℓ. By what we have just seen, it suffices to count the lattice points inside each of these triangles. The number of integers y in the interval $0 < y < kq/p$ is equal to $[kq/p]$. Thus, for $1 \le k \le (p-1)/2$, there are precisely $[kq/p]$ lattice points in T_1 directly above the point $(k, 0)$ and below ℓ; in other words, lying on the vertical line segment from $(k, 0)$ to $(k, kq/p)$. It follows that the total number of lattice points contained in T_1 is

$$\sum_{k=1}^{(p-1)/2} \left[\frac{kq}{p} \right].$$

A similar calculation, with the roles of p and q interchanged, shows that the number of lattice points within T_2 is

$$\sum_{j=1}^{(q-1)/2} \left[\frac{jp}{q} \right].$$

This accounts for all of the lattice points inside R, so

$$\frac{p-1}{2} \cdot \frac{q-1}{2} = \sum_{k=1}^{(p-1)/2} \left[\frac{kq}{p} \right] + \sum_{j=1}^{(q-1)/2} \left[\frac{jp}{q} \right].$$

From Gauss' lemma and the lemma just proven above,

$$\left(\frac{p}{q} \right) \left(\frac{q}{p} \right) = (-1)^{\sum_{j=1}^{(q-1)/2} [jp/q]} \cdot (-1)^{\sum_{k=1}^{(p-1)/2} [kq/p]}$$

$$= (-1)^{\sum_{j=1}^{(q-1)/2} [jp/q] + \sum_{k=1}^{(p-1)/2} [kq/p]}$$

$$= (-1)^{\frac{p-1}{2} \cdot \frac{q-1}{2}}.$$

Thus, the proof of the Quadratic Reciprocity Law is complete. $\qquad\square$

Regarding quadratic congruences with composite moduli, the following theorems gives the basic results:

Theorem V. *If p is an odd prime with $(a, p) = 1$, then $x^2 \equiv a \pmod{p^n}, n \ge 1$ has a solution if and only if* $\left(\dfrac{a}{p} \right) = 1$.

Proof.   If $x^2 \equiv a$ (mod p^n) has a solution, then clearly so does $x^2 \equiv a$ (mod p), so $\left(\dfrac{a}{p}\right) = 1$.

Conversely, suppose that $\left(\dfrac{a}{p}\right) = 1$, we prove that $x^2 \equiv a$ (mod p^n) has a solution by induction on n. For $n = 1$, nothing to prove since $\left(\dfrac{a}{p}\right) = 1$. Assume that the conclusion is true for $n = k$ $(k \geq 1)$, so if x_0 is a solution of $x^2 \equiv a$ (mod p^k), then $x_0^2 = a + bp^k$ for an appropriate choice of b. For $n = k + 1$, we can construct a solution of $x^2 \equiv a$ (mod p^{k+1}) by using x_0 and b as follows:

First of all, we solve the linear equation (in y) $2x_0 y \equiv -b$ (mod p) to obtain a unique solution y_0 modulo p. This is possible since $(2x_0, p) = 1$. Let $x_1 = x_0 + y_0 p^k$. Then

$$x_1^2 = x_0^2 + 2x_0 y_0 p^k + y_0^2 p^{2k} = a + (b + 2x_0 y_0) p^k + y_0^2 p^{2k} \equiv a \text{ (mod } p^{k+1}).$$

Thus, $x_1 = x_0 + y_0 p^k$ is a solution of $x^2 \equiv a$ (mod p^{k+1}). The inductive proof is completed. □

Theorem VI. *Let a be an integer and n a positive composite integer. When $n = p_1^{\alpha_1} p_2^{\alpha_2} \cdots p_r^{\alpha_r}$ is the prime factorization of n, then a is a quadratic residue modulo n if and only if a is a quadratic residue modulo p_i for all $i = 1, 2, \ldots, r$.*

Proof.   If there is integer x such that $x^2 \equiv a$ (mod n), then it is clear that $x^2 \equiv a$ (mod p_i) for $i = 1, 2, \ldots, r$.

When there is an integer x_i such that $x_i^2 \equiv a$ (mod p_i) for each of $i = 1, 2, \ldots, r$, then, by Theorem V, there is an integer x_i' such that $(x_i')^2 \equiv a$ (mod $p_i^{\alpha_i}$) for $i = 1, 2, \ldots, r$. By the Chinese Remainder Theorem, there is an integer x satisfying the system

$$x \equiv x_i' \pmod{p_i^{\alpha_i}}, \quad i = 1, 2, \ldots, r.$$

Hence

$$x^2 \equiv (x_i')^2 \equiv a \pmod{p_i^{\alpha_i}}, \quad i = 1, 2, \ldots, r,$$

so that

$$x^2 \equiv a \pmod{[p_1^{\alpha_1}, p_2^{\alpha_2}, \cdots, p_r^{\alpha_r}]}, \text{ i.e., } x^2 \equiv a \pmod{n}.$$

Thus, a is a quadratic residue modulo n. □

Appendix D

Proofs of Some Important Inequalities

Theorem I. *(Cauchy-Schwartz Inequality)* *For any real a_1, a_2, $\cdots$, a_n and b_1, b_2, $\cdots$, b_n,*

$$(a_1 b_1 + a_2 b_2 + \cdots + a_n b_n)^2 \le (a_1^2 + a_2^2 + \cdots + a_n^2)(b_1^2 + b_2^2 + \cdots + b_n^2).$$

Further, if a_1, a_2, $\cdots$, a_n are not all zeros, the equality holds if and only if there exists certain constant k such that $b_i = k a_i$ for all $i = 1, 2, \cdots, n$.

Proof. Without loss of generality we assume that $a_1^2 + a_2^2 + \cdots + a_n^2 > 0$. Then the function $f(x)$ given by

$$
\begin{aligned}
f(x) &= (a_1^2 + a_2^2 + \cdots + a_n^2)x^2 - 2(a_1 b_1 + a_2 b_2 + \cdots + a_n b_n)x \\
&\quad + (b_1^2 + b_2^2 + \cdots + b_n^2) \\
&= (a_1 x - b_1)^2 + (a_2 x - b_2)^2 + \cdots + (a_n x - b_n)^2
\end{aligned}
$$

is always non-negative, so its discriminant is less than or equal to zero, that is,

$$\Delta = 4(a_1 b_1 + a_2 b_2 + \cdots + a_n b_n)^2 - 4(a_1^2 + a_2^2 + \cdots + a_n^2)(b_1^2 + b_2^2 + \cdots + b_n^2) \le 0.$$

Thus, the inequality is proven. Further, since

$$\Delta = 0 \Leftrightarrow f(x) = 0 \text{ has two equal real roots } k \Leftrightarrow b_i = k a_i \text{ for all } i = 1, 2, \ldots, n,$$

we obtain the condition for holding the equality. $\qquad\qquad\square$

Theorem II. *(Schur's Inequality)* *For all nonnegative real numbers x, y, z and a positive number $r > 0$, the following inequality always holds:*

$$x^r(x - y)(x - z) + y^r(y - z)(y - x) + z^r(z - x)(z - y) \ge 0,$$

where the equality holds if and only if (i) $x = y = z$ or (ii) two of x, y, z are equal and the other is zero.

273

Proof.   Since x, y, z are symmetric in the inequality, we may assume that $x \geq y \geq z \geq 0$. Then

$$
\begin{aligned}
&x^r(x - y)(x - z) + y^r(y - z)(y - x) + z^r(z - x)(z - y) \\
&= (x - y)[x^r(x - z) - y^r(y - z)] + z^r(x - z)(y - z) \\
&\geq (x - y)y^r[(x - z) - (y - z)] + z^r(x - z)(y - z) \geq 0.
\end{aligned}
$$

When the equality holds, then $x - y = 0$ and $x - z = 0$ or $z = 0$, i.e., $x = y = z$ or $x = y, z = 0$. The inverse is clear. $\square$

Theorem III. (Rearrangement Inequality) *Let $a_1 \leq a_2 \leq \cdots \leq a_n$ and $b_1 \leq b_2 \leq \cdots \leq b_n$ be two groups of ordered real numbers. For a permutation $(j_1, j_2, \cdots, j_n)$ of $(1, 2, \cdots, n)$, the sums given by*

$$
\begin{aligned}
S_O &= a_1 b_1 + a_2 b_2 + \cdots + a_n b_n, \quad \textbf{(ordered sum)} \\
S_M &= a_1 b_{j_1} + a_2 b_{j_2} + \cdots + a_n b_{j_n}, \quad \textbf{(mixed sum)} \\
S_R &= a_1 b_n + a_2 b_{n-1} + \cdots + a_n b_1, \quad \textbf{(reverse sum)}
\end{aligned}
$$

must obey the inequalities : $S_R \leq S_M \leq S_O$.

Furthermore, $S_M = S_O$ for all S_M (or $S_M = S_R$ for all S_M) if and only if $a_1 = a_2 = \cdots = a_n$ or $b_1 = b_2 = \cdots = b_n$.

Proof.   (i)   First of all we show $S_M \leq S_O$, i.e.,

$$
a_1 b_{j_1} + a_2 b_{j_2} + \cdots + a_n b_{j_n} \leq a_1 b_1 + a_2 b_2 + \cdots + a_n b_n. \tag{D.1}
$$

If $j_n = n$, we can delete the terms $a_n b_n$ from both sides of (26.1), then change n to $n - 1$ and continue our discussion. Therefore, we assume that $j_n \neq n$. Let $j_k = n$ for some $1 \leq k \leq n - 1$. By interchanging b_{j_n} and b_n, another mixed sum S'_M is obtained:

$$
S'_M = a_1 b_{j_1} + a_2 b_{j_2} + \cdots + a_k b_{j_n} + \cdots + a_n b_n.
$$

Since

$$
S'_M - S_M = (a_k b_{j_n} + a_n b_n) - (a_k b_n + a_n b_{j_n}) = (a_n - a_k)(b_n - b_{j_n}) \geq 0,
$$

so it suffices to compare S'_M with S_O. Starting from any arrangement $(b_{j_1}, b_{j_2}, \ldots, b_{j_n})$, by continuing the above process it is always possible to obtain S_O by a finite number of exchanges as above without decreasing the value of S_M. This implies that $S_M \leq S_O$ for all S_M.

(ii)   We can prove $S_R \leq S_M$ based on the result of (i). Since $-b_n \leq -b_{n-1} \leq \cdots \leq -b_1$ and $(-b_{j_1}, -b_{j_2}, \cdots, -b_{j_n})$ is a permutation of $(-b_n, -b_{n-1}, \cdots, -b_1)$, by the result of (i),

$$
-S_R = \sum_{i=1}^{n} a_i(-b_{n+1-i}) \geq \sum_{i=1}^{n} a_i(-b_{j_i}) = -\sum_{i=1}^{n} a_i b_{j_i} = -S_M,
$$

therefore $S_R \leq S_M$.

(iii) "Regarding the necessary and sufficient conditions for holding the equality, it is clear that $S_M = S_O$ for all S_M when either $a_1 = a_2 = \cdots = a_n$ or $b_1 = b_2 = \cdots = b_n$. Conversely, when $S_M = S_O$ for all S_M, if it is neither $a_1 = a_2 = \cdots = a_n$ nor $b_1 = b_2 = \cdots = b_n$, then $a_1 < a_n$ and $b_1 < b_n$. Hence, for the special mixed sum given by $S_M = a_1 b_n + a_2 b_2 + \cdots + a_n b_1$, as shown above, we obtain that $S_O - S_M = (a_n - a_1)(b_n - b_1) > 0$, which contradicts $S_M = S_O$ for all S_M.

The discussion for $S_M = S_R$ is similar. Thus, the rearrangement inequality is proven. □

Theorem IV. (**Chebyshev's Inequality**) *Let $a_1 \leq a_2 \leq \cdots \leq a_n$ and $b_1 \leq b_2 \leq \cdots \leq b_n$ be two groups of ordered real numbers. Then*

$$\frac{1}{n}\sum_{i=1}^{n} a_i b_{n+1-i} \leq \left(\frac{1}{n}\sum_{i=1}^{n} a_i\right)\left(\frac{1}{n}\sum_{i=1}^{n} b_i\right) \leq \frac{1}{n}\sum_{i=1}^{n} a_i b_i.$$

Proof. First of all we show the right inequality. Since

$$\frac{1}{2n}\sum_{i=1}^{n}\sum_{j=1}^{n}(a_i b_i + a_j b_j) = \frac{1}{2n}\sum_{i=1}^{n}\sum_{j=1}^{n} a_i b_i + \frac{1}{2n}\sum_{i=1}^{n}\sum_{j=1}^{n} a_j b_j = \sum_{i=1}^{n} a_i b_i,$$

$$\frac{1}{2n}\sum_{i=1}^{n}\sum_{j=1}^{n}(a_i b_j + a_j b_i) = \frac{1}{n}\sum_{i=1}^{n}\sum_{j=1}^{n} a_i b_j = \frac{1}{n}\sum_{i=1}^{n} a_i \cdot \sum_{j=1}^{n} b_j,$$

therefore

$$\left(\frac{1}{n}\sum_{i=1}^{n} a_i\right)\left(\frac{1}{n}\sum_{i=1}^{n} b_i\right) \leq \frac{1}{n}\sum_{i=1}^{n} a_i b_i \Leftrightarrow \frac{1}{n}\sum_{i=1}^{n} a_i \cdot \sum_{i=j}^{n} b_j \leq \sum_{i=1}^{n} a_i b_i$$

$$\Leftrightarrow \frac{1}{2n}\sum_{i=1}^{n}\sum_{j=1}^{n}(a_i b_i + a_j b_j - a_i b_j - a_j b_i) \geq 0$$

$$\Leftrightarrow \frac{1}{2n}\sum_{i=1}^{n}\sum_{j=1}^{n}(a_i - a_j)(b_i - b_j) \geq 0,$$

and the last inequality is obvious since $(a_i - a_j)$ and $(b_i - b_j)$ have same signs for $1 \leq i, j \leq n$. □

Consequence 1. For $x_1, x_2, \ldots, x_n > 0$, when $pq > 0$,

$$\sum_{i=1}^{n} x_i^{p+q} \geq \frac{1}{n}\sum_{i=1}^{n} x_i^p \cdot \sum_{i=1}^{n} x_i^q; \text{ and } \sum_{i=1}^{n} x_i^{p+q} \leq \frac{1}{n}\sum_{i=1}^{n} x_i^p \cdot \sum_{i=1}^{n} x_i^q \text{ when } pq < 0.$$

Consequence 1 is a direct application of the Chebyshev's inequality.

Consequence 2. $x_1, x_2, \ldots, x_n > 0, r > s > 0$ and $\prod_{i=1}^{n} x_i = 1$, then

$$\sum_{i=1}^{n} x_i^r \geq \sum_{i=1}^{n} x_i^s.$$

In fact, $\sum_{i=1}^{n} x_i^r \geq \dfrac{1}{n} \sum_{i=1}^{n} x_i^{r-s} \cdot \sum_{i=1}^{n} x_i^s \geq \left(\prod_{i=1}^{n} x^{r-s} \right)^{\frac{1}{n}} \cdot \sum_{i=1}^{n} x_i^s = \sum_{i=1}^{n} x_i^s.$

Consequence 3. For $a_1 \leq a_2 \leq \cdots \leq a_n; b_1 \leq b_2 \leq \cdots \leq b_n$ and $m_1, m_2, \cdots m_n > 0$,

$$\sum_{i=1}^{n} m_i \cdot \sum_{i=1}^{n} a_i b_i \geq \sum_{i=1}^{n} \sqrt{m_i a_i} \cdot \sum_{i=1}^{n} \sqrt{m_i b_i}.$$

In fact,

$$\sum_{i=1}^{n} m_i \cdot \sum_{i=1}^{n} a_i b_i - \sum_{i=1}^{n} \sqrt{m_i a_i} \cdot \sum_{i=1}^{n} \sqrt{m_i b_i} = \frac{1}{2} \sum_{i,j=1}^{n} m_i m_j (a_i - a_j)(b_i - b_j) \geq 0.$$

Theorem V. (Jensen's Inequality) *For a strictly convex function $f(x)$ defined on $I = [a, b]$ (or $I = (a, b)$), the inequality*

$$f\left(\frac{x_1 + x_2 + \cdots + x_n}{n} \right) \leq \frac{1}{n}[f(x_1) + f(x_2) + \cdots + f(x_n)] \qquad (D.2)$$

holds for any $x_1, x_2, \cdots, x_n \in I$, and the equality holds if and only if $x_1 = x_2 = \cdots = x_n$.

Proof. We prove it by induction on n. For $n = 1, 2$, the conclusion is clear. Assume the conclusion is true for $n = k$ ($k \geq 2$), then for $n = k + 1$, let

$$A = \frac{1}{k+1}(x_1 + x_2 + \cdots + x_{k+1}),$$

$$B = \frac{1}{k}(x_1 + x_2 + \cdots + x_k),$$

$$C = \frac{x_{k+1} + (k-1)A}{k}.$$

We have $A = \dfrac{B+C}{2}$ and

$$f(A) = f\left(\frac{B+C}{2}\right) \le \frac{1}{2}[f(B) + f(C)]$$

$$\le \frac{1}{2}\left[\frac{1}{k}(f(x_1) + f(x_2) + \cdots + f(x_k)) + \frac{1}{k}(f(x_{k+1}) + (k-1)f(A))\right],$$

Therefore

$$f\left(\frac{x_1 + x_2 + \cdots + x_{k+1}}{k+1}\right) \le \frac{1}{k+1}[f(x_1) + f(x_2) + \cdots + f(x_{k+1})].$$

Above equalities hold if and only if $x_1 = x_2 = \cdots = x_k$ and $B = C$, which is equivalent to $x_1 = x_2 = \cdots = x_k = x_{k+1}$. $\square$

Theorem VI. (Weighted Jensen's Inequality) *Given that f is a continuous function defined on some interval $I = [a\,,\,b]$ (or $I = (a\,,\,b)$). Then f is strictly convex on I if and only if for any positive integer $n \in \mathbb{N}$ and $\lambda_1,\ \lambda_2,\ \cdots \lambda_n > 0$ satisfying $\sum_{i=1}^{N} \lambda_i = 1$,*

$$f(\lambda_1 x_1 + \lambda_2 x_2 + \cdots + \lambda_n x_n) \le \lambda_1 f(x_1) + \lambda_2 f(x_2) + \cdots + \lambda_n f(x_n), \quad \text{(D.3)}$$

for all $x_1, \cdots,\ x_n \in I$, and the equality holds if and only if $x_1 = x_2 = \cdots = x_n$.

Proof. The sufficiency is obvious. Below we only prove the necessity by induction on n. Let f be a continuous and strictly convex function.
The conclusion is clear for $n = 1$.

For $n = 2$, If the conclusion is not true, then there exist $x_1, x_2 \in I$ with $x_1 < x_2$ and $\lambda^* \in [0, 1]$, such that

$$f(\lambda^* x_1 + (1 - \lambda^*)x_2) > \lambda^* f(x_1) + (1 - \lambda^*)f(x_2).$$

Let $F(\lambda) = f(\lambda x_1 + (1 - \lambda)x_2) - \lambda f(x_1) - (1 - \lambda)f(x_2), 0 \le \lambda \le 1$. Then F is continuous on $[0, 1]$ and $F(0) = F(1) = 0, F(\lambda^*) > 0$. Therefore $M_0 = \max_{0 \le \lambda \le 1}\{F(\lambda)\} > 0$. Let

$$\lambda_0 = \inf\{\lambda | F(\lambda) = M_0, 0 \le \lambda \le 1\}.$$

Then $0 < \lambda_0 < 1$. Take $\delta > 0$ such that $[\lambda_0 - \delta, \lambda_0 + \delta] \subset [0, 1]$. Since

$$x_1 \le x_1^* = (\lambda_0 - \delta)x_1 + (1 - \lambda_0 + \delta)x_2, x_2^* = (\lambda_0 + \delta)x_1 + (1 - \lambda_0 - \delta)x_2 \le x_2,$$

we have

$$f\left(\frac{x_1^* + x_2^*}{2}\right) \le \frac{1}{2}\left[f(x_1^*) + f(x_2^*)\right],$$

i.e.,

$$f(\lambda_0 x_1 + (1 - \lambda_0)x_2)$$
$$\leq \tfrac{1}{2}\{f((\lambda_0 - \delta)x_1 + (1 - \lambda_0 + \delta)x_2) + f((\lambda_0 + \delta)x_1 + (1 - \lambda_0 - \delta)x_2))\},$$

which implies that

$$M_0 = F(\lambda_0) \leq \frac{1}{2}[F(\lambda_0 - \delta) + F(\lambda_0 + \delta)] < M_0,$$

a contradiction. Thus, the conclusion is true for $n = 2$.

Assume that (D.3) is true for $n \leq k$ ($k \geq 2$), then, for $x_1, x_2, \cdots, x_{k+1} \in I$ and the nonnegative numbers $\lambda_1, \lambda_2, \cdots, \lambda_{k+1}$ with sum 1, by considering

$$\lambda_k x_k + \lambda_{k+1} x_{k+1} = (\lambda_k + \lambda_{k+1}) \left(\frac{\lambda_k x_k}{\lambda_k + \lambda_{k+1}} + \frac{\lambda_{k+1} x_{k+1}}{\lambda_k + \lambda_{k+1}} \right) = (\lambda_k + \lambda_{k+1}) x_k',$$

where $x_k' = \dfrac{\lambda_k}{\lambda_k + \lambda_{k+1}} x_k + \dfrac{\lambda_{k+1}}{\lambda_k + \lambda_{k+1}} x_{k+1}$, then the induction assumption gives

$$f(x_k') \leq \frac{1}{\lambda_k + \lambda_{k+1}}[\lambda_k f(x_k) + \lambda_{k+1} f(x_{k+1})],$$

hence

$$f(\lambda_1 x_1 + \lambda_2 x_2 + \cdots + \lambda_{k+1} x_{k+1})$$
$$\leq \lambda_1 f(x_1) + \cdots + \lambda_{k-1} f(x_{k-1}) + (\lambda_k + \lambda_{k+1}) f(x_k')$$
$$\leq \lambda_1 f(x_1) + \lambda_2 f(x_2) + \cdots + \lambda_k f(x_k) + \lambda_{k+1} f(x_{k+1}).$$

Therefore, the proposition is proven for $n = k + 1$ also.   □

Theorem VII. (**Power Mean Inequality**)   *For any nonnegative real numbers* $x_1, x_2, \cdots, x_n$ *and* $\alpha, \beta > 0$ *with* $\alpha > \beta$,

$$\left(\frac{1}{n} \sum_{i=1}^{n} x_i^\alpha \right)^{\frac{1}{\alpha}} \geq \left(\frac{1}{n} \sum_{i=1}^{n} x_i^\beta \right)^{\frac{1}{\beta}}. \tag{D.4}$$

Proof.   In (D.4) let $y_i = x_i^\beta$ for $i = 1, 2, \cdots, n$, then (D.4) becomes

$$\frac{1}{n}(y_1^{\alpha/\beta} + y_2^{\alpha/\beta} + \cdots + y_n^{\alpha/\beta}) \geq \left(\frac{y_1 + \cdots + y_n}{n} \right)^{\alpha/\beta}. \tag{D.5}$$

Since $\alpha/\beta > 1$, the function $f(u) = u^{\alpha/\beta}, u > 0$ is convex, therefore (D.5) is true by Jensen's inequality.   □

Appendix E

Note On Cauchy's Problem in Functional Equations

Theorem I. *When the function $f(x)$ satisfies equation (30.1) in Lecture 30 and satisfies one of the following conditions:*

    (A)   *$f(x)$ is bounded in an interval (a, b) where $a < b$;*
    (B)   *$f(x) \geq 0$ on the interval $[0, \epsilon]$ or $f(x) \leq 0$ on $[0, \epsilon]$, where $\epsilon > 0$,*
    (C)   *$f(x)$ is continuous at some point x_0,*
then $f(x) = Cx$, where C is a constant.

Proof of (A).   From the proof in Example 1 in Lecture 30, we find that equation (30.1) forces f to have the following properties:

$$f(0) = 0; \qquad f(r) = Cr \text{ for any } r \in \mathbb{Q}.$$

Define $g(x) = f(x) - Cx, x \in \mathbb{R}$. Then $g(r) = 0$ on $\mathbb{Q}$, and for $x, y \in \mathbb{R}$

$$g(x+y) = f(x+y) - C(x+y) = (f(x) - Cx) + (f(y) - Cy) = g(x) + g(y).$$

    For any given real number x, take an arbitrary rational number r in the interval $(x-b, x-a)$, and let $x_1 = x-r$, then $x_1 \in (a, b)$, and $g(x) = g(x_1+r) = g(x_1)$. Hence g is a bounded function.

    We now show $g(x) = 0$ for any $x \in \mathbb{R}$. Suppose that there is $x_0 \in \mathbb{Q}^c$ such that $g(x_0) = d \neq 0$, then $g(nx_0) = ng(x_0) = nd$, therefore $|g(nx_0)| = n|d| \to +\infty$ as $n \to +\infty$, a contradiction. Thus, $g(x)$ is equal to zero identically, i.e. $f(x) = Cx$ for all $x \in \mathbb{R}$.

Proof of (B).   It suffices to show that under condition (ii), $f(x)$ is monotone on $[0, \epsilon]$. Let $x, y \in [0, \epsilon]$ with $x \leq y$, then $0 \leq y - x \leq \epsilon$, therefore

$$f(y) = f(x + y - x) = f(x) + f(y - x) \geq f(x) \qquad \text{if } f(u) \geq 0 \text{ on } [0, \epsilon],$$

279

or

$$f(y) = f(x + y - x) = f(x) + f(y - x) \leq f(x) \qquad \text{if } f(u) \leq 0 \text{ on } [0, \epsilon].$$

Since $f(-x) = -f(x)$, so f is monotone on $[-\epsilon, \epsilon]$. By the proof of the Example 1 in Lecture 30, we have $f(x) = Cx$ for all $x \in \mathbb{R}$.

Proof of (C).   Since f is continuous at x_0, there must be $a < b$ such that $x_0 \in (a, b)$ and f is bounded on (a, b), so f satisfies condition (A).

Index